COMPUTATIONAL MODELING IN BIOENGINEERING AND BIOINFORMATICS

COMPUTATIONAL MODELING IN BIOENGINEERING AND BIOINFORMATICS

NENAD FILIPOVIC

ELSEVIER

ACADEMIC PRESS
An imprint of Elsevier

Academic Press is an imprint of Elsevier
125 London Wall, London EC2Y 5AS, United Kingdom
525 B Street, Suite 1650, San Diego, CA 92101, United States
50 Hampshire Street, 5th Floor, Cambridge, MA 02139, United States
The Boulevard, Langford Lane, Kidlington, Oxford OX5 1GB, United Kingdom

Library of Congress Cataloging-in-Publication Data
A catalog record for this book is available from the Library of Congress

British Library Cataloguing-in-Publication Data
A catalogue record for this book is available from the British Library

ISBN: 978-0-12-819583-3

For information on all Academic Press publications visit our website at https://www.elsevier.com/books-and-journals

Publisher: Mara Conner
Acquisition Editor: Chris Katsaropoulos
Editorial Project Manager: Isabella C. Silva
Production Project Manager: R.Vijay Bharath
Cover Designer: Matthew Limbert

Typeset by SPi Global, India

Contents

About the author

Dr. Nenad Filipovic is a rector of the University of Kragujevac, full professor in the Faculty of Engineering, and head of the Center for Bioengineering at the University of Kragujevac, Serbia. He was a research associate at the Harvard School of Public Health in Boston, the United States. His research interests are in the area of biomedical engineering, cardiovascular disease, fluid-structure interaction, biomechanics, multiscale modeling, data mining, software engineering, parallel computing, computational chemistry, and bioprocess modeling. He is an author and coauthor of 15 textbooks and 7 monographs, over 300 publications in peer reviewed journals and over 15 software programs for modeling with finite element methods and discrete methods from fluid mechanics and multiphysics. He also leads a number of national and international projects in EU and the United States in the area of bioengineering and bioinformatics. He is the director of the Center for Bioengineering at the University of Kragujevac and leads joint research projects with Harvard University and the University of Texas in the area of bionanomedicine computer simulation. He also leads a number of national and international projects in the area of bioengineering and bioinformatics. He is a managing editor for Journal of Serbian Society for Computational Mechanics and member of the European Society of Biomechanics (ESB) and the European Society for Artificial Organs (ESAO).

Preface

The current medicine treatment relies exclusively on diagnostic imaging data to define the present state of the patient, biomarkers, and experience of the medical doctors to evaluate the efficacy of prior treatments for similar patients. Computational methods may give opportunity for a patient-specific model in order to improve the prediction for the disease progression.

The book *Computational Modeling in Bioengineering and Bioinformatics* promotes complementary disciplines that hold great promise for the advancement of research and development in complex medical and biological systems, environment, public health, drug design, and so on. It provides a common platform for the cross-fertilization of ideas and shaping knowledge and scientific achievements by bridging these two very important and complementary disciplines into an interactive and attractive forum. The chapters includes 12 chapters varying from modeling of atherosclerosis, data mining, cancer modeling, stent design, lab-on-a-chip modeling, aortic dissection modeling, bone biomechanics, computational chemistry, modeling of DPIs for pulmonary drug delivery, cochlea biomechanics, microfluidic chips, abdominal aortic aneurysm, and sport biomechanics.

The book is also prepared to be useful for researchers in various fields related to bioengineering and other scientific fields, including medical applications. The book provides basic information about how a bioengineering (or medical) problem can be modeled, which computational models can be used, and what is the background of the applied computer models. Each of the bioengineering problems treated in this book has been analyzed elsewhere from different aspects, with more details and particular theoretical considerations. We have referred to these analyses to a certain extent, but this referring is far from being complete, since the field of computer modeling in bioengineering is vast and consistently expanding.

The chapters are results of many international projects of the authors over a number of years, in area of bioengineering and bioinformatics of the Faculty of Engineering of the University of Kragujevac, and Research and Development Center for Bioengineering BioIRC from Serbia.

Nenad Filipovic
Full Professor and Head of Center for Bioengineering, University of Kragujevac, Kragujevac, Serbia

CHAPTER 1

Computational modeling of atherosclerosis

Contents

1 Theoretical background

Atherosclerosis is becoming the number one cause of death worldwide as a progressive disease characterized in particular by the accumulation of lipids and fibrous elements in artery walls. It is characterized by dysfunction of endothelium; vasculitis; and accumulation of lipid, cholesterol, and cell elements inside blood vessel wall. The process of atherosclerosis develops in arterial walls (Loscalzo and Schafer, 2003). It starts to develop from oxidized low-density lipoprotein (LDL) molecules. When oxidized LDL evolves in plaque formations within an artery wall, a series of reactions occur to repair the damage to the artery wall caused by oxidized LDL (Libby, 2002). The body's immune system responds to the damage to the artery wall caused by oxidized LDL by sending specialized white blood cells—macrophages

Computational Modeling in Bioengineering and Bioinformatics
https://doi.org/10.1016/B978-0-12-819583-3.00001-1

(MPHs)—to absorb the oxidized LDL forming specialized foam cells (Meyer et al., 1996). Macrophages accumulate inside arterial intima. Also, smooth muscle cell (SMC) accumulates in the atherosclerotic arterial intima, where they proliferate and secrete extracellular matrix to form a fibrous cap (Begent and Born, 1970). Unfortunately, macrophages are not able to process the oxidized LDL and ultimately grow and rupture, depositing a larger amount of oxidized cholesterol into the artery wall. The schematic process of atherosclerosis has been shown in Fig. 1.1.

Formerly focused on luminal narrowing due to the bulk of atheroma, the current concepts recognize the biological attributes of the atheroma as key determinants of its clinical significance (Libby, 2002). Inflammatory process starts with penetration of low-density lipoproteins (LDL) in the intima. This penetration, if too high, is followed by leucocyte recruitment in the intima. This process may participate in the formation of the fatty streak, in the initial lesion of atherosclerosis, and then in formation of a plaque (Meyer et al., 1996).

Several mathematical models have described the transport of macromolecules, such as low-density lipoproteins, from the arterial lumen to the arterial wall and inside the wall (e.g., Tarbell, 2003; Zunino, 2002; Prosi et al., 2005; Quarteroni and Valli, 1999; Quarteroni et al., 2002). It is now well known that the early stage of the inflammatory disease is the result of interaction between plasma low-density lipoproteins that filtrate through endothelium into the intima, cellular components (monocytes/macrophages, endothelial cells, and smooth muscle cells), and the extracellular matrix of the arterial wall (Libby, 2002; Ross, 1993).

Low endothelial wall shear stress (WSS) promotes the development of early fibroatheromas, which evolution follows an individualized natural history of progression. In vivo assessment of the local WSS, extent of vascular inflammation, and arterial remodeling response, all responsible for individual plaque evolution, in combination with systemic biomarkers of vascular inflammation, may all together improve risk stratification of individual early atherosclerotic plaques, thereby personalizing the therapeutic strategies.

The purpose of this study was to determine wall shear stress with computational fluid dynamics in the coronary arteries using patient-specific data obtained from computed tomography. Also, plaque concentration in the arterial wall was calculated. Coronary geometries of several patients were used. Two time periods were analyzed: baseline and follow-up. Plaque progression was performed using numerical approach. Mass transport of low-density lipoprotein through the arterial wall was firstly described. Fluid

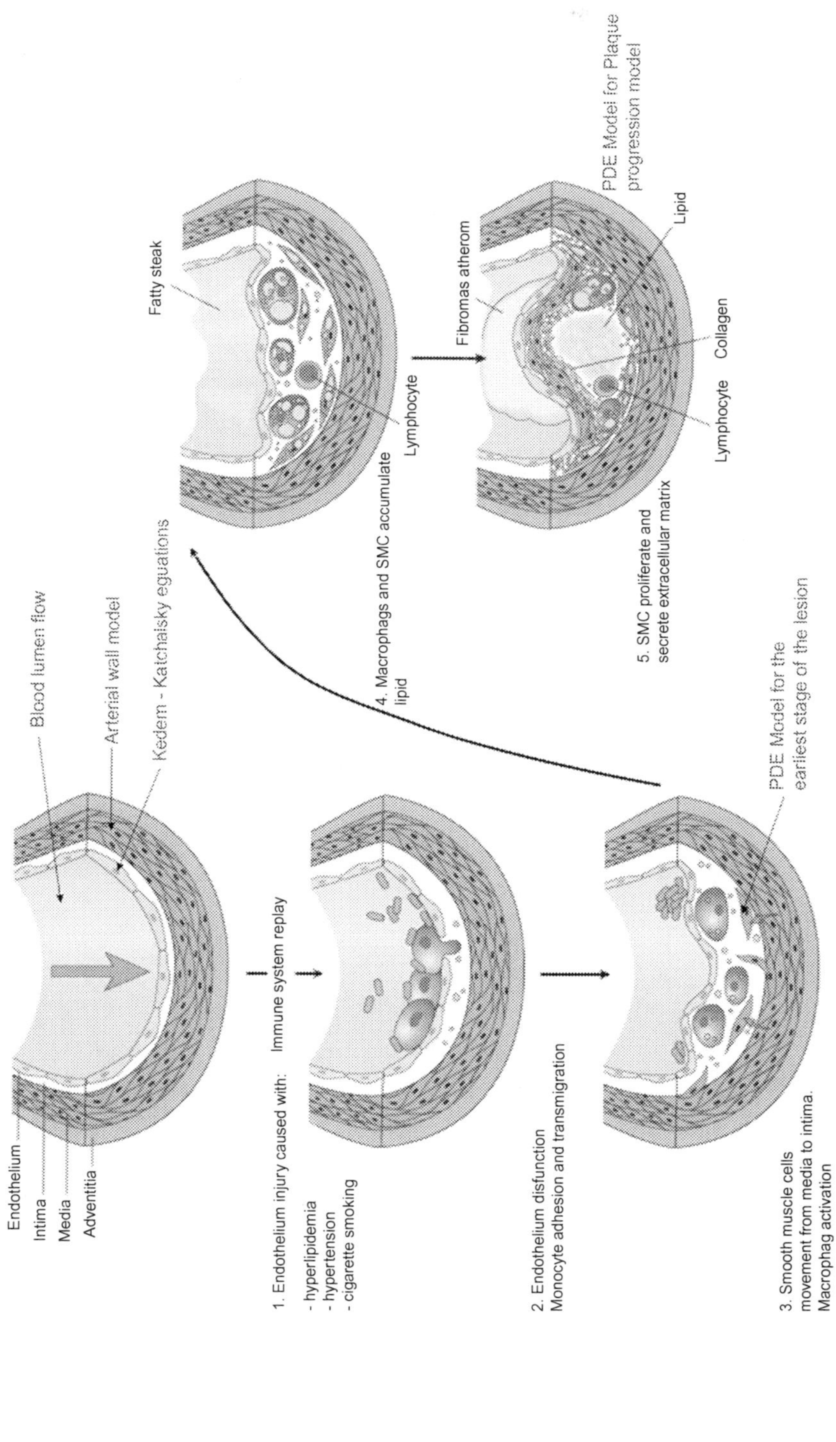

Fig. 1.1 Atherosclerotic plaque development. *Adapted from Loscalzo J., Schafer A. I., 2003, Thrombosis and Hemorrhage, third ed. Lippincott Williams & Wilkins, Philadelphia.*

motion in the lumen domain is described with Navier-Stokes equations, the fluid filtration with the Darcy law, while the Kedem-Katchalsky equations (Kedem and Katchalsky, 1958, 1961) is used for the solute flow between the lumen domain, endothelium, and the first layer of the vessel wall—intima. Also, we used coupled approach for modeling of blood flow with the collection of dissipative particle dynamic (DPD) particles (Boryczko et al., 2003; Haber et al., 2006; Filipovic et al., 2006, 2008a, 2008b, 2010). Motion of each DPD particle is governed by Newton's law equation. At the end, some discussion related to shear stress distribution and plaque growth was given.

2 Methods

2.1 Blood flow simulation

The blood can be considered as an incompressible homogenous viscous fluid for flow in large blood vessels. Also, the laminar flow is dominant in physiological flow environment. Therefore, the fundamental laws of physics that include balance of mass and balance of linear momentum are applicable here. These laws are expressed by continuity equation and the Navier-Stokes equations.

We here present the final form of these equations to emphasize some specifics related to blood flow. The incremental-iterative balance equation of a finite element for a time step "n" and equilibrium iteration "i" has a form

$$\begin{bmatrix} \frac{1}{\Delta t}\mathbf{M} + {}^{n+1}\tilde{\mathbf{K}}_{vv}^{(i-1)} & \mathbf{K}_{vp} \\ \mathbf{K}_{vp}^{T} & \mathbf{0} \end{bmatrix} \begin{Bmatrix} \Delta\mathbf{V}^{(i)} \\ \Delta\mathbf{P}^{(i)} \end{Bmatrix}_{blood} =$$

$$\begin{Bmatrix} {}^{n+1}\mathbf{F}_{ext}^{(i-1)} \\ 0 \end{Bmatrix} - \begin{bmatrix} \frac{1}{\Delta t}\mathbf{M} + {}^{n+1}\mathbf{K}^{(i-1)} & \mathbf{K}_{vp} \\ \mathbf{K}_{vp}^{T} & 0 \end{bmatrix} \begin{Bmatrix} {}^{n+1}\mathbf{V}^{(i-1)} \\ {}^{n+1}\mathbf{P}^{(i-1)} \end{Bmatrix} + \begin{Bmatrix} \frac{1}{\Delta t}\mathbf{M}^{n}\mathbf{V} \\ 0 \end{Bmatrix} \tag{1.1}$$

where ${}^{n+1}\mathbf{V}^{(i-1)}\ {}^{n+1}\mathbf{P}^{(i-1)}$ are the nodal vectors of blood velocity and pressure, with the increments in time step $\Delta\mathbf{V}^{(i)}$ and $\Delta\mathbf{P}^{(i)}$ (the index "blood" is used to emphasize that we are considering blood as the fluid), Δt is the time step size and the left upper indices "n" and "$n+1$" denote start and end of time step, and the matrices and vectors are defined in (Kojic et al., 2008; Filipovic et al., 2006). Note that the vector ${}^{n+1}\mathbf{F}_{ext}^{(i-1)}$ of external forces includes the volumetric and surface forces. In the assembling of these equations, the system of equations of the form (Eq. 1.1) is obtained, with the volumetric external forces and the surface forces acting only on the fluid domain boundary (the surface forces among the internal element boundaries cancel).

The specifics for the blood flow are that the matrix ${}^{n+1}\mathbf{K}^{(i-1)}$ may include variability of the viscosity if non-Newtonian behavior of the blood is considered. We have that

$$\left[K_{KJ}^{(i-1)}\right]_{mk} = \left[\hat{K}_{KJ}^{(i-1)}\right]_{mk} + \int_V \mu^{(i-1)} N_{K,j} N_{J,j} dV \tag{1.2}$$

where $\mu^{(i-1)}$ corresponds to the constitutive law for the last known conditions (at iteration "$i-1$"). In case of use of the Cason relation (1.2), the second invariant of the strain rate $D_{II}^{(i-1)}$ is to be evaluated when computing $\mu^{(i-1)}$.

We note here that the penalty method can also be used, as well as the ALE formulation in case of large displacements of blood vessel walls (Kojic et al., 2008; Filipovic et al., 2006; Filipovic et al., 2010).

In addition to the velocity and pressure fields of the blood, the distribution of stresses within the blood can be evaluated. The stresses ${}^t\sigma_{ij}$ at time "t" follow from

$${}^t\sigma_{ij} = -{}^t p\delta_{ij} + {}^t\sigma_{ij}^{\mu} \tag{1.3}$$

where

$${}^t\sigma_{ij}^{\mu} = {}^t\mu^t\left(v_{i,j} + v_{j,i}\right) \tag{1.4}$$

is the viscous stress. Here, ${}^t\mu$ is viscosity corresponding to the velocity vector ${}^t\mathbf{v}$ at a spatial point within the blood domain. The field of the viscous stresses is given by Eq. (1.4).

Further, the wall shear stress at the blood vessel wall is calculated as

$${}^t\tau = {}^t\mu \frac{\partial^t v_t}{\partial n} \tag{1.5}$$

where ${}^t v_t$ denotes the tangential velocity and n is the normal direction at the vessel wall. Practically, we first calculate the tangential velocity at the integration points near the wall surface and then numerically evaluate the velocity gradient $\partial^t v_t/\partial n$; finally, we determine the viscosity coefficient ${}^t\mu$ using the average velocity at these integration points. In essence, the wall shear stress is proportional to the shear rate γ at the wall and the blood dynamic viscosity μ.

For a pulsatile flow, the mean wall shear stress within a time interval T can be calculated as (Taylor et al., 1998)

$$^{T}\tau_{mean} = \left| \frac{1}{T} \int_{0}^{T} {}^{t}\tau_{n} dt \right| \tag{1.6}$$

Another scalar quantity is a time-averaged magnitude of the surface traction vector, calculated as

$$^{T}\tau_{mag} = \frac{1}{T} \int_{0}^{T} |{}^{t}\mathbf{t}| dt \tag{1.7}$$

where the vector ${}^{t}\mathbf{t}$ is given by the Cauchy formula.

2.2 Plaque formation and progression modeling—Continuum approach

Continuum-based methods are an efficient way for modeling the evolution of plaque. In our model, LDL concentration is first introduced into the system of partial differential equations as a boundary condition. The model simulates the inflammatory response formed at the initial stages of plaque formation.

Regarding the particle dynamics, the model is based on the involvement of LDL/oxidized LDL, monocytes and macrophages, and foam cells and extracellular matrix. Reaction-diffusion differential equations are used to model these particle dynamics. The adhesion rate of the molecules depends on the local hemodynamics, which is described by solving the Navier-Stokes equations. Intima LDL concentration is a function of the wall shear stress, while the adhesion of monocytes is a function of shear stress and VCAM. Finally, the alterations of the arterial wall are simulated. A finite element solver is used to solve the system of the equations. The lumen is defined as a 2D domain, while the intima is simplified as 1-D model due to its thin geometry. First, the LDL penetration to the arterial wall and the wall shear stress is calculated. Then, the concentration of the various components of the model is calculated to simulate the intima fattening in the final step.

The LDL penetration is defined by the convection-diffusion equation, while the endothelial permeability is shear stress dependent. This model produces results about the initial stages of the atherosclerotic plaque formation. More specifically, concentration of LDL is calculated on the artery wall and in the next step the oxidized LDL. Furthermore, monocytes and their modified form (macrophages) are also counted. Solution to the system provides to the user the concentration of foam cells created when a threshold on LDL concentration is reached.

The previous model describes the initial stages of atherosclerosis. However, atherosclerosis is characterized by the proliferation of SMCs. A medical user needs a prediction for the plaque formation that is based on the concentration of SMCs, the necrotic core, and the extracellular matrix. In this respect, a new approach to count the concentration of SMCs is being developed. The user is also provided with results regarding the formation of plaque in an overall manner.

The fluid is assumed to be steady, incompressible, and laminar for modeling fluid dynamics in the lumen. Navier-Stokes equations were used (Eqs. 1.8 and 1.9):

$$-\mu\nabla^2 u_l + \rho(u_l \cdot \nabla)u_l + \nabla p_l = 0 \tag{1.8}$$

$$\nabla u_l = 0 \tag{1.9}$$

where u_l is blood velocity, p_l is pressure, μ is blood dynamic viscosity, and ρ is blood density.

Darcy's law was used to model mass transfer across the wall (transmural flow) of the blood vessel:

$$u_w - \nabla\left(\frac{k}{\mu_p} p_w\right) = 0 \tag{1.10}$$

$$\nabla u_w = 0 \tag{1.11}$$

where u_w is transmural velocity, p_w is pressure in the arterial wall, μ_p is viscosity of blood plasma, and k is the Darcian permeability coefficient of the arterial wall (Eqs. 1.10 and 1.11). Convective diffusion equations were occupied for modeling mass transfer in the lumen (Eq. 1.12):

$$\nabla \cdot (-D_l \nabla c_l + c_l u_l) = 0 \tag{1.12}$$

where c_l represents blood concentration in the lumen and D_l is diffusion coefficient of the lumen.

Convective diffusion reactive equations (1.13) were used for modeling mass transfer in the wall, which are related to transmural flow:

$$\nabla \cdot (-D_w \nabla c_w + K c_w u_w) = r_w c_w \tag{1.13}$$

where c_w is solute concentration in the arterial wall, D_w is diffusive coefficient of solution in the wall, K is solute lag coefficient, and r_w is consumption rate constant.

The coupling of fluid dynamics and solute dynamics at the endothelium was achieved by the Kedem-Katchalsky equations (Kedem and Katchalsky, 1958, 1961) (Eqs. 1.14 and 1.15):

$$J_v = L_p(\Delta p - \delta_d \Delta \pi) \tag{1.14}$$

$$J_s = P\Delta c + \left(1 - \delta_f\right) J_v \bar{c} \tag{1.15}$$

where L_p is the hydraulic conductivity of the endothelium, Δc is the solute concentration difference across the endothelium, Δp is the pressure drop across the endothelium, $\Delta \pi$ is the oncotic pressure difference across the endothelium, σ_d is the osmotic reflection coefficient, σ_f is the solvent reflection coefficient, P is the solute endothelial permeability, and $\bar{c}$ is the mean endothelial concentration (Nanfeng et al., 2006).

The inflammatory process is modeled using three additional reaction-diffusion partial differential equations (Filipovic et al., 2011, 2012, 2013):

$$\begin{aligned} &\partial_t O = d_1 \Delta O - k_1 O \cdot M \\ &\partial_t M + \operatorname{div}(v_w M) = d_2 \Delta M - k_1 O \cdot M + S/(1+S) \\ &\partial_t S = d_3 \Delta S - \lambda S + k_1 O \cdot M + \gamma(O - O^{thr}) \end{aligned} \tag{1.16}$$

where O is the oxidized LDL in the wall; M and S are concentrations in the intima of macrophages and cytokines, respectively; d_1, d_2, d_3 are the corresponding diffusion coefficients; λ and γ are degradation and LDL oxidized detection coefficients; and v_w is the inflammatory velocity of plaque growth (Filipovic et al., 2011, 2013).

2.3 Discrete approach

Blood flow in the small coupled domain within finite element mesh is viewed as a motion of the collection of DPD particles. Motion of each DPD particle is described by Newton's law equation:

$$m_i \dot{\mathbf{v}}_i = \sum_j \left(\mathbf{F}_{ij}^C + \mathbf{F}_{ij}^D + \mathbf{F}_{ij}^R \right) + \mathbf{F}_i^{ext} \tag{1.17}$$

where m_i is the mass of particle "i"; $\dot{\mathbf{v}}_i$ is the particle acceleration as the time derivative of velocity; $\mathbf{F}_{ij}^C$, $\mathbf{F}_{ij}^D$, and $\mathbf{F}_{ij}^R$ are the conservative (repulsive), dissipative, and random (Brownian) interaction forces, which particle "j" exerts on particle "i," respectively, provided particle "j" is within the radius of influence r_c of particle "i"; and $\mathbf{F}_i^{ext}$ is the external force exerted on particle "i," which usually represents gradient of pressure or gravity force as a driving

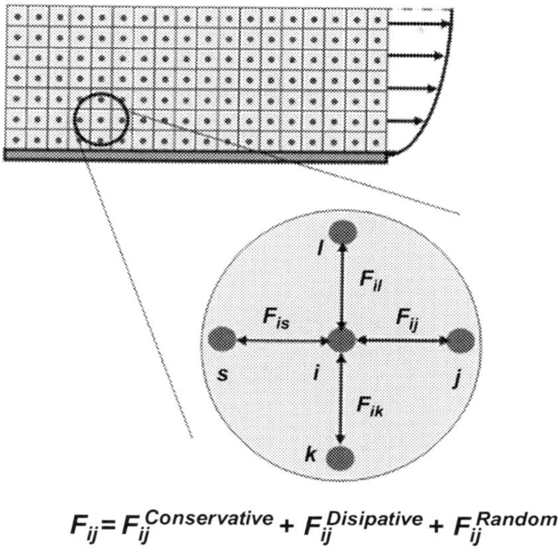

Fig. 1.2 Interaction forces in the DPD approach.

force for the fluid domain (Boryczko et al., 2003). Hence the total interaction force $\mathbf{F}_{ij}$ (Fig. 1.2) between the two particles is

$$\mathbf{F}_{ij} = \mathbf{F}_{ij}^{C} + \mathbf{F}_{ij}^{D} + \mathbf{F}_{ij}^{R} \tag{1.18}$$

2.4 DPD modeling of oxidized LDL particle adhesion to the wall

When an oxidized LDL comes close to the wall and if shear rate allows, it binds to the wall. However, when adhered LDL particles are exposed simultaneously to other forces stronger than the binding forces, the bonds break. To incorporate LDL adhesion to vessel walls, we introduce an attractive (bonding) force, ($\mathbf{F}_{ij}^{a}$), in addition to the conservative, viscous, and random interaction forces. As an approximation, we model the attractive force with a linear spring attached to the LDL's surface. The spring is attached to the vessel wall or to an already adhered LDL particle. The effective spring constant for LDL adhesion on the vessel wall, or to another stationary LDL particle, is denoted by k_{bw}.

The additional parameter involved in the model is the size of the domain from the LDL coated wall ($L_{\max}^{wall}$) for which the action of attractive force needs to be considered. We take the attractive force as

$$F_{w}^{a} = k_{bw}\left(1 - L_{w}/L_{max}^{wall}\right) \tag{1.19}$$

where L_w is the distance of the oxidized LDL from the wall.

3 Results

3.1 Animal pig experiments

In this section, a model of plaque formation on the pig left anterior descending coronary artery (LAD) is simulated numerically using a specific animal data obtained from IVUS and histological recordings. The 3D blood flow is described by the Navier-Stokes equations, together with the continuity equation. Mass transfer within the blood lumen and through the arterial wall is coupled with the blood flow and is modeled by a convection-diffusion equation. The LDL transports in lumen of the vessel and through the vessel tissue (which has a mass consumption term) are coupled by Kedem-Katchalsky equations (Kedem and Katchalsky, 1958, 1961). The inflammatory process is modeled using three additional reaction-diffusion partial differential equations. A full three-dimensional model was created that includes blood flow and LDL concentration, as well as plaque formation. Matching of IVUS and histological animal data is performed using a 3D histological image reconstruction and 3D deformation of elastic body.

We used experimental data from pigs submitted to a high-cholesterol diet for 2 months. Specific software for 3D reconstruction of lumen domain and wall artery (coronary artery) was developed. Matching of histological data and IVUS slices has been applied. A 3D reconstruction was performed from standard IVUS and angiography images. After that, a full three-dimensional finite element analysis was performed using our in-house finite element code to find low and oscillatory WSS zones. The LAD was selected for this analysis. The process of matching with IVUS images was achieved by 2D modeling of tissue deformation for a number of cross sections recorded by histological analysis; those cross sections are deformed until the internal lumen circumferential lengths in IVUS images are reached. Volume of the plaque obtained from histological analysis (after 2 months of high-fat diet for plaque formation) was fitted by employing a nonlinear least-square analysis (Chavent, 2010), to determine material parameters in Section 2.

We examined experimental data obtained for the LAD artery of a pig after 2 months of high-fat diet, to determine material parameters of the computer model. Matching computed plaque location and progression in time with experimental observations demonstrates a potential benefit for future prediction of this vascular disease by using computer simulation.

The results for shear stress distribution for pig are shown in Fig. 1.3. It can be observed that there are low wall shear stress zones <5 dyne/cm^2 that are indicated in Fig. 1.3 in the proximal zones of the coronary arteries which is in good agreement with histological measurement from CNR.

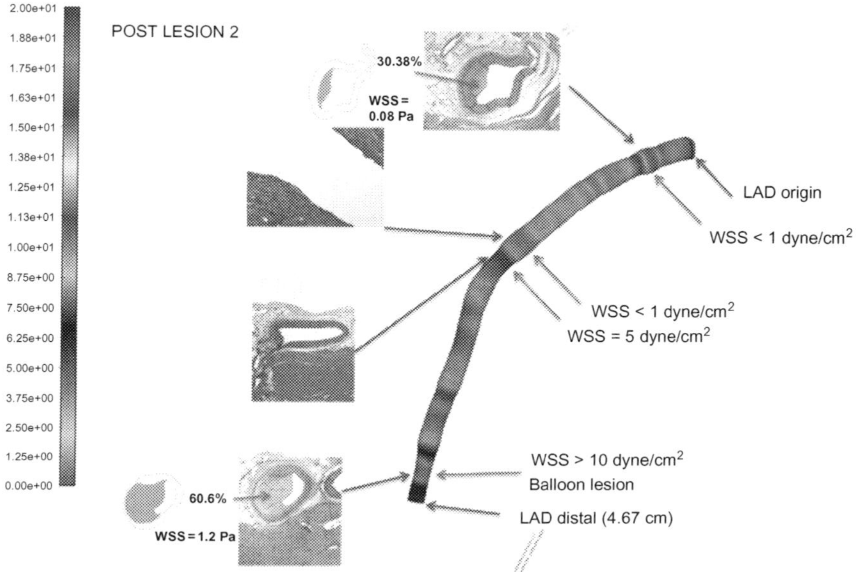

Fig 1.3 Shear stress distribution for pig.

3.2 Comparison experimental and numerical results for pigs data

In this section, the comparison of numerical and experimental results for high-fat diet pig is presented. We tried to match first of all lesion and foam cell (FC) lipids. The initiator for plaque formation is shear stress function.

We choose four characteristic pigs to fit model parameters and compare it with experimental measurement from CNR described in WP4. These data for four pigs are HF6, HF9, HF11, and LIHF2, and their experimental data for flow, pressure, and blood analysis are given in Tables 1.1 and 1.2.

Experimental histological data for these four pigs together with WSS are presented in Table 1.2. We concentrated on the FC lipids, lesion area, CXCR4, and MIF signals.

For a specific pig HF9, we tried to fit lesion area function with WSS obtained from simulation. The function has four parameters (a,b,c,d)

$$\text{Lesion area} = a \cdot \log\left(b + \frac{c}{wss + d}\right) \tag{1.20}$$

The fitted values for a, b, c, d are 100.95, 1.0, 0.00773, and −0.8849, respectively (Fig. 1.4).

The comparison of numerical and experimental data for pig HF9 is presented in Figs. 1.5, 1.6, and 1.7. CXCR4 is considered to be initiator

Table 1.1 Experimental data for boundary conditions, flow, pressure, and blood analysis for pigs HF6, HF8, and HF9

Type	Trivella 06		Trivella 08		Trivella 09	
Position	Distal	Proximal	Distal	Proximal	Distal	Proximal
Date	27/04/2010		11/05/2010		18/05/2010	
Basal flow: min (cm/s)	5	6	3	5	5	3
Basal flow: max (cm/s)	25	30	23	33	21	24
Basal flow: average (cm/s)	14	16	13	17	10	11
Adenosine flow: min (cm/s)	14	5	5	6	8	17
Adenosine flow: max (cm/s)	66	42	24	38	36	53
Adenosine flow: average (cm/s)	30	28	16	20	16	19
Coronary reserve	2.14	1.75	1.23	1.18	1.60	1.73
Pressure: basal (sistolic) (mmHg)	128	137	120	96	121	126
Pressure: basal (diastolic) (mmHg)	106	117	113	91	116	115
Pressure: basal (average) (mmHg)	113	124	115	93	118	119
Pressure: adenosine (sistolic) (mmHg)	129	127	116	106	112	116
Pressure: adenosine (diastolic) (mmHg)	104	100	102	100	107	107
Pressure: adenosine (average) (mmHg)	112	109	107	102	109	110
WBC (*1000 n/μL)	12.6		17.3		17.3	
HCT (%)	29.5		37.1		34.2	
Total cholesterol (mg/dL)	669		522		680	
HDL (mg/dL)	49		25		40	
LDL (mg/dL)	604.8		491		629.2	
TG (mg/dL)	76		32		54	
Albumin (g/dL)	2.95		2.07		2.44	
GOT (IU/L)	114		76		49	
GPT (IU/L)	41		53		58	
GGT (IU/L)	90		81		103	
Apoprotein A-I (mg/dL)	75.1		40.4		67.5	

Table 1.2 Experimental data for boundary conditions, flow, pressure, and blood analysis for pig LIHF2

	Post_Lesion 2	
Position	**Distal**	**Proximal**
Date	**09/07/2010**	
Basal flow: min (cm/s)	7	5
Basal flow: max (cm/s)	40	34
Basal flow: average (cm/s)	20	14
Adenosine flow: min (cm/s)	16	15
Adenosine flow: max (cm/s)	71	67
Adenosine flow: average (cm/s)	29	35
Coronary reserve	1.45	2.50
Pressure: basal (sistolic) (mmHg)	88	89
Pressure: basal (diastolic) (mmHg)	74	82
Pressure: basal (average) (mmHg)	79	84
Pressure: adenosine (sistolic) (mmHg)	87	97
Pressure: adenosine (diastolic) (mmHg)	81	87
Pressure: adenosine (average) (mmHg)	83	90
WBC (*1000 n/μL)	18.2	
HCT (%)	28.7	
Total cholesterol (mg/dL)	288	
HDL (mg/dL)	86	
LDL (mg/dL)	191	
TG (mg/dL)	55	
Albumin (g/dL)	2.58	
GOT (IU/L)	57	
GPT (IU/L)	43	
GGT (IU/L)	79	
Apoprotein A-I (mg/dL)	57.5	

and plaque formation together with WSS distribution. The system of reaction-diffusion Eq. (3.7) (Parodi et al., 2012) is now

$$\begin{aligned} &\partial_t CXCR4 = d_1 \Delta CXCR4 - k_1 CXCR4 \cdot M \\ &\partial_t FC + \mathrm{div}(v_w FC) = d_2 \Delta FC - k_1 CXCR4 \cdot FC \\ &\partial_t MIF = d_3 \Delta MIF - \lambda MIF + k_1 CXCR4 \cdot FC + \gamma(CXCR4 - CXCR4^{thr}) \end{aligned} \tag{1.21}$$

Boundary condition for first equation is WSS function from Fig. 1.4. and MIF signal function for the second equation. For fluid domain, we used around 100,000 eight-node finite elements. Wall domain was modeled with around 80,000 eight-node finite elements. Boundary conditions and fitted

Shear stress [Pa]	Lesion area (mm^2)
2.36	0.40
2.12	0.50
2.39	0.30
2.12	0.10
3.77	0.00
4.31	0.22
5.12	0.20

Fitted coefficients	
a	100.95
b	1.00
c	0.00773
d	−0.8849

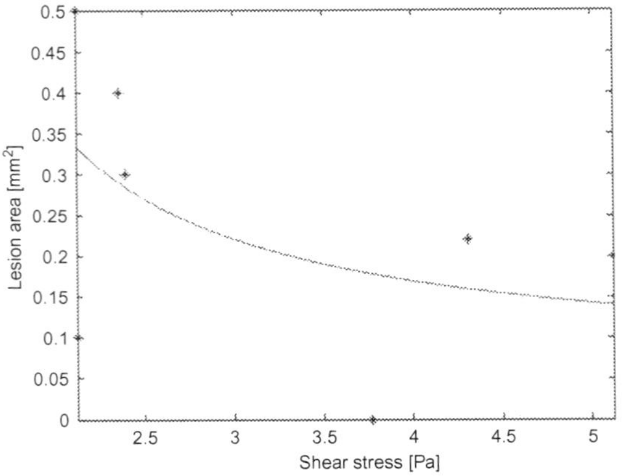

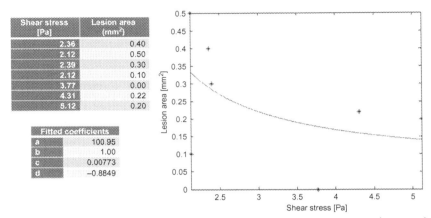

Fig. 1.4 Fitted data for WSS function from lesion area. Lesion area $= a \cdot \log\left(b + \frac{c}{wss + d}\right)$. Four parameters *a, b, c, d* were fitted.

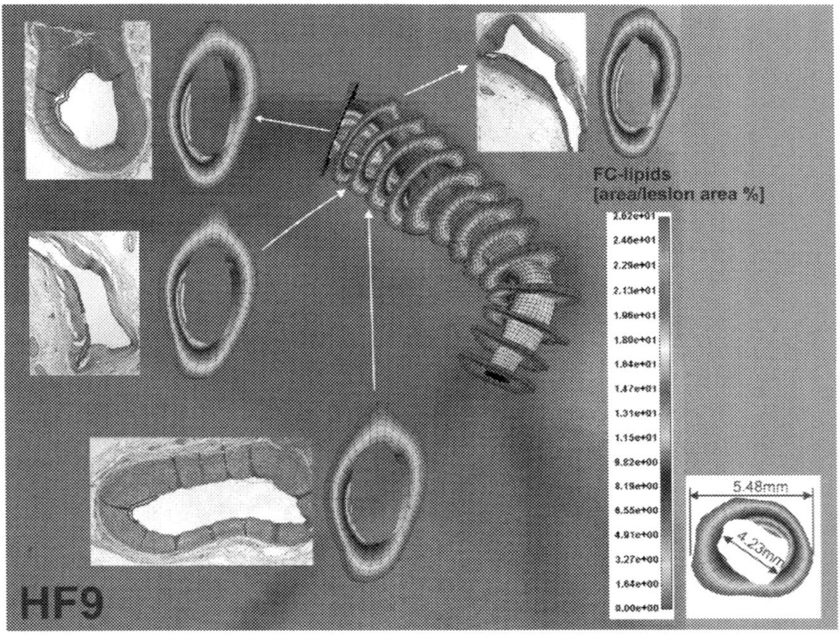

Fig. 1.5 Numerical results for HP9. Foam cell (FC) lipid percentage area inside lesion area 26.2 %. Histological data per some characteristic cross sections and comparison with numerical data. Numerical cross-sectional dimension.

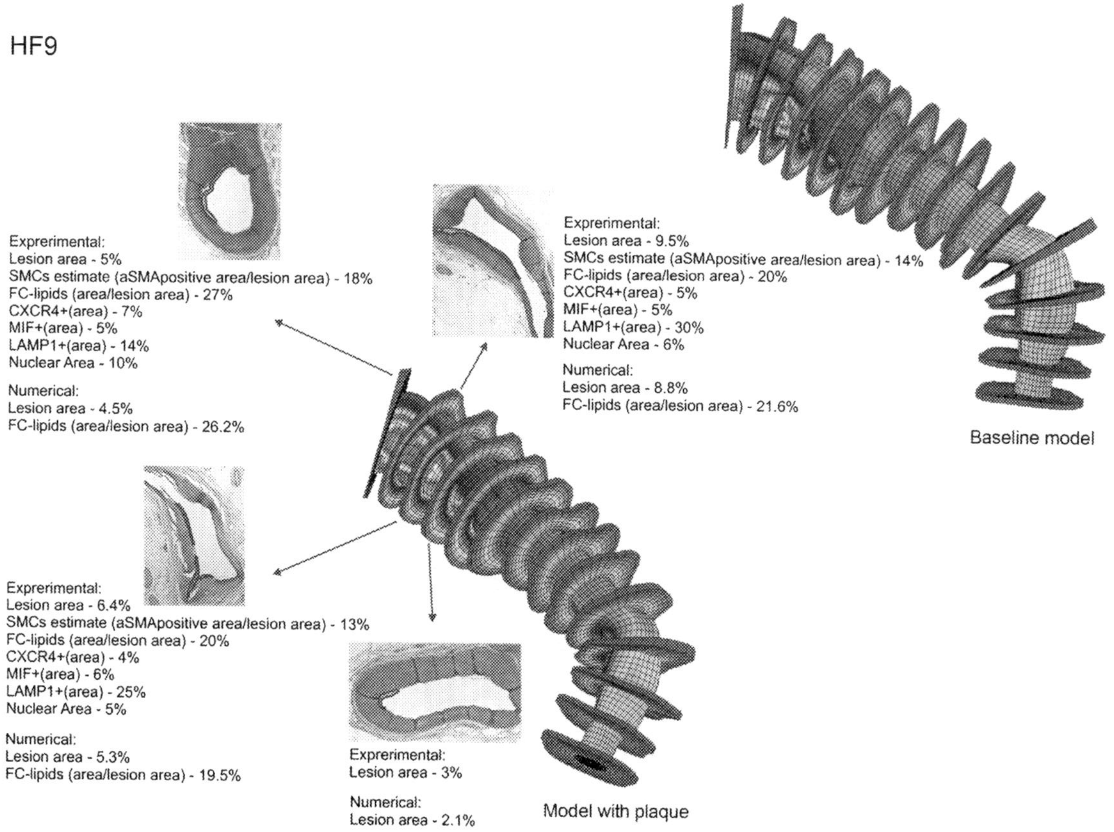

Fig. 1.6 Comparison between experimental and numerical data for pig HF9. Baseline and model with plaque. Cross-sectional presentation of histological and numerical data.

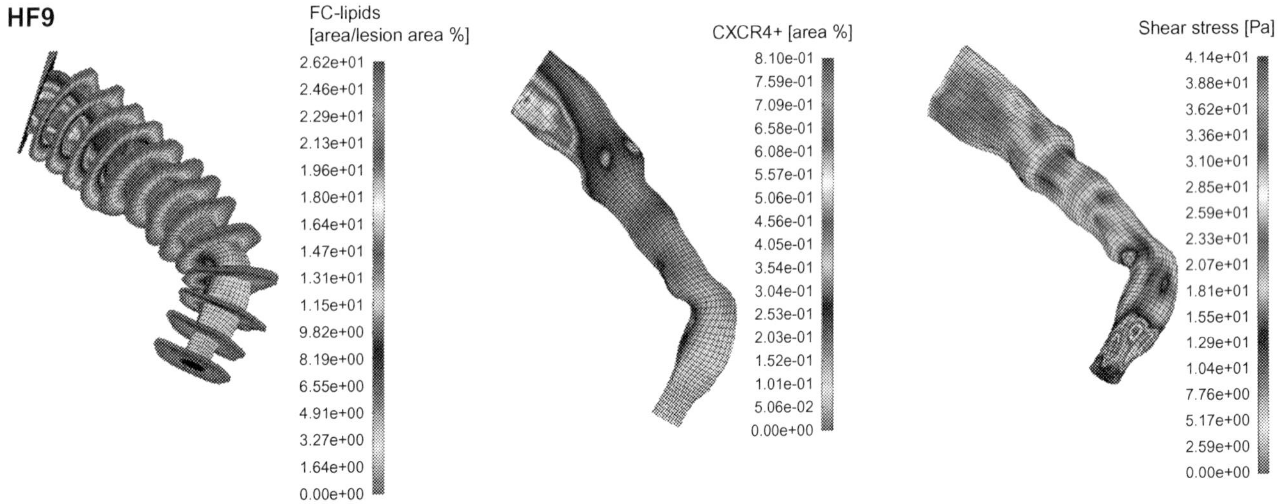

Fig. 1.7 Numerical results for pig HF9. Percentage of the FC lipids inside lesion area. Percentage of CXCR4 inside wall area. Shear stress distribution [Pa].

parameters are given in Table 1.3. ALE formulation was applied for the moving mesh domains. Time domain of 2 months was achieved for each pig.

Numerical and experimental results for FC lipid percentage inside lesion area are presented in Fig. 1.5. It can be observed that good accuracy was achieved for lesion area, FC lipids, and CXCR4.

Comparison of these experimental and numerical data is given in Table 1.4.

3.3 Fitting parameters of ODE model for pigs

In this section, we tried to fit ODE system for plaque formation by using experimental data from two HF pigs HF9 and HF11 and two HHF pigs HHF4 and HHF10. As ODE system is in the time domain, we assumed that different types of plaque lesion correspond to different time even it is on the same pigs.

3.3.1 Hybrid genetic algorithm

We firstly described a hybrid genetic algorithm for fitting parameters of ODE model for oxidized LDL, macrophage, smooth muscle cell, and foam cell concentration evolution in time. The goal of the fitting procedure is to determine parameters of ODE model, so it agrees with experimental data as much as possible. ODE model is given with the following equations:

$$\begin{aligned} C' &= \Phi - \alpha C\phi - \lambda_1 C \\ \phi' &= \beta_0 C\phi(1-\beta_1\phi) - \alpha C\phi \\ \rho' &= \frac{\phi}{1+\phi}\gamma_0\rho(1-\gamma_1\rho) - \varphi_F\rho \\ \varphi_F' &= \alpha C\phi - \lambda_2\varphi_F \end{aligned} \quad 1.22$$

where C, ϕ, ρ, and φ are concentrations of oxidized LDL, macrophages, smooth muscle cells, and foam cells, respectively; β_0, β_1, γ_0, γ_1, α, λ_1, and λ_2 are real positive numbers; and Φ is a wall shear stress function:

$$\Phi(WSS) = \alpha_0 \log\left(1 + \frac{\alpha_1}{WSS + \alpha_2}\right) \quad (1.23)$$

where α_0, α_1, and α_2 are also real positive numbers.

A hybrid genetic algorithm combines the power of the genetic algorithm (GA) with the speed of a local optimizer. The GA excels at gravitating toward the global minimum. It is not especially fast at finding the minimum when in a locally quadratic region. Thus, the GA finds the region of the

Table 1.3 Model parameter values for HF9 pig experiment

Blood domain	$\rho = 1000\,kg/m^3$	$\mu = 0.035$ [P]	$D_l = 1.0 \times 10^{-11}\ m^2/s$	$Vsr = 11\,cm/s$	$Pout = 116\,mmHg$	$C_{LDL} = 629.2\,mg/dL$
Intima			$D_w = 1.0 \times 10^{-11}\ m^2/s$	$r_w = -2.6 \times 10^{-4}$	$Pmed = 100\,mmHg$	
Inflammation	$d_1 = 10^{-8}\ m^2/s$	$d_2 = 10^{-8}\ m^2/s$	$d_3 = 10^{-8}\ m^2/s$	$k_1 = 10^{-4}\ m^3/kg\,s$	$\lambda = 1.2\,s^{-1}$	$\gamma = 1.1e-^{5}\ s^{-1}$
WSS function	$a = 100.95$	$b = 1.0$	$c = 0.0073$	$d = -0.8849$		

Table 1.4 Comparison of experimental and numerical data for pig HF9

	Cross section 0 mm, WSS = 2.37 Pa		Cross section 2 mm, WSS = 2.14 Pa		Cross section 3 mm, WSS = 2.20 Pa		Cross section 5 mm, WSS = 2.15 Pa	
HF9	**Experimental**	**Numerical**	**Experimental**	**Numerical**	**Experimental**	**Numerical**	**Experimental**	**Numerical**
Lesion area [%]	5	4.5	9.5	8.8	6.4	5.3	3	2.1
FC lipids (area/lesion area) [%]	27	26.2	20	21.6	20	19.5	–	11
CXCR4 (area) [%]	7	6.4	5	4.3	4	4.6	–	0.1

optimum, and then the local optimizer takes over to find the minimum. Hybrid GA can take one of the following forms:

1. Running a GA until it slows down, then letting a local optimizer take over. Hopefully the GA is very close to the global minimum.
2. Seeding the GA population with some local minima found from random starting points in the population.
3. After a few iterations, running a local optimizer on the best solution or the best few solutions and adding the resulting chromosomes to the population.

The continuous GA will easily couple to a local optimizer, since local optimizers use continuous variables. Generally, the local optimizer begins its job when the GA shows little improvement after many generations. The local optimizer uses the best chromosome in the population as its starting point. We used Nelder-Mead simplex optimization algorithm as a local optimizer (Nelder and Mead, 1965).

3.3.2 Results for fitting model

For fitting ODE model, we used HF and HHF pig data. Unfortunately, we do not have data about concentration evolution in time for LDL, macrophages (MPH), smooth muscle cell (SMC), and foam cell (FC). Instead, we used different lesion types as different time steps for each pig. Instead of macrophage concentration, we used MIF concentration. Oxidized LDL concentration is not available, so we fitted our ODE model according to MIF, SMC, and FC concentrations. Available data are represented in Tables 1.5 and 1.6 for HF pigs and Tables 1.10 and 1.11 for HHF pigs.

Available data for pigs HF9 and HF11 are given in Tables 1.5 and 1.6.

The best fit minimizes the sum of squared residuals, where residuals represent differences between experimental and calculated concentration of MIF, smooth muscle cells, and foam cells. Function to be minimized is in the following form:

Table 1.5 HF9 pig data

HF9					
Time	**Lesion type**	**WSS**	**MIF**	**SMC**	**FC**
T_0	Intact	3.77	0.001	0.150	0.00
T_1	Type II b	4.61	0.070	0.130	0.20
T_2	Type IIa-III	2.17	0.055	0.135	0.20

Table 1.6 HF11 pig data

		HF11			
Time	**Lesion type**	**WSS**	**MIF**	**SMC**	**FC**
T_0	Intact	5.02	0.001	0.24	0.00
T_1	Type I-II	3	0.100	0.24	0.12
T_2	Type IIb	1.29	0.130	0.22	0.18
T_3	Type IIa advanced	1.865	0.140	0.17	0.20

$$ERROR = \sum_{i=1}^{N_{HF9}} \left(\left(\phi_i^{exp} - \phi_i^{calc}\right)^2 + \left(\rho_i^{exp} - \rho_i^{calc}\right)^2 + \left(\varphi_i^{exp} - \varphi_i^{calc}\right)^2 \right)$$
$$+ \sum_{i=1}^{N_{HF11}} \left(\left(\phi_i^{exp} - \phi_i^{calc}\right)^2 + \left(\rho_i^{exp} - \rho_i^{calc}\right)^2 + \left(\varphi_i^{exp} - \varphi_i^{calc}\right)^2 \right)$$

where N_{HF9} is the number of time steps for pig HF9; N_{HF11} is the number of time steps for pig HF11; ϕ_i^{exp}, ρ_i^{exp}; and φ_i^{exp} are experimental MIF, SMC, and FC concentrations at time step i; and ϕ_i^{calc}, ρ_i^{calc}, and φ_i^{calc} are calculated MIF, SMC, and FC concentrations at time step i.

In this case, we are fitting α_0, α_1, and α_2 (coefficients in $\Phi(WSS)$) that are common for both pigs and $\beta_0, \beta_1, \gamma_0, \gamma_1, \alpha, \lambda_1$,and λ_2 that are separate for pigs HF9 and HF11. That makes total 17 parameters to fit.

For fitting, we used hybrid GA with parameters given in Table 1.7.

Optimization progress (minimization of ERROR function) is depicted in Fig. 1.8.

Figs. 1.9 and 1.10 show final results of fitting procedure.

Table 1.8 contains fitted values of $\alpha_0, \alpha_1, \alpha_2, \beta_0, \beta_1, \gamma_0, \gamma_1, \alpha, \lambda_1$,and $\lambda_2$for pigs HF9 and HF11.

HHF pigs

Available data for pigs HHF4 and HHF10 are given in Tables 1.9 and 1.10.

Table 1.7 Genetic algorithm parameters

Population size	**Number of generations**	lb **(Vector of lower bounds)**	ub **(Vector of upper bounds)**
200	100	[0 0 0 0 0 0 0 0 0 0 0 0 0 0 0 0 0]	[inf inf inf inf inf inf inf inf inf inf inf inf inf inf inf inf inf]

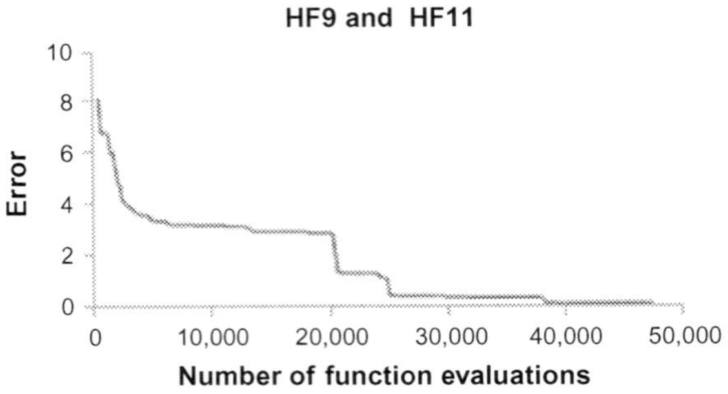

Fig. 1.8 Optimization progress for pigs HF9 and HF11.

The best fit minimizes the sum of squared residuals, where residuals represent differences between experimental and calculated concentration of macrophages, smooth muscle cells, and foam cells. Function to be minimized is in the following form:

$$ERROR = \sum_{i=1}^{N} \left(\left(\phi_i^{exp} - \phi_i^{calc}\right)^2 + \left(\rho_i^{exp} - \rho_i^{calc}\right)^2 + \left(\varphi_i^{exp} - \varphi_i^{calc}\right)^2 \right)$$

where N is the number of time steps; ϕ_i^{exp}, ρ_i^{exp}, and φ_i^{exp} are experimental MIF, SMC, and FC concentrations at time step i; and ϕ_i^{calc}, ρ_i^{calc}, and φ_i^{calc} are calculated MIF, SMC, and FC concentrations at time step i.

In this case, we do not have available WSS data, so we are fitting β_0, β_1, γ_0, γ_1, α, and λ_2 separate for pigs HHF4 and HHF10. We have two different optimization processes, both minimizing the ERROR function given earlier.

We used hybrid GA with parameters given in Table 1.11.

Optimization progress (minimization of ERROR function) for pigs HHF4 and HHF10 is depicted in Fig. 1.11.

Final results of the fitting procedures are represented in Figs. 1.12 and 1.13.

Table 1.12 contains fitted values of α, β_0, β_1, γ_0, γ_1, and λ_2 for pigs HHF4 and HHF10.

3.4 Coupled model of atherosclerosis

Coupled method for modeling of atherosclerosis assumed combination of continuum (finite element method) and discrete (DPD method) described in the previous section (Fig. 1.14). Atherosclerosis development for specific patient at coronary artery was simulated. Upstream of the bifurcation level,

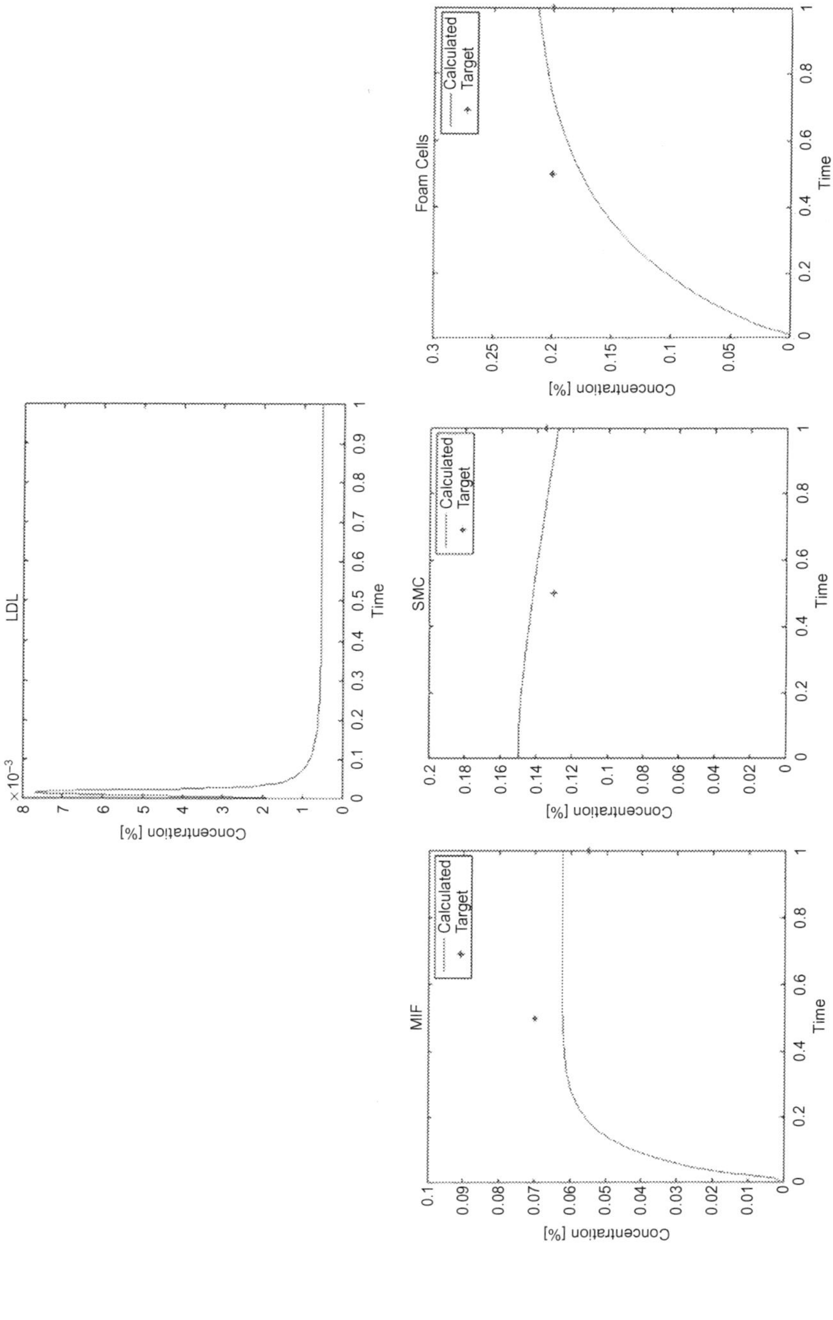

Fig. 1.9 Fitting results for pig HF9.

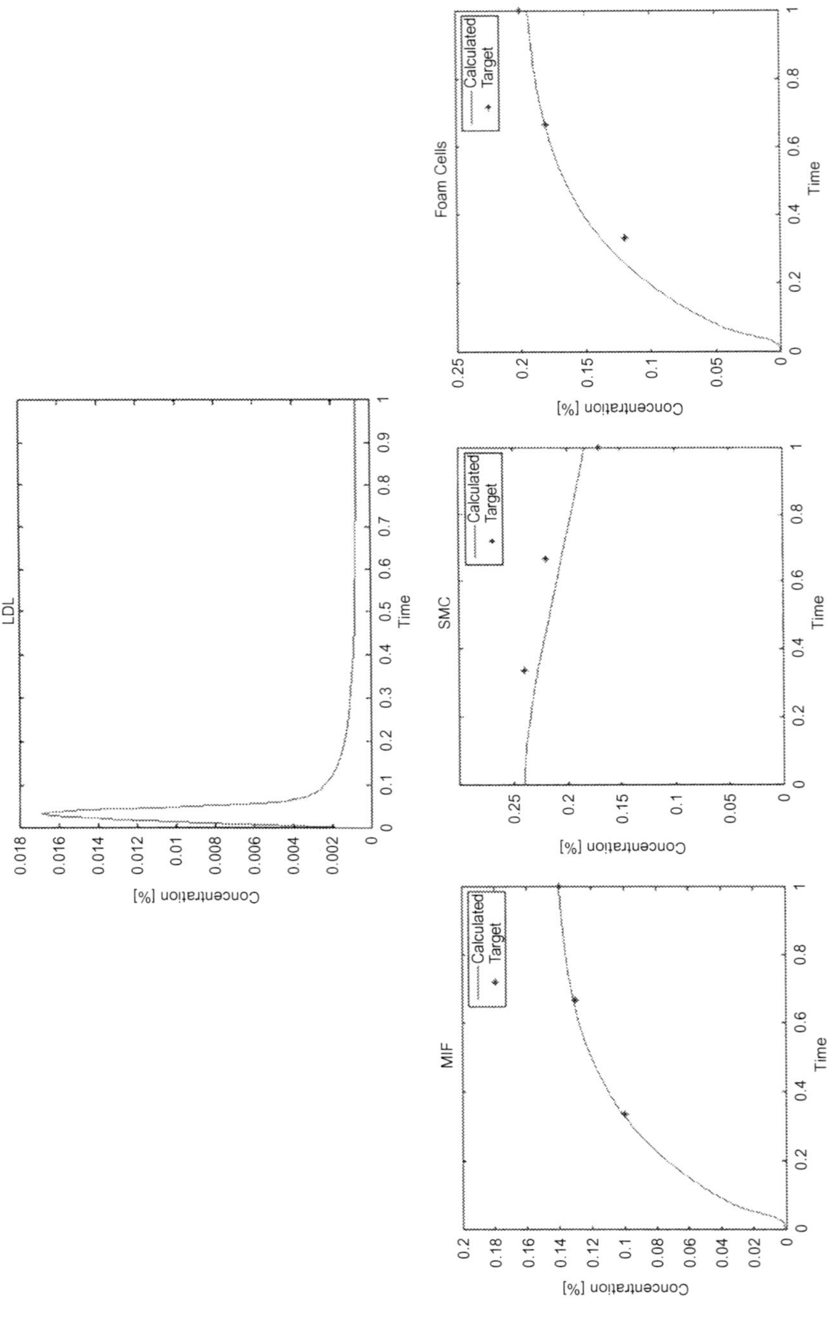

Fig. 1.10 Fitting results for pig HF11.

Table 1.8 Fitted parameter values for pigs HF9 and HF11

	Fitted parameters									
Pig	α_0	α_1	α_2	α	λ_1	β_0	β_1	γ_0	γ_1	λ_2
HF9	2.56	223.61	702.02	20583.7	9.41E−05	40977.5	7.96	0.00407	59.33	3.18
HF11	2.56	223.61	702.02	7151.4	1.83E−09	12415.9	2.94	0.00177	3519.06	3.54

Table 1.9 HHF4 pig data

HHF4				
Time	**Lesion type**	**MIF**	**SMC**	**FC**
T_0	Intact	0.001	0.28	0.00
T_1	Type IIA	0.020	0.28	0.06
T_2	Type III	0.065	0.11	0.18
T_3	Type IV	0.020	0.25	0.10
T_4	Type V	0.020	0.16	0.06

Table 1.10 HHF10 pig data

HHF10				
Time	**Lesion type**	**MIF**	**SMC**	**FC**
T_0	Type I	0.0010	0.3500	0.000
T_1	Type IIA	0.0800	0.3500	0.100
T_2	Type IIB	0.0343	0.1800	0.170
T_3	Type IV	0.0095	0.2750	0.155
T_4	Type V	0.0213	0.2725	0.095

Table 1.11 Genetic algorithm parameters

Population size	Number of generations	lb (Vector of lower bounds)	ub (Vector of upper bounds)
100	120	[0 0 0 0 0 0]	[inf inf inf inf inf inf]

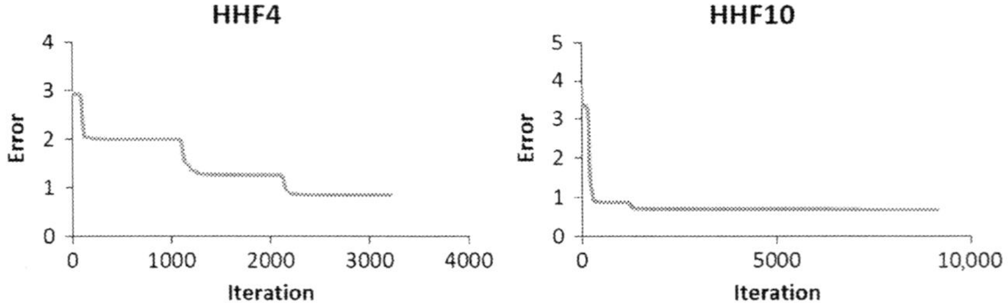

Fig. 1.11 Optimization process for pigs HHF4 and HHF10.

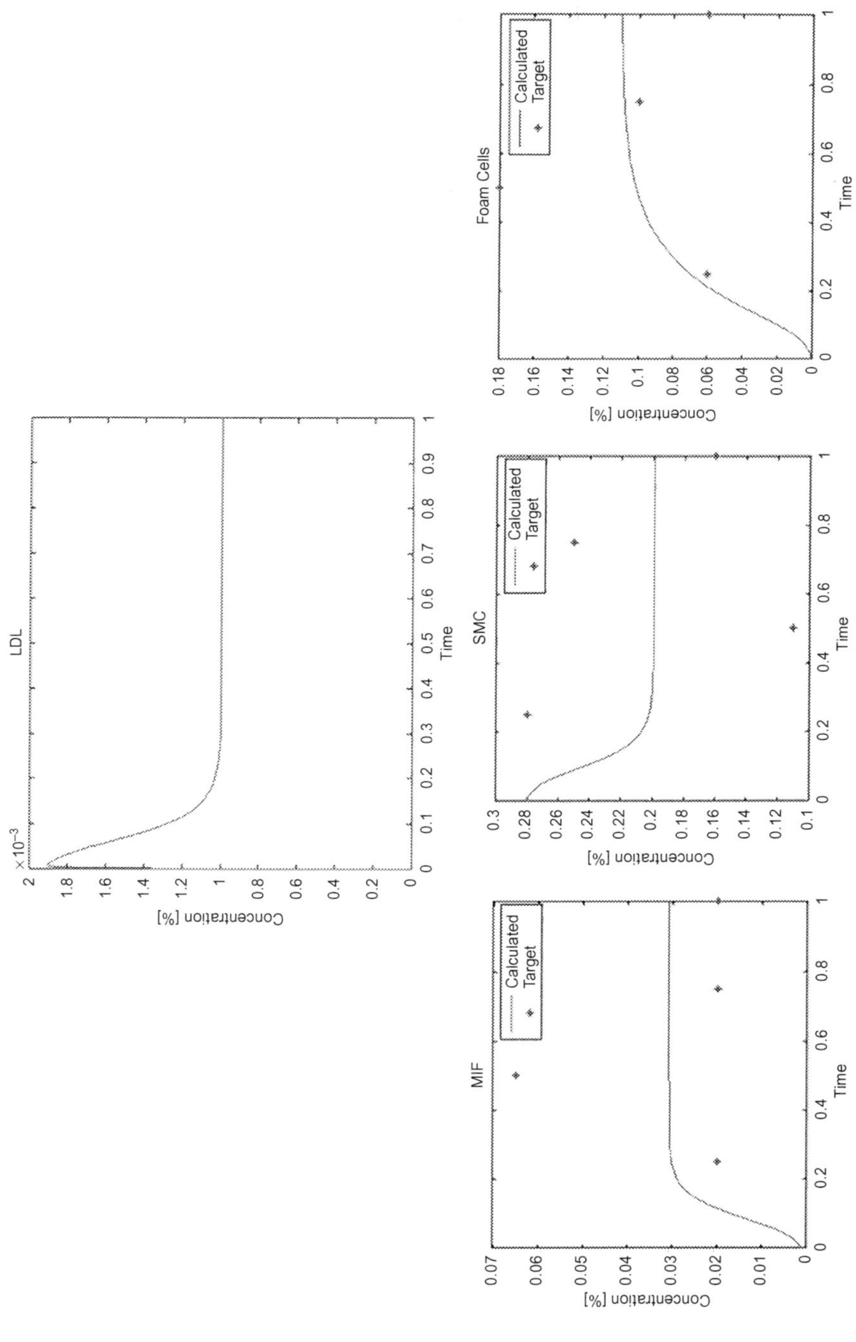

Fig. 1.12 Fitting results for pig HHF4.

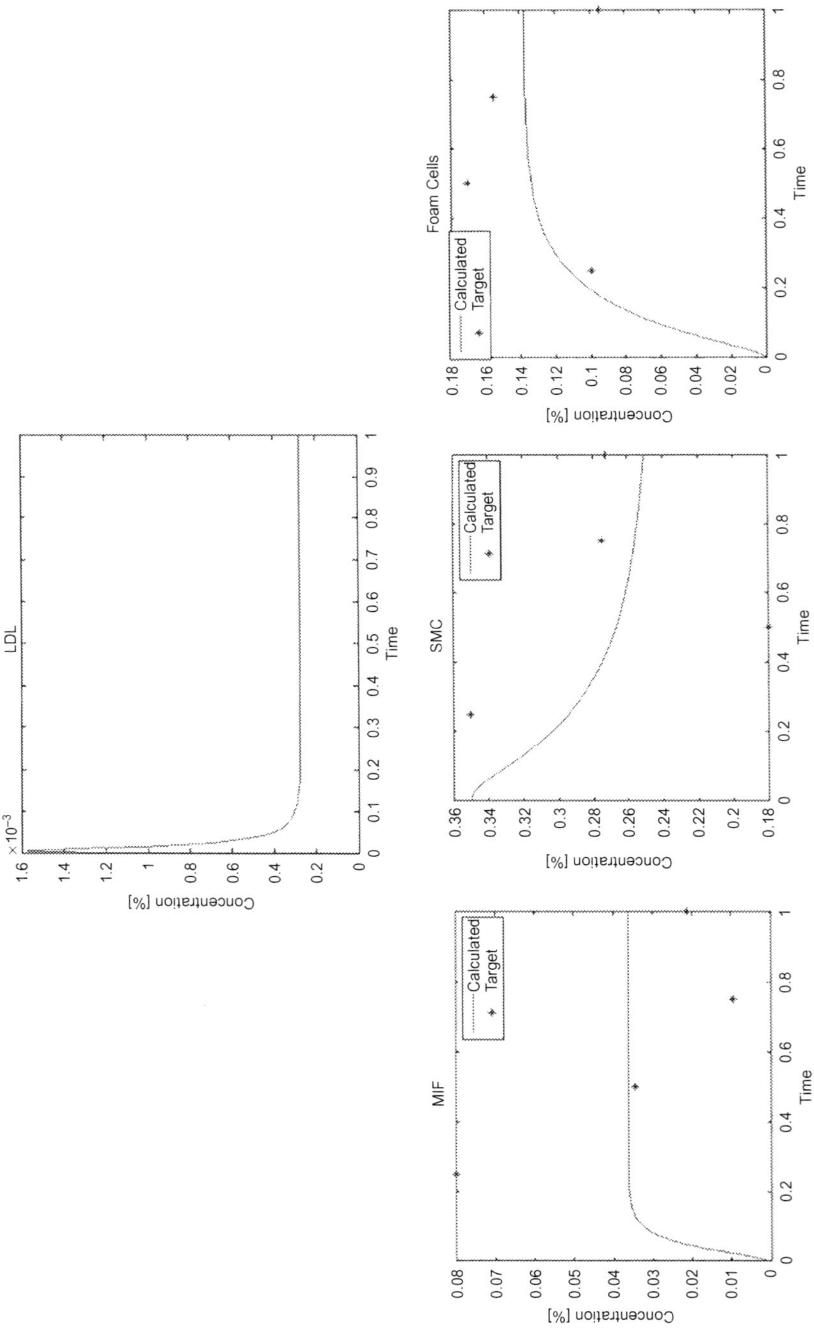

Fig. 1.13 Fitting results for pig HHF10.

Table 1.12 Fitted parameter values for pigs HHF4 and HHF10

	Fitted parameters					
Pig	α	β_0	β_1	γ_0	γ_1	λ_2
HHF4	21947.99	44458.6	16.51	672.73	4.999	6.067
HHF10	102083.19	200602.45	13.542	95.126	3.878	7.421

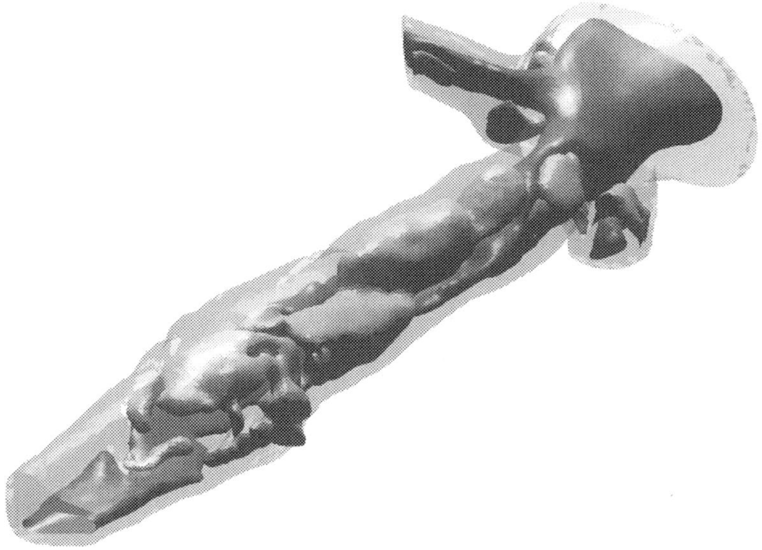

Fig. 1.14 Three-dimensional reconstruction for HHF4 pigs from "ex vivo" MicroCT acquisition. Calcified plaque domain inside the wall domain is denoted with *yellow* color.

there was plaque progression (Fig. 1.15). The results give better insight in the microlevel approach for plaque progression.

3.5 Plaque concentration distribution in the coronary artery

Computer models are generated from DICOM CT medical images. A three-dimensional finite element mesh model was obtained using an automatic segmentation algorithm within the Materialize Mimics 10.01 software. Blood flow through the right coronary arteries was simulated using PAK solver (Kojic et al., 2008). Blood was considered as a Newtonian fluid with a dynamic viscosity of $\mu = 0.00365$ Pas and incompressible with a density of $\rho = 1050\,\text{kg/m}^3$. Pulsatile coronary inlet velocity waveform was used.

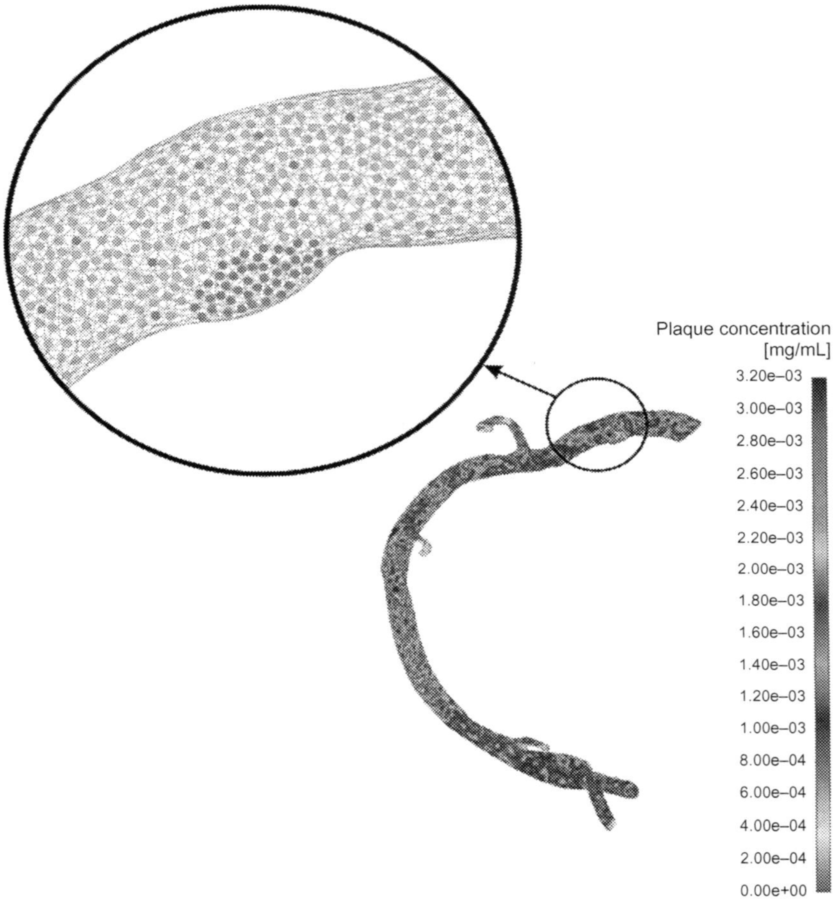

Fig. 1.15 The right coronary artery for specific patient. Coupled simulation finite element and DPD method.

A three-dimensional simulation of blood flow through lumen and plaque progression in the vessel wall was simulated. The biomolecular parameters such as LDL, HDL, and triglycerides used for the computer simulation are presented in Table 1.13.

Patient #01 is a 69-year-old male, which of cardiovascular risks has past smoking and dyslipidemia. From the current therapy receives the statins. Statins, also known as HMG-CoA reeducate inhibitors, are a class of lipid-lowering medications. Statins have been found to reduce cardiovascular disease (CVD) and mortality to those who are at high risk of cardiovascular disease. The evidence is strong that statins are effective for

Table 1.13 Biomolecular parameters and adhesion molecules

		Biomolecular parameters		
		LDL	**HDL**	**Triglycerides**
Patient #01	Baseline	108	62	179
	Follow-up	99.7	73	157
Patient #02	Baseline	114	58	125
	Follow-up	111	64	107

treating CVD in the early stages of the disease (secondary prevention) and to those at elevated risk but without CVD (primary prevention). Shear stress distribution and plaque concentration for patient #01 have been presented in the Fig. 1.16.

Three characteristic sites (red circles) are marked by a doctor. The first location is 4 mm from the beginning of the artery; the second 32 mm and the third 51 mm are located from the inlet cross section. The value of wall shear stress was observed. Low shear stress values are associated with increased plaque concentration, as it has been shown in the Fig. 1.16 (right; red color). In the right panel of Fig. 1.16, a cross section of the wall is shown, to better see the presence of plaque in the wall.

Average values range from 0.32 and 0.19 to 0.79 Pa, measured at cross sections, respectively. The second time moment, follow-up (Fig. 1.17), shows higher values of wall shear stress in the observed zone and higher plaque concentration values that indicating there has been further plaque progression.

Patient #02 is a 69-year-old male, which of cardiovascular risks has hypertension. From the current therapy receives the ACE inhibitors and aspirin. An angiotensin-converting enzyme inhibitor (ACE inhibitor) is a pharmaceutical drug used primarily for the treatment of hypertension (elevated blood pressure) and congestive heart failure. This group of drugs causes relaxation of blood vessels and a decrease in the volume of blood, which leads to lower blood pressure and reduced demand from the heart. Aspirin, also known as acetylsalicylic acid (ASA), not only is a nonsteroidal antiinflammatory drug (NSAID) and works similar to other NSAIDs but also suppresses the normal functioning of platelets. Results of shear stress distribution and plaque concentration for patient #02 are shown on Fig. 1.18.

In the case of this patient, two characteristic sites were observed.

As can be seen from Fig. 1.18, at the inlet of the coronary artery, from 5 mm and from 29 mm, there is a low wall shear stress. The average values

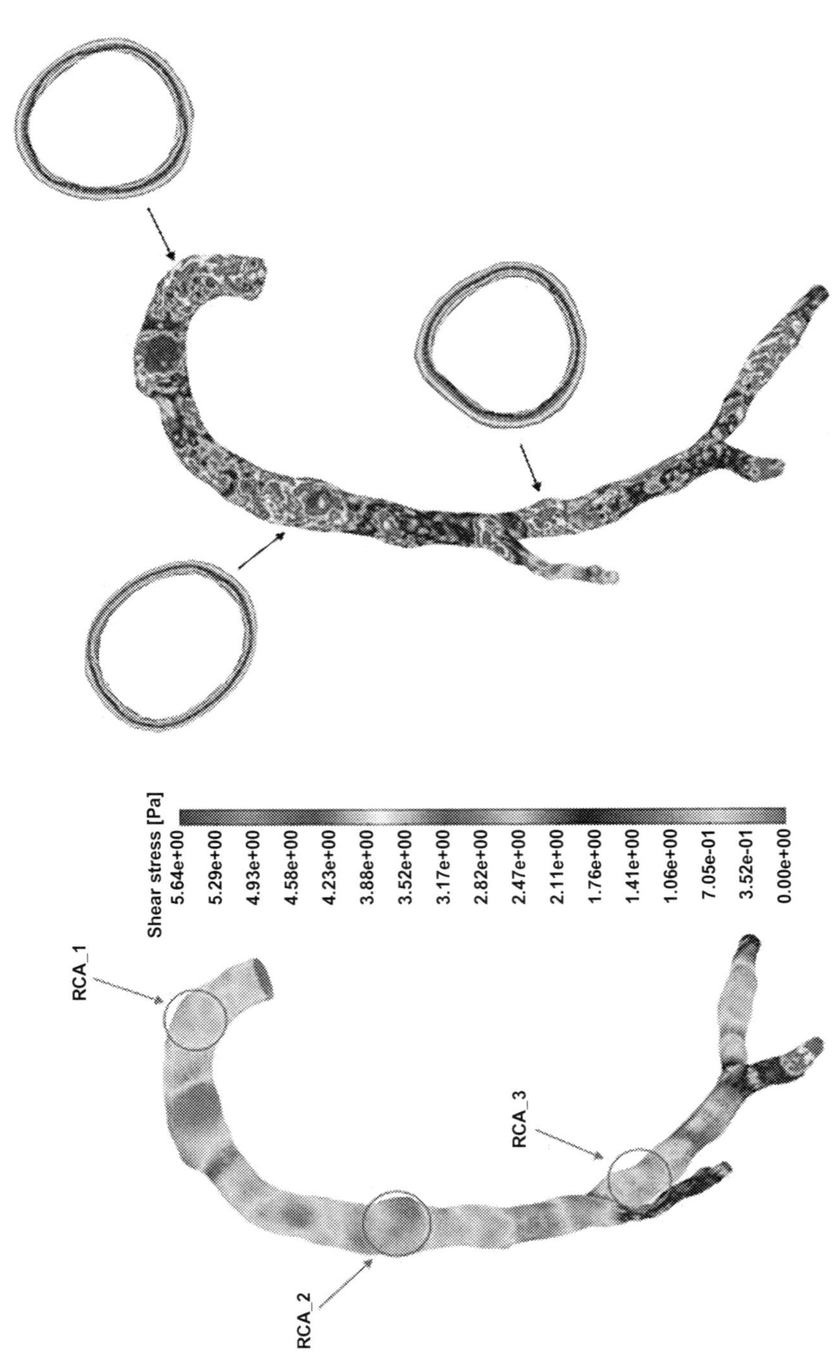

Fig. 1.16 Shear stress distribution and plaque concentration for patient #01 (baseline).

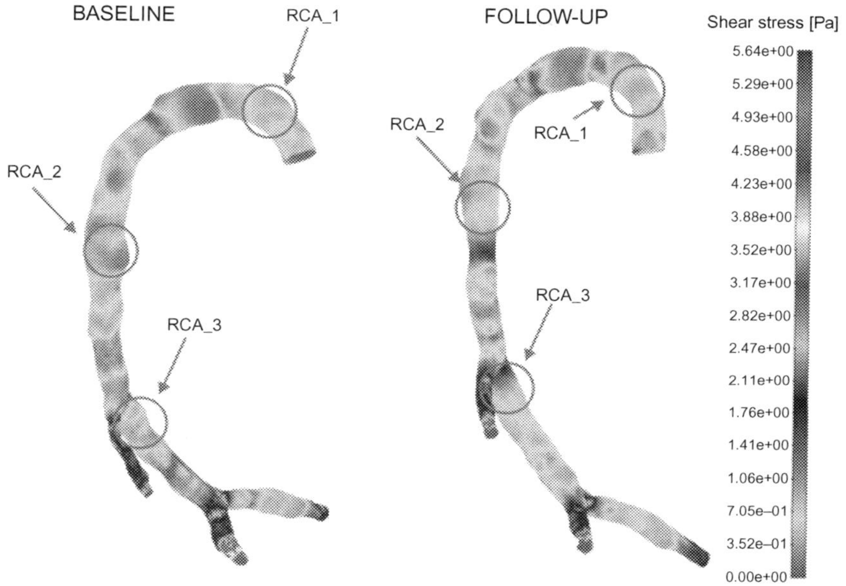

Fig. 1.17 Shear stress distribution for patient #01 (baseline and follow-up).

range from 0.32 to 0.37 Pa, measured at cross sections. At the same time, on the mentioned part of the coronary artery, an increased plaque concentration with a maximum value of 2.2E − 3 mg/mL was observed. The second time moment, follow-up (Fig. 1.19, right), shows lower values of wall shear stress in the observed zone, so it can be concluded that due to further progress of this disease, there was a decreased diameter of the artery and increased plaque concentrations.

The biomolecular parameters such as LDL, HDL, and triglycerides are used for the computer simulation and adhesion molecules ICAM1, VCAM1, and E-selectin (Table 1.14).

The results of first numerical simulation (baseline) are presented in Fig. 1.20.

At a distance of 1–4 mm, from the inlet of the coronary artery, a low value of wall shear stress was observed. Average values range from 0.23 to 0.36 Pa, measured at cross sections, at a distance of 0.5 mm. A low shear stress was also noticed at 15.8 mm distance from the beginning of the artery. Low shear stress values are associated with increased plaque concentration, as shown in the Fig. 1.20 (right). The highest values of plaque concentration in these zones are 1.7E − 3 mg/mL.

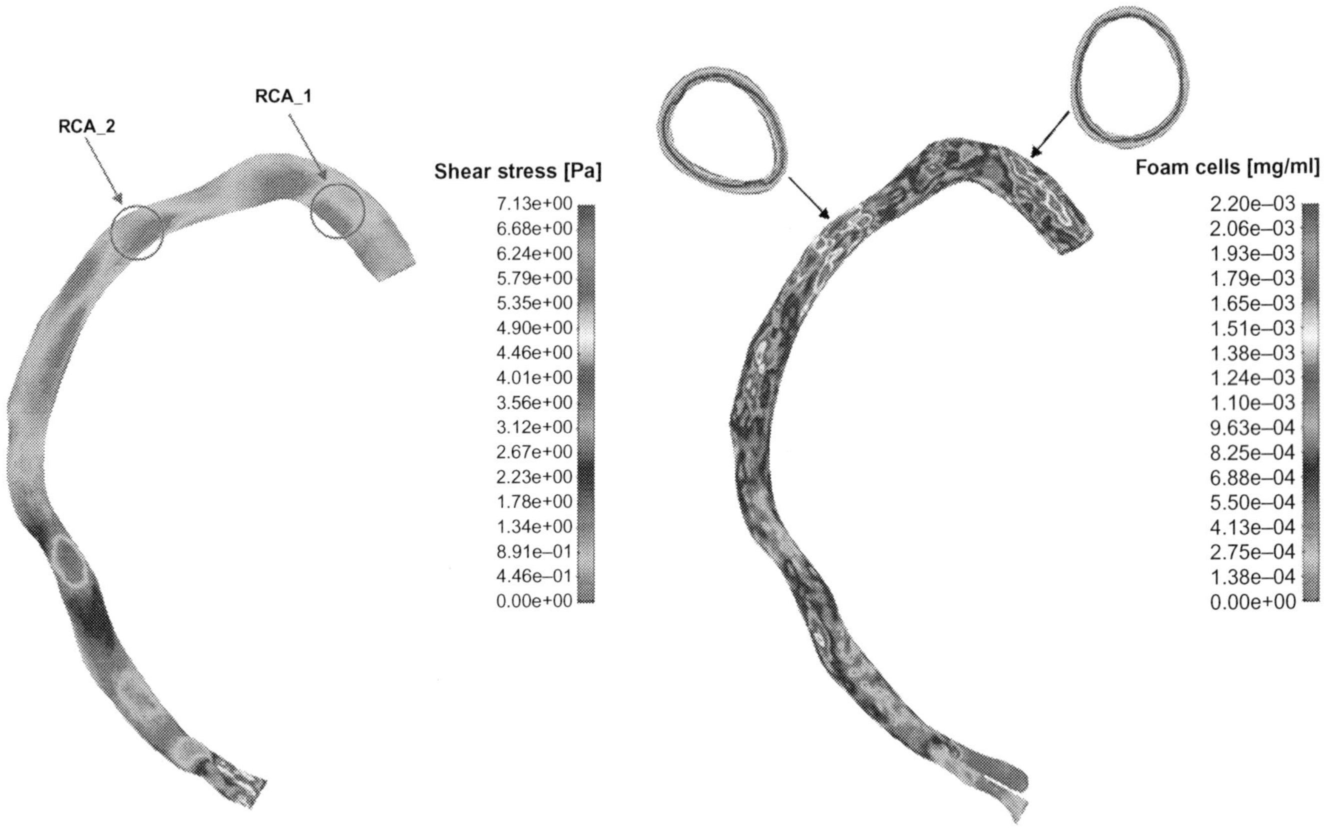

Fig. 1.18 Shear stress distribution and plaque concentration for patient #02 (baseline).

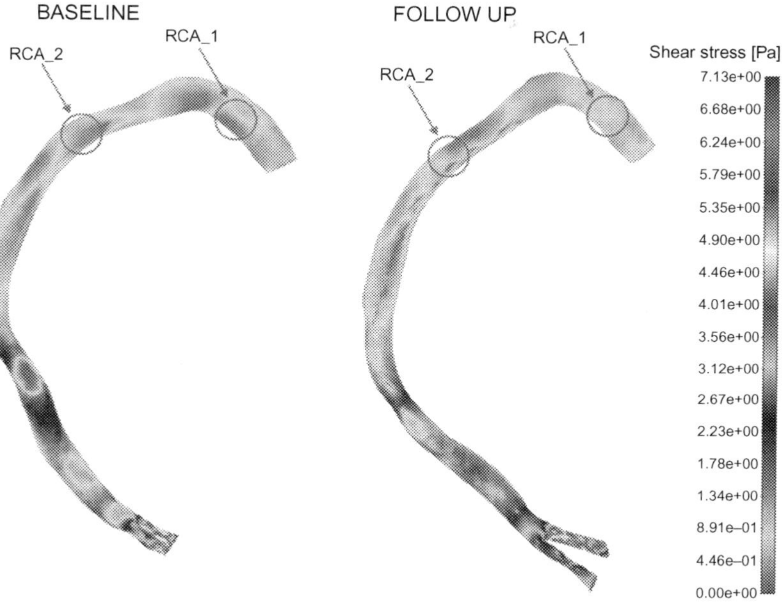

Fig. 1.19 Shear stress distribution for patient #02 (baseline and follow-up).

Table 1.14 The biomolecular parameters and adhesion molecules

Time	LDL	HDL	Triglycerides	ICAM1	VCAM1	E-selectin
Baseline	118	37	172	166.19	557.17	26
Follow-up 1	/	/	/	128.93	483.37	22.62
Follow-up 2	55	57	178	142.23	587.72	/

The result of numerical simulation for the second time—follow-up 1—is presented in the Fig. 1.21.

Further development of plaque concentrations was observed in the areas where low shear stress appears, there is an increase in these values, and the range of these values is from 0.31 to 0.39 Pa. It can be seen that the area of plaque concentration is expanding (Fig 1.21, right).

The third numerical simulation presents the condition of the artery after 6 years (Fig. 1.22). Average values of shear stress in the observed zones of the artery range from 0.43 to 0.47 Pa. Also, it is clear from Fig. 1.22 (right) that there has been a significant increase in the plaque development in the observed zones.

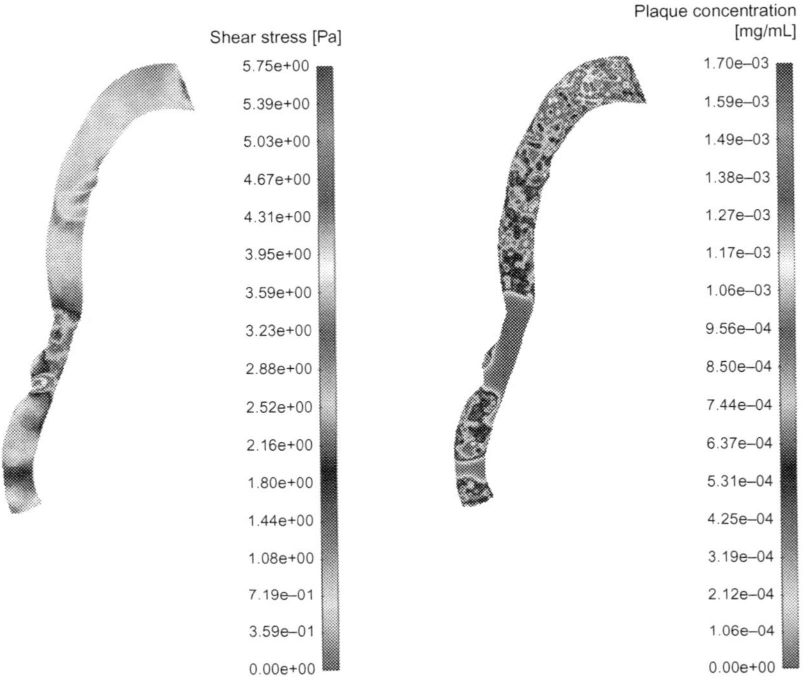

Fig. 1.20 Shear stress distribution (*left*) and plaque concentration (*right*) for baseline.

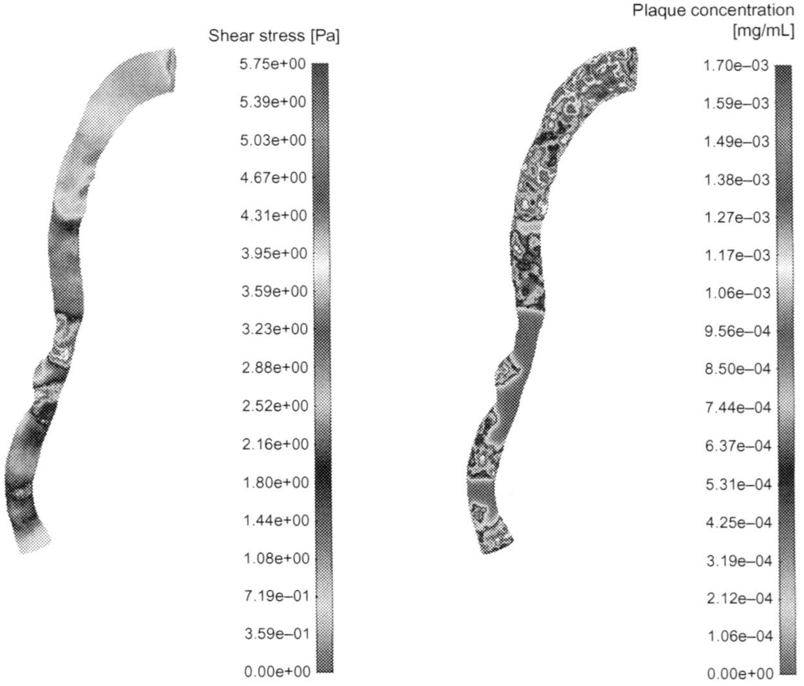

Fig. 1.21 Shear stress distribution (*left*) and plaque concentration (*right*) for follow-up 1.

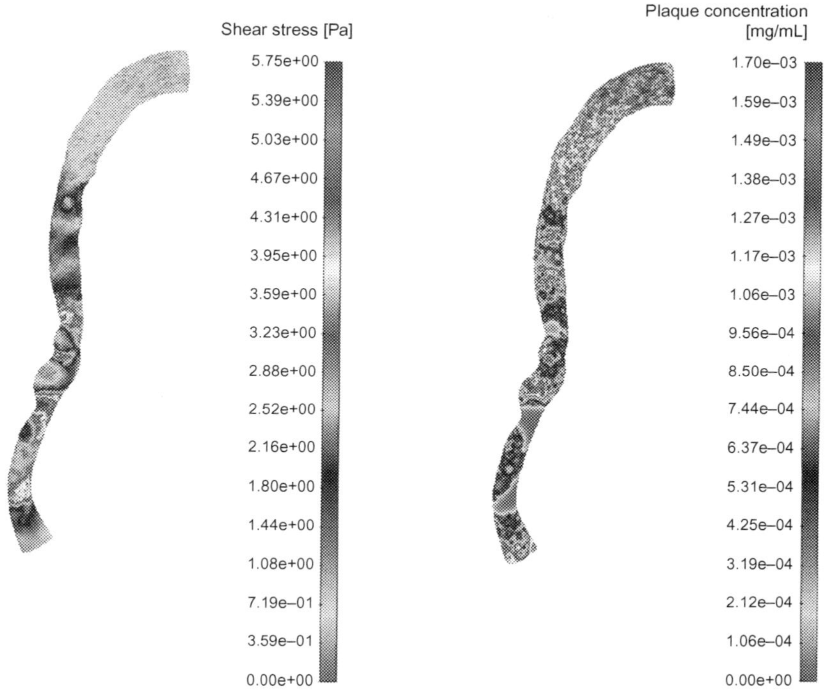

Fig. 1.22 Shear stress distribution (*left*) and plaque concentration (*right*) for follow-up 2.

4 Conclusions

Cardiovascular disease is responsible for an increasing number of mortality in all developed countries. Great attention is focused on studying this disease to reduce the mortality rate. In this study, three-dimensional simulations were investigated to determine hemodynamic parameter such as wall shear stress with computational fluid dynamics in the coronary artery for pig experiment and patient-specific clinical data from computed tomography. Also, plaque concentration in the arterial wall was calculated. We used continuum finite element and coupled discrete DPD approach modeling. Two different times for patients were observed: baseline and follow-up. After first time point, we analyzed the shear stress distribution. Our model incorporates specific patient biomolecular parameters such as LDL, HDL, and triglycerides. The locations with low shear stress are more prominent for further development of atherosclerosis. Then we compared our predictive model with follow-up with particular patients. The results of the performed analyses have shown that sites with lower shear stress values were correlated with the sites of plaque accumulation measurements. Coupled method finite element and DPD give more detailed accumulation of the plaque taken into account discrete nature of the biological process. It is possible to predict the sites of plaque occurrence and concentration in certain places of the artery using computer simulation.

Acknowledgments

Serbian Ministry of Education, Science and Technological Development III41007, ON174028 and EC HORIZON2020 689068 SMARTool project.

References

Begent, N., Born, G.V., 1970. Growth rate in vivo of plate let thrombi, produced by iontophoresis of ADP, as a function of mean blood flow velocity. Nature 227, 926–930.

Boryczko, K., Dzwinel, W., Yuen, D., 2003. Dynamical clustering of red blood cells in capillary vessels. J. Mol. Model. 9, 16–33.

Chavent, G., 2010. Nonlinear Least Squares for Inverse Problems: Theoretical Foundations and Step-by-Step Guide for Applications. second print, Springer, New York.

Filipovic, N., Haber, S., Kojic, M., Tsuda, A., 2008a. Dissipative particle dynamics simulation of flow generated by two rotating concentric cylinders: II. Lateral dissipative and random forces. J. Phys. D. Appl. Phys. 41 (3), 6–16.

Filipovic, N., Isailovic, V., Djukic, T., Ferrari, M., Kojic, M., 2011. Multi-scale modeling of circular and elliptical particles in laminar shear flow. IEEE Trans. Biomed. Eng. https://doi.org/10.1109/TBME.2011.2166264. PMID: 21878403, ISSN 0018-9294.

Filipovic, N., Kojic, M., Ferrari, M., 2010. Dissipative particle dynamics simulation of circular and elliptical particles motion in 2D laminar shear flow. Microfluid. Nanofluid. 10 (5), 1127–1134.

Filipovic, N., Kojic, M., Tsuda, A., 2008b. Modeling thrombosis using dissipative particle dynamics method. Phil. Trans. R. Soc. A 366 (1879), 3265–3279.

Filipovic, N., Mijailovic, S., Tsuda, A., Kojic, M., 2006. An implicit algorithm within the Arbitrary Lagrangian-Eulerian formulation for solving incompressible fluid flow with large boundary motions. Comput. Methods Appl. Mech. Eng. 195, 6347–6361.

Filipovic, N., Rosic, M., Tanaskovic, I., Milosevic, Z., Nikolic, D., Zdravkovic, N., Peulic, A., Kojic, M., Fotiadis, D., Parodi, O., 2012. ARTreat project: three-dimensional numerical simulation of plaque formation and development in the arteries. Inform. Technol. BioMed. 16 (2), 272–278.

Filipovic, N., Teng, Z., Radovic, M., Saveljic, I., Fotiadis, D., Parodi, O., 2013. Computer simulation of three dimensional plaque formation and progression in the carotid artery. Med. Biol. Eng. Comput. 51 (6), 607–616.

Haber, S., Filipovic, N., Kojic, M., Tsuda, A., 2006. Dissipative particle dynamics simulation of flow generated by two rotating concentric cylinders. Part I: boundary conditions. Phys. Rev. E 74, 1–8.

Kedem, O., Katchalsky, A., 1958. Thermodynamic analysis of the permeability of biological membranes to non-electrolytes. Biochim. Biophys. 27, 229–246.

Kedem, O., Katchalsky, A., 1961. A physical interpretation of the phenomenological coefficients of membrane permeability. J. Gen. Physiol. 45, 143–179.

Kojic, M., Filipovic, N., Stojanovic, B., Kojic, N., 2008. Computer Modeling in Bioengineering: Theoretical Background, Examples and Software. John Wiley and Sons, Chichester, England.

Libby, P., 2002. Inflammation in atherosclerosis. Nature 420 (6917), 868–874.

Loscalzo, J., Schafer, A.I., 2003. Thrombosis and Hemorrhage, third ed. Lippincott Williams & Wilkins, Philadelphia.

Meyer, G., Merval, R., Tedgui, A., 1996. Effects of pressure-induced stretch and convection on low-Density Lipoprotein and Albumin uptake in the rabbit aortic wall. Circ. Res. 79, 532–540.

Nanfeng, S., Nigel, W., Alun, H., Simon, T.X., Yun, X., 2006. Fluid-wall modelling of mass transfer in an axisymmetric stenosis: effects of shear-dependent transport properties. Ann. Biomed. Eng. 34, 1119–1128.

Nelder, J., Mead, R., 1965. A simplex method for function minimization. Comput. J. 7 (4), 308–313.

Parodi, O., Exarchos, T., Marraccini, P., Vozzi, F., Milosevic, Z., Nikolic, D., Sakellarios, A., Siogkas, P., Fotiadis, D.I., Filipovic, N., 2012. Patient-specific prediction of coronary plaque growth from CTA angiography: a multiscale model for plaque formation and progression. IEEE Trans. Inf. Technol. Biomed. 16 (5), 952–965.

Prosi, M., Zunino, P., Perktold, K., Quarteroni, A., 2005. Mathematical and numerical models for transfer of low-density lipoproteins through the arterial walls: a new methodology for the model set up with applications to the study of disturbed luminal flow. J. Biomech. 38, 903–917.

Quarteroni, A., Valli, A., 1999. Domain Decomposition Methods for Partial Differential Equations. Oxford University Press.

Quarteroni, A., Veneziani, A., Zunino, P., 2002. Mathematical and numerical modeling of the solute dynamics in blood flow and arterial walls. SIAM J. Numer. Anal. 39, 1488–1511.

Ross, R., 1993. Atherosclerosis: a defense mechanism gone awry. Am. J. Pathol. 143, 987–1002.

Tarbell, J.M., 2003. Mass transport in arteries and the localization of atherosclerosis. Annu. Rev. Biomed. Eng. 5, 79–118.

Taylor, C.A., Hughes, T.J.R., Zarins, C.K., 1998. Finite element modeling of blood flow in arteries. Comp. Methods Appl. Mech. Eng. 158, 155–196.

Zunino, P., 2002, Mathematical and Numerical Modeling of Mass Transfer in the Vascular System, PhD Thesis.

Further reading

Boynard, M., Calvez, V., Hamraoui, A., Meunier, N., Raoult, A., 2009. Mathematical modelling of earliest stage of atherosclerosis. In: COMPDYN 2009—SEECCM 2009, 22–24 June 2009, Island of Rhodes, Greece.

Calvez, V., Ebde, A., Meunier, N., Raoult, A., 2008. Mathematical modelling of the atherosclerotic plaque formation. ESAIM Proc. 28, 1–12.

Caro, C.G., Fitz-Gerald, J.M., Schroter, R.C., 1971. Atheroma and arterial wall shear observation, correlation and proposal of a shear-dependent mass transfer Mechanism for Atherogenesis. Proc. Roy. Soc. London 177, 109–159.

Giannogolou, G.D., Soulis, J.V., Farmakis, T.M., Farmakis, D.M., Louridas, G.E., 2002. Haemodynamic factors and the important role of local low static pressure in coronary wall thickening. Int. J. Cardiol. 86, 27–40.

CHAPTER 2

Machine learning approach for breast cancer prognosis prediction

Contents

1 Introduction

As the leading cause of cancer mortality among women, breast cancer is a global public health problem accounting for 23% of the total number of diagnosed cancer (McGuire et al., 2015). Although scientists revealed that among important risk factors for disease occurrence are aging, menstrual period, genetic factors and family history, late first pregnancy or not having children, exposure to stress, and poor diet, it is still unknown how these factors (in combination or individually) trigger the cells to become cancerous. Generally, we could say that genetic, hormonal, or dietary risk factors are numerous and these factors are often intertwined so it is not possible to isolate the specific role of each one. A lot of emphasis has been put on the

Computational Modeling in Bioengineering and Bioinformatics
https://doi.org/10.1016/B978-0-12-819583-3.00002-3

understanding of how these risk factors cause certain changes in DNA structure and, consequently, its relation to the disease occurrence.

Modern oncological examination, surgical examination, and examinations in the field of molecular biology have changed the way the breast cancer is perceived in the past years. The results impacted the understanding of premalignant processes and induced the new criteria and approaches in disease classification, prognosis, and treatment. However, due to heterogeneity of cancer disease, the area of cancer research is still under constant monitoring. It is well known that malign tumors could be classified into several groups according to invasiveness, histological, and biological characteristics, etc. Tumor biology is specific for every patient, which causes the great variability of mechanisms defined in the process of cancer development and progression. Also, different tumor biology induces different behavior of the disease, and, thus, different treatment regime should be tailored.

During the history of breast surgery, numerous classification systems for optimal cancer treatment choices have been made. One of the widely used classification systems is the so-called TNM (tumor-node-metastasis) system (Leslie et al., 2009). It is a standardized system used to determine the disease stage based on the clinical description of tumor, where the letter T followed by a number from 0 to 4 describes the tumor's size and spread to the skin or chest wall under the breast, the letter N followed by a number from 0 to 3 indicates whether the cancer has spread to lymph nodes near the breast, and the letter M followed by a 0 or 1 indicates whether or not the cancer has spread to distant organs. Recently, molecular profiling classification scheme has started to be used in medical practice (Kittaneh et al., 2013), according to which the breast cancer is considered a very heterogeneous group, divided into subtypes (normal like, luminal A, luminal B, HER2-enriched, low claudin, and basal like). The classification is based on the expression of hormonal and growth factor receptors—estrogen (*ER*), progesterone (*PR*), and human epidermal growth factor receptor 2 (*HER2*)—and on gene expression profiling. These classifications have implications for the choice of therapeutic modalities and consequently the outcome (Koren and Bentires-Alj, 2015; De Abreu et al., 2014).

According to the last revision of TNM classification by the American Joint Committee on Cancer (AJCC), which is effective as of January 1, 2018, the manual brings together all currently available knowledge on staging of cancer at various anatomic sites (Amin et al., 2017). Briefly, the evidence-based TNM system incorporates not only prognostic biomarkers (such as HER2, ER, and PR) and the use of multigene panels in specific situations, but also includes clinical findings and findings from surgery and resected tissue.

Obviously, in line with the scientific and technological progress, advanced imaging techniques and treatments evolve and constantly move the survival period in positive direction. Clinicians now have a wide choice of therapy modalities, ranging from breast-conserving surgery to mutilating surgical procedures followed by aggressive chemotherapy and radiotherapy. On the other hand, current oncology protocols include relatively small number of parameters and therefore inevitably lead to generalizations, so that overly aggressive treatment and underestimation of disease continue to be a common problem. Although aggressive treatments are considered to reduce the spread of breast cancer by decreasing the distant metastases, studies have shown that 70% of patients receiving these therapies would have survived without them (Sun et al., 2007). For every breast cancer death prevented, three women would be unnecessarily overtreated (Lee et al., 2015). Also, one of the major clinical problems of breast cancer is the recurrence of therapeutically resistant disseminated disease. Despite the progress made, the apparent high morbidity and mortality of breast cancer patients require new approaches to the treatment of this disease.

Modern oncology defined a personalized medicine approach and individualization of therapy as the ultimate goal (Toss and Cristofanilli, 2015; Amin et al., 2017). Clinician's aim is not only to predict the emergence of newly diagnosed disease but also to predict response to therapy. Reliable prognostic estimations remain one of the basic issues in treatment decision-making. It is sufficient not only to simply identify patients who, because of their excellent prognosis, should be spared any adjuvant treatment-induced toxicity but also to identify those high-risk patients who should be given the opportunity of very aggressive regime (De Laurentiis et al., 1999). The ability to predict disease outcomes more accurately would allow physicians to make informed decisions on the extent of surgical procedure and potential necessity of adjuvant treatment (Park et al., 2013).

It is very difficult for clinicians, even for the most skilled ones, to objectively assess the impact of a large number of parameters that may be important in the development of diseases, to analyze them with objectivity and make adequate decisions on the choice of therapy. Computer-aided systems based on different artificial intelligence techniques for classification can help experts substantially (Kourou et al., 2015; Cruz and Wishart, 2007). However, the use of these models in clinical practice is strongly based on proving the reliability of predictions and demonstration of acquired knowledge.

In this chapter, we propose a work which aim is to discuss the predictive ability of machine learning (ML) techniques to estimate the survival period and relapse of breast cancer based on real patient data. Various parameters ranging from clinical, histological, imaging, and molecular domains were selected based on literature according to greatest significance in the diagnosis and monitoring of patients with breast cancer (Kittaneh et al., 2013; Toss and Cristofanilli, 2015; Gail et al., 1989). Our hypotheses are that the current oncological protocols are too rigid to approximate and neglect both characteristics of the individual patients and their tumor biology and that knowledge about patient-specific manifestation of disease could be sought in the appropriate combination of these parameters. Specifically, two types of problems are examined (5-year survival period and the recurrence of disease within this period) with the objective to highlight the ability of practical use of predictive models including the development of minimum data sets.

2 Machine learning applications for prediction of breast cancer prognosis

Usage of prognostic calculators aimed at classifying patients into groups with increased risk of recurrence and metastases is not new. One of the first predictive systems that have been introduced in clinical practice has been the so-called Gail model, and it was used to predict the risk of developing an invasive breast cancer. It has been confirmed in several studies (Jacobi et al., 2009; Gail et al., 1989; Mealiffe et al., 2010; Costantino et al., 1999). Gail model is designed as an interactive tool that takes into account five factors including current age, menarche, and previous breast biopsy, also taking into account the tissue atypia in these biopsies, age at first childbirth, and family history of breast cancer, but only among first-degree relatives (mother, sisters, and daughters). Revised Gail model is used at the University of Texas Southwestern Medical Center in Dallas, and it is called the NSABP model 2 (Owens et al., 2011). Gail model is handled with only five parameters, and one of the major limitations is underestimating risk in 50% of families with cancer in the paternal lineage (Evans and Howell, 2007). Therefore, the need for more precise calculators that would process more data became evident (Hippisley-Cox and Coupland, 2015).

Growing trend of application of machine learning techniques for the prediction of cancer prognosis is mainly due to the fact that large databases composed of complex data from various domains could be efficiently explored in short time and key features could be detected with remarkable degree of accuracy (Kourou et al., 2015). Also, the relationships among features are often nonlinear, which forms major difficulty for experts to make correct decisions or limitation for the employment of classical statistical analysis.

In the area of breast cancer, three main tasks where ML techniques could be efficiently employed are (1) risk assessment for disease occurrence, (2) the prediction of disease recurrence, and (3) the prediction of survival period for patients who suffer from this cancer. In this chapter, we focus only on the last two tasks, since they could be explored only when the disease is diagnosed. Although all the tasks could be perceived from both classification and regression aspect, we designed the prediction on survivability as a classification problem that predicts whether the patient is going to survive a specified period of time and the prediction of disease recurrence as a classification that predicts whether the primary tumor is going to be relapsed.

Predictions of 5-year survival and recurrence for breast cancer patients are very important in clinical practice. These two parameters are one of the most significant outputs of the disease outcome and success of treatment. But having the system that provides the likelihood that the patient will (not) survive for 5 years or that he/she will suffer the recurrence of cancer enables the clinicians with possibility to evaluate and compare expected benefits or risks of specific therapies in the initial stage of treatment. This could significantly improve patients' quality of life and cost-effectiveness of treatment. Moreover, classification of a patient into low- or high-risk group may also be useful for communication among physicians. A key requirement for this type of application of a prediction model is that it could be proved that the models are validated in representative cohort of real patients' data; in other words, that predictions are reliable (Steyerberg, 2008).

The literature shows that a variety of prediction models for breast cancer prognosis and databases have been developed and utilized. Park et al. (2013) used SEER database (SEER, 2010) to compare artificial neural networks (ANN) (accuracy $=0.65 \pm 0.02$), support vector machines (SVM) (accuracy $=0.51 \pm 0.01$), and semisupervised learning (SSL) (accuracy $=0.71 \pm 0.02$) with the aim to predict 5-year survival for breast cancer patients. Kim et al. (2012) developed models for breast cancer recurrence comparing SVM (Accuracy $=0.846$) and ANN (Accuracy $=0.812$) classifiers using the novel database of the Breast Cancer Center of a Korean tertiary teaching hospital. Maglogiannis et al. (2009) developed an intelligent system for breast cancer diagnosis and prognosis using SVM-based classifier, but additionally, Naïve Bayes (NB) and

ANN models were trained and tested using WPBC database (Newman et al., 1998). Using the American College of Surgeons' Patient Care Evaluation (PCE) data set and SEER breast carcinoma data set, Burke et al. (1997) concluded that ANN predictions of 5-year survival rate were significantly more accurate than those of the TNM staging system (AccuracyTNM = 0.720, AccuracyANN = 0.770 and ACTNM = 0.692, ACANN = 0.730, for PCE and SEER databases respectively). Additionally, Štrumbelj et al. (2010) presented a detailed study on using standard machine learning approaches for the prediction of breast cancer recurrence. They used the database provided by the Institute of Oncology, Ljubljana, and achieved results comparable with those of oncologists.

The quality of databases that forms the basic component of the predictive models is very important. With the advent of genetics and proteomics, many data mining research has grown in this direction (Paik et al., 2004; Chu and Wang, 2005; Alexe et al., 2006; Parker et al., 2009; Cuzick et al., 2011; Bilal et al., 2013; Martin et al., 2016). However, although several gene signatures are found to have significant power to address prognosis outcome, their biological meaning is not sufficiently described and should be explored in more detail regarding molecular cancer subgroups (Desmedt et al., 2008). Also, these studies are quite expensive and are not available to smaller health centers. On the other hand, it has been shown that the level of the most important protein and genomic expression in tumor and peritumoral tissue (MMP-9, VEGF-A, HIF-1, CXCL-12, and iNOS) depends on several clinico-histological features and biomolecular markers (ER, PR, and HER2) (Wu et al., 2012; Li et al., 2017). Furthermore, studies that analyzed multiple clinical, biological, and histological factors improved the understanding of the biology of the cancer and thus can contribute to control cancer progression (Rakha et al., 2010).

3 Methodological framework for machine learning techniques

Machine learning, as a branch of artificial intelligence, represents the set of methods that have the ability to learn from data, that is, to detect hidden patterns in databases with the aim to use generated knowledge to predict new outputs of the system (Kononenko and Kukar, 2007). Various mathematical models are implemented to approximate complex relationships in data for descriptive or predictive aims. Therefore, ML is a very useful approach for various domains where collections of relevant real data could not be explored in deterministic way due to the presence of uncertainty or noise and nonlinear dependences among features.

As we have stated in the previous section, various ML methods are applied for the development of predictive models for breast cancer prognosis—NB, DT, SVM, ANN, etc. These algorithms belong to the group of supervised ML methods. Namely, in most cases, we desire to train ML algorithms on labeled data sets to find best hypothesis that map the input data to the wanted output. The other group of predictive algorithms that could also be efficiently employed belongs to the group of semisupervised ML methods which characteristics are the ability to deal with data sets where some examples are labeled and others are not. Methods from this group are applied when the database is composed of greater number of unlabeled samples. And, finally, the third large group of ML algorithms is the group of unsupervised ML methods, where all samples in the database are unlabeled and without any notion of output values. More specifically, the unsupervised algorithms are designed to find the patterns among data and distinguish samples according to the discovered groups of input data. However, the most common tasks in predictions of prognosis related to breast cancer are classification and regression, with the emphasis on classification, so the role of supervised ML algorithms in the development of predictive systems is dominant.

There is a standardized process to perform classification—Knowledge Discovery in Databases (KDD) (Maimon and Rokach, 2005). The main purpose is to adequately organize data so that ML methods could be applied in a proper manner and to evaluate the developed model that could lead to its further deployment phase. The most important steps for the development of ML models are presented in Fig. 2.1.

In the following sections, we will describe a methodological framework for ML that is followed in the study presented by Andjelkovic-Cirkovic et al. (2015) and which results will be specifically discussed. The paper describes a research work which aim was to perform ML techniques to examine a real patients' database that was created in a comprehensive manner within the 5-year study. Also, some of the issues that represent common problems in the process of development of predictive models will be noted and discussed.

3.1 Preprocessing

A series of preprocessing steps are envisaged in data preparation phase to ensure clean data and to reduce the number of variables (features).

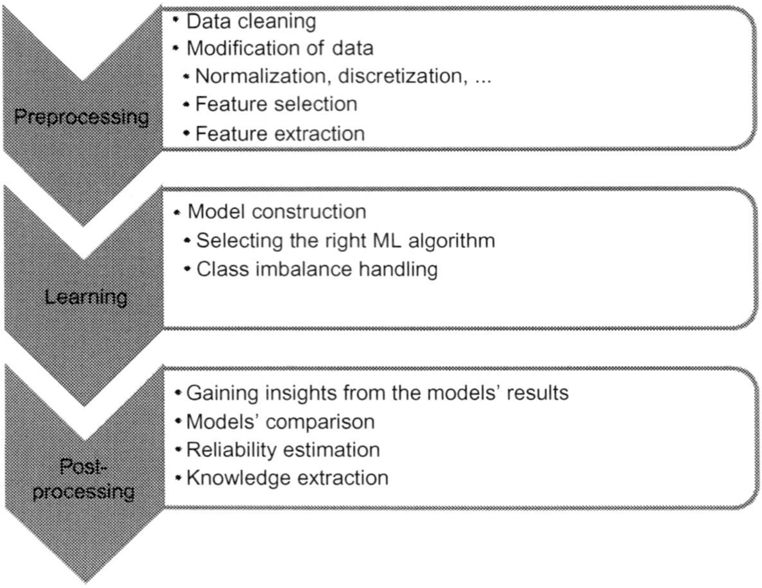

Fig. 2.1 Three main steps in the process of ML model development.

Specifically, variables with high percentage of missing values should be omitted from further analysis, while the missing (empty) fields of variables with low percentage of missing values could be treated with some of widely proposed procedures. The standard way to deal with this issue is to replace the missing values with the feature means (for the numerical variables) and modes (for nominal variables).

One of the characteristic situations for medical databases is the presence of multivalued discrete features since the variables are discretized by clinicians based on how they use it in everyday medical practice. This is mainly due to the bilateral nature of the disease, which means that cancer can occur in both breasts. For patients with such diagnosis, it is not uncommon for the disease at different sites to be of different histological type and subtype, etc. This problem could be solved by applying binarization technique (Kononenko and Kukar, 2007). This method increases the overall number of input features and could be briefly described in the following manner: for each value of the original feature A, a new binary feature is generated by joining all other values into one. For example, the feature is localization of tumor—left or right breast. If the tumor has bilateral nature, the attribute (feature) will have two different values in one field, which is the problem for

ML algorithms. The binarization technique resolves this issue by creating two attributes related to the localization of tumor in comparison with the first case—localization_left and localization_right. Both features are binary, which means that for the mentioned case, both features will have values 1. In any other case, one of them will have value 1 and other 0 and vice versa.

Improvement of data quality is without any doubt the base step in preprocessing. Some other methods belong to a so-called strategy approach that assumes the transformation or modification of input data to better fit a particular ML algorithm.

3.2 Imbalanced data sets

Considering the actual estimates about 5-year survival and disease recurrence rate among women with breast cancer (Brewster et al., 2008), imbalanced distribution of patients for both classification problems is expected. Generally, events with lower frequency could negatively affect the ML models, and its misclassifying could lead to unfavorable decisions and results in heavy costs. Thus, for model evaluation, performance metrics that takes class distribution into account should be adopted (this will be described in the next subsection). Many ML approaches are developed to cope with this problem (Haixiang et al., 2017). These approaches are mainly based on resampling techniques for data transformation, cost-sensitive learning based on algorithm modifications, and ensemble methods. According to Haixiang et al. (2017), resampling methods are the most common approach for dealing with imbalanced data. SMOTE algorithm (Chawla et al., 2002) is shown as a good choice when data set consists of a small number of minority instances. The method is based on k-NN approach and oversamples the instances from minority class by creating synthetic minority class examples similar to the already available ones. Although the main effect is to identify more specific decision regions in the feature space for the minority class, without spreading into majority class region, the question is whether these artificially created samples can really represent real cases.

3.3 Feature selection

After data organization, feature vectors can be fed directly as an input to the classification algorithms, or the feature selection (FS) procedures can be employed to maintain the feature subset that contributes most to the classification accuracy and facilitates the classification task. Moreover, the removal of the irrelevant features could significantly reduce the risk of minority class samples to become discarded in the case of imbalanced classification problems (Yijing et al., 2016). Feature selection techniques can be broadly categorized into three subcategories: the filter, wrapper, and embedded methods. The embedded and wrapper methods incorporate a machine learning algorithm in the process of the selection of attributes. The difference is that embedded methods learn which features best contribute to the accuracy of the model while the model is being created, while wrapper methods consider the feature subset selection as a search problem and evaluate and compare combinations by estimation of trained classifier, mostly by estimation of accuracy. In the filter approach, feature selection is done independently of the learning algorithm of a classifier. The filter algorithms use statistical and information theory measures to rank the individual features according to their importance to a target class. Therefore, these algorithms may also be used to evaluate the significance of individual features and perform comparison between them.

Each filter method assesses the features using different measures and algorithms; thus, they may produce different rankings. Some research works stated that better results may be achieved by combining different rankings (Prati, 2012; Bolón-Canedo et al., 2015).

The standard way in ML development phase is to use FS methods within the cross-validation procedure, where FS should be applied only on training set. In that way, test data will never be statistically evaluated before the final evaluation of classifier's performances and biased results will be avoided.

In our study, we tested three filter algorithms and wrapper approach for feature selection. Among filter methods, we selected the mRMR (Peng et al., 2005), ReliefF (Robnik-Sikonja and Kononenko, 1997), and information gain (Cover and Thomas, 1991) according to their different mathematics in the background and frequency of their application in literature:

1. mRMR (minimum redundancy-maximum relevance) is a feature selection algorithm that tends to select the subset of features having the highest relevance with the output and the least correlation among the features themselves.
2. ReliefF algorithm was firstly developed by Kira and Rendell (1992) and improved by Kononenko (1994) and Robnik-Sikonja and Kononenko (1997). The algorithm estimates the usefulness of features according to their different/similar values for neighbor's instances belonging to different/same classes. The algorithm is able to correctly estimate the quality of attributes in problems with strong dependencies between the attributes (Robnik-Sikonja and Kononenko, 2003).

3. Information gain (IG) algorithm considers a single feature at a time. This filter splits a set of instances into disjoint subsets according to the values of the attribute and measures the gain in information entropy of that attribute.
4. The wrapper algorithm (Kohavi and John, 1997) was combined with a best-first search algorithm (Russell and Norvig, 2003; Korf, 1999) and IWSS algorithm (Bermejo et al., 2009) following forward selection. The evaluation of feature subset was achieved by performing estimation of classifier's AUC parameters within the inner 10-fold cross-validation procedure.

In the case of wrapper approach, different search approaches based on heuristics could be applied with the aim to explore the feature subsets in a comprehensive manner. Besides a best-first search algorithm, the most popular algorithms are hill climbing, beam search, genetic algorithms (GA), etc. Many other search algorithms were developed recently, and each of them could be discussed from positive and negative perspective.

IWSS algorithm is designed with the purpose to reduce computational costs. It firstly employs a filter measure to obtain a ranking of the attribute relevance with respect to the class, and then, a greedy algorithm is used to run over the ranking by adding those variables that are relevant to the classification process. The relevance of including a new variable is measured in a wrapper way. On the other hand, best-first search algorithm performs higher number of evaluations of feature subset, and, according to empirical results in many research papers, this algorithm is positioned on the top of the list of the most sophisticated search algorithms.

3.4 Classification models and evaluation of performances

Data preprocessing step is followed by learning task. Here, we briefly describe the most commonly used ML algorithms for breast cancer prognosis:

1. Naive Bayes (NB) is the classification algorithm based on Bayes's theorem that often performs excellently in real problems, especially in medicine (Kononenko, 1993).
2. Logistic regression (LR) is used exclusively for two-class problems. It uses maximum likelihood estimation to evaluate the probability of class membership.
3. Decision tree (DT), based on C4.5 algorithm, selects the best features for the root node using the concept of information gain ratio and then builds its subtrees in a recursive manner.

4. Random forest (RF) is an ensemble classifier that consists of many decision trees and outputs the class using the "voting" principle. Not only the bootstrap aggregating technique is applied to provide different training subsets to decision trees with the goal of reducing the variance, but also the tree learning algorithm is modified to select a random subset of features for each splitting.
5. Support vector machine (SVM) belongs to the group of the most successful algorithms for binary classification. In contrast to other algorithms, SVM tends to use all available features, even if they are not of much importance. The method places class-separating hyperplanes in the original or transformed feature space, and the new sample is labeled with the class label that maximizes decision function—the distance between support vectors (examples of different classes closest to the hyperplane).
6. Artificial neural network (ANN)—a multilayer perception with a single hidden layer—is most suitable for the purposes of binary classification and survival prediction (Zhang, 2000). Inspired by the human brain, the algorithm consists of a number of highly interconnected nodes organized into layers making the architecture for information processing capable of identifying complex and nonlinear patterns between input data and output variables.

Within the process of development of classification models, it is desirable to try more ML techniques to find the one with the best predictive value (Robnik-Sikonja et al., 2018). A good classification model should provide minimal training (resubstitution error) and minimal generalization error, that is, fit the training set well and classify test samples with the minimal possible error. The predictive performance of the model is computed according to the generalization ability measured on test set, so it is of vital importance to provide a comprehensive estimation of classifier's performances. When the data set contains a vast number of samples, the simplest way is to use a large training data set for model development and large test set for its evaluation. Just one thing in this situation should be taken into consideration—to provide the quality of training and testing data sets. This could be achieved by ensuring that both data cohorts contain the same proportion of classes.

In medical practice the situation is more complicated, especially for the evaluation of disease progress or any other prognostic task. In most cases, data sets are composed of limited number of patients' records with complete prognostic information. One of the most popular approaches to cope with this issue is cross validation. This statistical technique randomly split data samples into n disjunctive folds, where n is previously defined. Then, in n-iteration procedure steps, $n-1$ folds are used for training, and the remaining one is used for testing. Final classifier's performance is the average value of performances achieved in every step of iteration. It is important to enable equal proportion of classes in every fold, and such cross-validation procedure is called stratified. Additionally, the variance of classifier's performance could further be reduced by repeating the cross validation with different random sampling of data. The standard way of predicting the error rate of a learning technique is to use stratified 10-fold cross validation.

Accuracy, sensitivity, specificity, precision, kappa statistic, and the area under the ROC curve (AUC) are standard metrics used to measure the classifiers' performances. These values are calculated on the basis of the entries of the confusion matrix that contains information about the actual and predicted classification (Table 2.1).

Accuracy (AC) is the proportion of the total number of predictions that were correct. It is determined using (1), where TP and TN present the number of positive and negative instances that are correctly classified while FN and FP represent the number of misclassifications of positive and negative instances, respectively. Sensitivity (SENS) is defined as the proportion of true positives that are correctly identified by the classifier (2). It represents an ability to correctly detect patients who do have a condition. Specificity is a measure of how accurate a test is against false positives (3). Precision (PREC), as the ratio of predicted positive examples that are really positive, reflects the percentage of patients who actually have the disease among all tested positive (4). The AUC is an effective and combined measure of sensitivity and specificity that describes the inherent validity of prognostic tests (Majnik and Bosnić, 2013). The area measures discrimination, that is, the ability of the test to correctly perform classification. The Kappa statistic denotes the agreement between the predicted results obtained by the model and the actual values calculated according to Eq. (2.5), where p_o is the percentage of the observed agreement between the predictions and actual values

Table 2.1 Confusion matrix

		Predicted	
		Positive	**Negative**
Actual	Positive	TP	FN
	Negative	FP	TN

and p_e the percentage of chance agreement between the predictions and actual values. Although accuracy is the most general measure that evaluates the overall correctness of the model, in the imbalanced models, it is biased toward the majority class. In this case, other measures give better evaluations of models, and they are used for model comparison:

$$AC = \frac{(TP + TN)}{(TP + FN + FP + TN)} \tag{2.1}$$

$$SENS = \frac{TP}{(TP + FN)} \tag{2.2}$$

$$SPEC = \frac{TN}{(TN + FP)} \tag{2.3}$$

$$PREC = \frac{TP}{(TP + FN)} \tag{2.4}$$

$$kappa = \frac{p_o - p_e}{1 - p_e} \tag{2.5}$$

Through the process of designing, training, and testing the classifiers, it is important to investigate the potential role of developed models to improve the accuracy of prognosis. The learning ability of classifiers is determined by its parameters, and, thus, the task of designing includes identification of the optimal architecture and parameters of classifiers. Many classifiers provide similar results for a large number of different parameter settings making the search space difficult to explore (Reif et al., 2012). The optimization of classifier's parameters should be performed in controlled manner to avoid overfitting.

It is of particular interest to optimize ANN and SVM classifiers due to the fact that success of these algorithms is based on the combination of large number of parameters. The optimal ANN configuration could be set automatically by using the GA as presented in Vukicevic et al. (2016). Specifically, this paper provides the approach to automatically select the optimal classifier's parameters simultaneously with the selection of best feature subset. Although the method is very computationally expensive, it has been proved that it outperforms other standard classification and feature selection algorithms, which is especially valuable for the exploration of real data set.

It should be mentioned that the only regular way to perform parameter tuning is within the process of prediction model development. Especially, within the process of cross validation, the standard is to apply this procedure within inner cross validation and only on training data.

4 To what extent can we trust a prediction model?

The evaluation of model accuracy or other model measures provides us with the estimates of the average model performances based on training/validation data set. But how much the average accuracy of a model could convince us about the accuracy of computed predictions for unseen data? Often, users of ML models need some additional sophisticated parameter that can provide the information of reliability of computed prediction accuracy. An estimation of reliability for classification predictions could be very helpful in fields with risk-sensitive decisions such as medicine. Based on this parameter, clinicians can decide to what extent to trust a particular prediction of decision support system. Thus, extending the prediction framework with the confidence estimates of the single prediction is very valuable and empowers the prediction model with clinical usability.

Two interesting papers focused on this idea are Costa et al. (2013) and Pevec et al. (2011). The former represented method for estimating certainty in DT individual classifications, while the latter provided the overview of several methods developed for the estimation of reliability of ANN individual classification. Both research papers relied on similar procedures provided by Kukar and Kononenko (2002).

Three reliability estimates proposed by Pevec et al. (2011) for individual classifications are described in the following:

1. CNK (reliability estimation using local modeling of prediction error)—approach based on the nearest neighbors' labels. If developed prediction model predicts *K* as class probability distribution for unlabeled example *X*, the estimate CNK (CNeighbors—*K*), for example, *X*, is defined as the average distance between the predictions based on *k*-nearest neighbors and the example's prediction *K* (Eq. 2.6):

$$CNK = 1 - \frac{\sum_{i=1}^{n} \|C_i, K\|}{k} \tag{2.6}$$

   where *Ci* is the true label of the *i*th nearest neighbor, $i = 1,\ldots, k$.
2. LCV (local cross-validation reliability estimation)—approach based on the nearest neighbors' labels and leave-one-out cross validation (LOOCV). For unlabeled example *X*, in the first step, the algorithm forms the subspace of *k*-nearest neighbors. Then, following the principles of LOOCV, *k* local models are generated, each of them excluding one of the *k*-nearest neighbors, and leave-one-out predictions K_i, $i = 1,\ldots, k$ are computed for each of the nearest neighbors. Also, for

each of the nearest neighbors, true labels *Ci*, $i = 1, \ldots, k$ are known, and absolute local leave-one-out prediction errors $E_i = \|(C_i, K_i)\|$ can be calculated. The LCV estimation is then calculated using Eq. (2.7):

$$LCV = \frac{1 - \sum_{i=1}^{k} E_i}{k} \tag{2.7}$$

3. DENS (density-based reliability estimator)—approach based on the density of problem space around the unlabeled example for which we want to know prediction. The density estimate *p(X)* for unlabeled example *X* was calculated by using the Parzen windows with the Gaussian kernels, which were applied on distances between unseen example and all of the examples from learning data set (Eq. 2.8):

$$p(X) = \frac{\sum_{i=1}^{n} K(\|X, X_i\|)}{n} \tag{2.8}$$

where $\|.,.\|$ denotes a Euclidean distance function between two samples, *K* denotes Gaussian kernel function, and X_i, $1 = 1, \ldots, n$ denotes the learning samples. In that way, the problem of computing the multidimensional Gaussian kernel was reduced to computing the two-dimensional kernel by using a distance function.

The assumptions are that the error is lower for predictions that are made in denser training problem subspaces and higher for predictions that are made in sparser subspaces. Therefore, the reliability estimate is given by (Eq. 2.9):

$$DENS = \max_{x_i, i=1,\ldots,n} (p(X_i) - p(X)) \tag{2.9}$$

These reliable estimates belong to the group of model-independent approaches since they do not focus on classifier's parameter influence on the reliability value, but only on learning set. For the calculation of distances between probability distributions in CNK and LCV, Hellinger's distance is used.

The reliability values and prediction accuracy (predicted probability for the class that the example actually belongs to) can be computed for all examples within the cross-validation procedure. In other words, simultaneously with the testing, the classifier on test data, reliability values, and prediction accuracy for every example from test set can be calculated. The expectation is that a positive correlation between these two metrics could be found in case of reliable predictions. So, any correlation coefficients (e.g., Spearman's correlation coefficients) between these two can be computed and discussed in regard to significance level.

5 Findings

Researching the literature regarding the application of ML techniques for the prediction of breast cancer survival period and recurrence of primary tumor, we found a plethora of studies that demonstrate successful approach. Somehow, direct comparison cannot be completely performed since most studies use different databases or have specific objectives. General conclusion is that there is a constant tendency in scientific literature to reach a new highly specific and sensitive parameter that would be a marker of cancer progress.

Here, we will see the results of ML application on data set uniquely designed to include a vast number of prognostic and predictive factors that are available in oncological practice. Although all these features are well studied, only few of them are actively used in clinics according to standardized classification schemes. The aim was to develop intelligent predictive model able to capture links between vast number of features that previously eluded researchers and that will provide promising results in comparison with scientific literature.

5.1 Data source

The database we analyzed was designed as a combination of clinical and laboratory examination for the research project purposes (Andjelkovic-Cirkovic et al., 2015). Pathological and consequent immunohistochemically analyses are part of routine praxis after surgical intervention. Thus, the preoperative and postoperative data including the follow-up data were collected within the 5-year study. Numerous anamnestic, clinical, pathohistological, and biological parameters such as PH, age, menstrual status, histological type, histological grade, nuclear grade, mitotic index, vascular invasion, perineural invasion, lymphatic invasion, HER-2 status, estrogen receptor, progesterone receptor, stage, quadrant, and lymph nodes were assembled with the aim to identify the most important ones that characterize the patients' profiles regarding the survival period and recurrence of breast cancer disease. The patient population with complete prognostic information consisted of 146 women operated on breast cancer and 80 features that also included indicators about recurrence of disease locally, metastasis, and survival. The type of features contained in the data set varied from binary and discrete to continuous value.

The general overview about the distribution of patients according to 5-year survival rate and relapse of disease is presented in Table 2.2.

5.2 Prognosis on survivability

Due to imbalanced nature of data set, the classifiers (described in Section 3) demonstrated the higher ability to detect patients from major class in comparison with the minority samples (specificity is greater than 0.9, while sensitivity is lower than 0.7). Besides these two parameters, we proposed AUC value as relevant measure for comparing the classifiers (Chawla et al., 2004).

Additionally, FS significantly contributed to the improvement of classifiers' performances. In Table 2.2, we summarized the best achieved results for all classification algorithms after employment FS procedure, SMOTE, and procedure for tuning the classifiers (ANN and SVM) within the stratified 10-fold cross-validation procedure. For ANN algorithm, we optimized number of hidden nodes and momentum constant and for SVM parameters C and kernel (polynomial, RBF), as well as degree of kernel (1 or 2) and gamma according to previously selected kernel.

From Table 2.3, we concluded that reducing the numbers of features notably impacted the improvement of NB classifier. Actually, in all experiment settings, NB achieved the AUC value greater than 0.9 that qualifies it as the classifier able to perform the best differentiation between samples of different classes. All other classifiers achieved lower values of AUC in all settings. Although LR, ANN, and SVM demonstrated the better SENS values, taking into account all other parameters, we selected NB classifier as the best option for this problem. Moreover, NB classifier has inherent ability of explaining its decisions, which is the main reason for using it in medical practice.

Regarding the FS methods, mRMR algorithm demonstrated the best results (NB_mRMR with 14 selected features: AC=0.897, SENS=0.711, SPEC=0.963, PREC=0.871, kappa=0.716, and AUC=0.924). mRMR

Table 2.2 Distribution of patients according to 5-year survival period and relapse of disease

The distribution of class values	Class1	Class2	Number of attributes
Five-year survival period (Class1, more than 5 years; Class2, less than 5 years)	109	37	79
Relapse od disease (Class1, no relapse; Class2, with relapse)	110	36	77

Table 2.3 Average values of parameters that measure the classifiers' performances for the prediction of 5-year survival

Classifier	AC	SENS	SPEC	PREC	Kappa	AUC
NB	0.875 ± 0.017	0.711 ± 0	0.923 ± 0.023	0.792 ± 0.059	0.665 ± 0.039	0.912 ± 0.012
LR	0.844 ± 0.019	0.757 ± 0.045	0.875 ± 0.029	0.684 ± 0.051	0.610 ± 0.044	0.846 ± 0.03
DT	0.846 ± 0.028	0.685 ± 0.037	0.903 ± 0.025	0.714 ± 0.062	0.595 ± 0.068	0.789 ± 0.039
RF	0.84 ± 0.014	0.625 ± 0.025	0.917 ± 0.013	0.726 ± 0.036	0.567 ± 0.036	0.858 ± 0.023
ANN	0.841 ± 0.025	0.744 ± 0.05	0.875 ± 0.041	0.684 ± 0.074	0.600 ± 0.051	0.838 ± 0.016
SVM	0.853 ± 0.033	0.743 ± 0.032	0.891 ± 0.048	0.716 ± 0.087	0.626 ± 0.065	0.817 ± 0.022

selected the following features that have the highest correlation with class variable but demonstrated the smallest correlation among themselves: histological type, medular; secondary filed, classic; lymphatic invasion; angioinvasion; tumor size; bilateral tumor; perinodal inflation of adipose tissue; second cycle chemotherapy; second cycle radiotherapy; metastasis; ER; NPI; tertiary radiotherapy; and cause of death.

It was not surprising that NB achieved the best performance, since the main assumption for this algorithm was that attributes do not depend on each other.

5.3 Prognosis on breast cancer recurrence

Similarly as with previous problem, for testing the classifiers, stratified 10-fold cross-validation procedure was used. FS procedure was applied on the training set in every iteration step, as well as SMOTE algorithm, which was applied after FS. Summarized results are described in Table 2.4.

From Table 2.4, we concluded that all classifiers achieved very good results. We especially highlight RF, SVM, and ANN classifiers which results after performing FS were remarkable. Indeed, RF_ReliefF with 14 selected features, SVM_mRMR with 10 selected features, and ANN_IG with 10 selected features achieved same results: AC=0.979, SENS=0.944, SPEC=0.991, PREC=0.958, kappa=0.944, and AUC=0.924. Optimal parameters for ANN_IG classifier were learning rate=0.1, momentum constraint=0.2, and number of hidden nodes=5 and for SVM_mRMR C=10 and kernel – RBF with gamma=0.1.

Selected features are presented in Table 2.5.

5.4 Estimation of reliability for individual predictions

As we stated previously, all algorithms for the estimation of individual classifications, which we presented here, are designed to provide local information about the expected error of individual prediction for a given unseen example in a probabilistic way. This means that the calculated reliability values belong to interval [0, 1]. For CNK and LCV measures, 0 represents the confidence of the most inaccurate prediction and 1 the confidence of the most accurate one. For DENS, it is the opposite case.

These algorithms were tested as described in Section 4 specifically targeting the ML methods that manifested best performances. Namely, in Table 2.6 and Fig. 2.2, we summarized results related to the estimation of

Table 2.4 Average values of parameters that measure the classifiers' performances for the prediction of cancer recurrence

Classifier	AC	SENS	SPEC	PREC	Kappa	AUC
NB	0.950 ± 0.011	0.937 ± 0.013	0.955 ± 0.017	0.873 ± 0.04	0.870 ± 0.024	0.969 ± 0.005
BN	0.954 ± 0.02	0.896 ± 0.042	0.973 ± 0.02	0.917 ± 0.057	0.875 ± 0.052	0.95 ± 0.009
LR	0.956 ± 0.009	0.938 ± 0.026	0.962 ± 0.012	0.889 ± 0.028	0.882 ± 0.023	0.963 ± 0.012
DT	0.971 ± 0.006	0.917 ± 0.022	0.989 ± 0.005	0.964 ± 0.014	0.920 ± 0.018	0.941 ± 0.007
RF	0.974 ± 0.006	0.931 ± 0.016	0.989 ± 0.005	0.964 ± 0.014	0.930 ± 0.018	0.964 ± 0.008
ANN	0.966 ± 0.015	0.93 ± 0.027	0.978 ± 0.012	0.931 ± 0.035	0.908 ± 0.04	0.966 ± 0.023
SVM	0.976 ± 0.007	0.93 ± 0.027	0.991 ± 0	0.971 ± 0.001	0.934 ± 0.02	0.961 ± 0.014

Table 2.5 List of selected features for breast cancer recurrence problem

FS algorithm	Selected features
IG	Second cycle radiotherapy, second cycle hormonal therapy, second cycle chemotherapy, tertiary radiotherapy, perinodal inflation of adipose tissue, number of malign nodes, NPI, disease stage, angioinvasion, T4
ReliefF	Second cycle radiotherapy, second cycle hormonal therapy, second cycle chemotherapy, perinodal inflation of adipose tissue, tertiary radiotherapy, angioinvasion, multicentricity, lymphatic invasion, histological type_lobular, in situ kom_cribriform, HG, histological type_ductal, disease stage, ER
mRMR	Second cycle radiotherapy, second cycle hormonal therapy, second cycle chemotherapy, tertiary radiotherapy, perinodal inflation of adipose tissue, multicentricity, age, secondary field_classic, secondary field_solid, secondary field_DFCM

Table 2.6 Correlation coefficients between calculated predictions and reliability estimations for the problem of 5-year survival period

Algorithm	Spearman correlation coefficient	*P*-value
CNK	0.5481	<0.001
LCV	0.5747	<0.001
DENS	0.2239	<0.01

reliability of NB_mRMR classifier and 5-year survival data set and in Table 2.7 and Fig. 2.3 results for the estimation of reliability of RF_ReliefF classifier and recurrence data set.

From Tables 2.6 and 2.7, we can conclude that high correlations were achieved for both problems. More specifically, for survival problem, CNK and LCV estimations could be used according to correlation coefficients and their significance, while for problem of recurrence, CNK and DENS achieved significant results, with emphasis on CNK. From Figs. 2.2 and 2.3, we can observe separation power of correct and false predictions by reliability estimates.

Summarizing, for ML models presented in this work, these estimators are in line with the average models' statistics and could be used to provide information about certainty of predictions on unseen data.

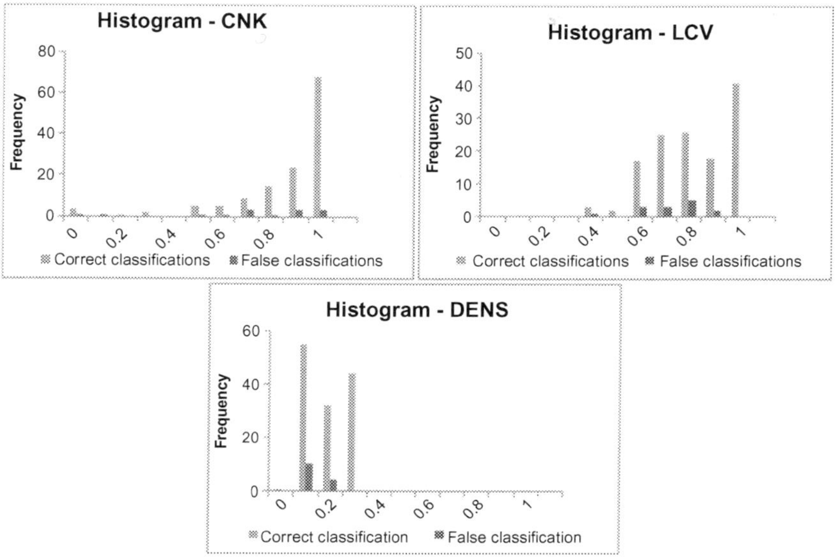

Fig. 2.2 Separation power of correct and false predictions by the reliability estimates for the NB classifier—case of 5-year survival prediction.

Table 2.7 Correlation coefficients between calculated predictions and reliability estimations for the problem of cancer recurrence

Algorithm	Spearman correlation coefficient	*P*-value
CNK	0.8894	<0.001
LCV	0.0643	0.4406
DENS	0.5838	<0.001

6 Conclusions and future trends

In this chapter, we presented a brief description of ML approach capability to face the problems in the area of breast cancer prognosis—the prediction of survival period and recurrence of primary breast cancer. Besides the importance to exploit relationships and structural information in data, we focused our intention on methods that can complement classical ML approach in terms of providing additional metrics to measure uncertainty in predictions on new unseen data. These properties are very constructive in the practical applications on medical data analysis.

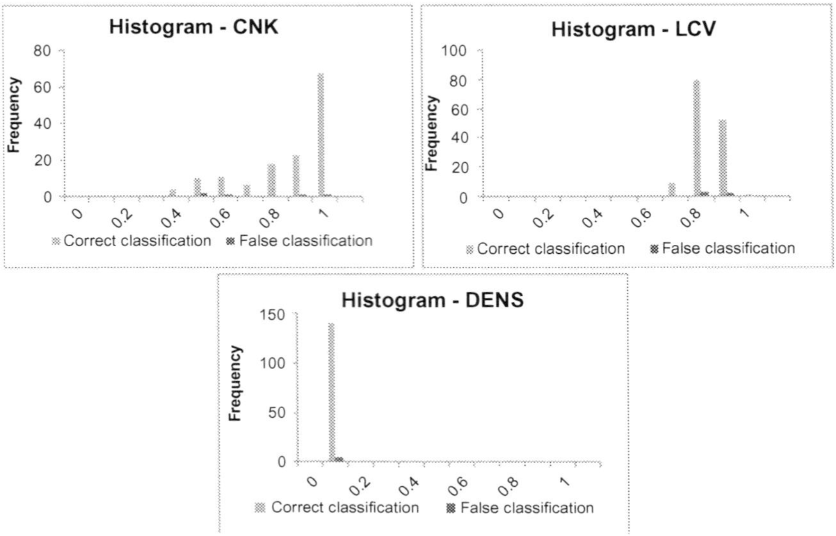

Fig. 2.3 Separation power of correct and false predictions by the reliability estimates for the RF classifier—case of cancer recurrence prediction.

Also, from the aspect of deployment of ML models for clinical purposes, it is especially valuable to provide an explanation framework for so-called black box algorithms. For instance, in the problem of the prediction of cancer recurrence, ANN, SVM, and RF demonstrated the highest model performances, but neither of these techniques provided any deduction mechanism that could help clinicians make informed decision. This problem is of particular interest for our future research.

We also provided a brief literature review on the application of ML methods for breast cancer disease. Regarding the amount of data used for the development of ML models, as already mentioned, most studies were performed on limited data sets, and, thus, further validation on additional test sets is needed to prove model performances and use them in clinics. Therefore, it is very important to design a database in a fashionable manner that can enable good calibration of data from variety of populations and under different conditions.

Significant contribution on more intensive use of predictive systems based on artificial intelligence techniques could be provided by joint initiatives between more institutions. These collaborations could empower research infrastructure and enable collection of large number of data from various domains.

Using prediction systems like we presented in this chapter could help physicians to segregate from a very heterogeneous group of patients who are at increased risk of developing metastasis or recurrence and poor outcome, which we could not estimate on the basis of classical, ordinary used decision-making parameters. Such patients should be treated with more aggressive oncology treatment, with the aim of preventing the development of systemic disease.

References

Alexe, G., Alexe, S., Axelrod, D.E., Bonates, T.O., Lozina, I.I., Reiss, M., Hamme, P.L., 2006. Breast cancer prognosis by combinatorial analysis of gene expression data. Breast Cancer Res. 8 (4), 41. https://doi.org/10.1186/bcr1512.

Amin, M.B., Greene, F.L., Edge, S.B., Compton, C.C., Gershenwald, J.E., Brookland, R.K., Meyer, L., Gress, D.M., Byrd, D.R., Winchester, D.P., 2017. The Eighth Edition AJCC Cancer Staging Manual: continuing to build a bridge from apopulation-based to a more "personalized" approach to cancer staging. CA Cancer J. Clin. 67 (2), 93–99.

Andjelkovic-Cirkovic, B., Cvetkovic, A., Ninkovic, S., Filipovic, N., 2015. An intelligent system for estimation of survival rate, relapse and metastasis—a role in individualization of breast cancer therapy. In: IEEE 15th International Conference on Bioinformatics and Bioengineering, BIBE 2015, Belgrade, November 2–4, 2015, pp. 171–176. https://doi.org/10.1109/BIBE.2015.7367658.

Bermejo, P., Gamez, J.A., Puerta, J.M., 2009. Incremental Wrapper-based subset Selection with replacement: an advantageous alternative to sequential forward selection. In: IEEE Symposium on Computational Intelligence and Data Mining, CIDM 2009, March 30–April 2, pp. 367–374. https://doi.org/10.1109/CIDM.2009.4938673.

Bilal, E., Dutkowski, J., Guinney, J., Jang, I.S., Logsdon, B.A., Pandey, G., et al., 2013. Improving breast cancer survival analysis through competition-based multidimensional modeling. PLoS Comput. Biol. 9, e1003047. https://doi.org/10.1371/journal.pcbi.1003047.

Bolón-Canedo, V., Sanchez-Marono, N., Alonso-Betanzos, A., 2015. Feature Selection for High-Dimensional Data. Springer, https://doi.org/10.1007/978-3-319-21858-8_2 (Chapter 2).

Brewster, A.M., Hortobagyi, G.N., Broglio, K.R., Kau, S.W., Santa-Maria, C.A., Arun, B., et al., 2008. Residual risk of breast cancer recurrence 5 years after adjuvant therapy. J. Natl. Cancer Inst. 1179–1183. https://doi.org/10.1093/jnci/djn233.

Burke, H.B., Goodman, P.H., Rosen, D.B., Henson, D.E., Weinstein, J.N., Harrell, F.E., Marks, J.R., Winchester, D.P., Bostwick, D.G., 1997. Artificial neural networks improve the accuracy of cancer survival prediction. Cancer 79, 857–862.

Chawla, N., Bowyer, K., Hall, L., Kegelmeyer, W., 2002. SMOTE: syntheticminority oversampling technique. J. Artif. Intell. Res. 16, 321–357.

Chawla, N., Japkowicz, N., Kolcz, A., 2004. Editorial: special issue on learning from imbalanced data sets. ACM SIGKDD Explor. Newsl. 6 (1), 1–6.

Chu, F., Wang, L., 2005. Applications of support vector machines to cancer classification with microarray data. Int. J. Neural Syst. 15, 475. https://doi.org/10.1142/S0129065705000396.

Costa, E.P., Verwer, S., Blockeel, H., 2013. Estimating prediction certainty in decision trees. In: Tucker, A., Höppner, F., Siebes, A., Swift, S. (Eds.), IDA 2013, LNCS. In: vol. 8207. Springer, Heidelberg, pp. 138–149.

Costantino, J.P., Gail, M.H., Pee, D., Anderson, S., Redmond, C.K., Benichou, J., Wieand, H.S., 1999. Validation studies for models projecting the risk of invasive and total breast cancer incidence. J. Natl. Cancer Inst. 91 (18), 1541–1548.

Cover, T.M., Thomas, J.A., 1991. Elements of Information Theory. Wiley.

Cruz, J.A., Wishart, D.S., 2007. Applications of machine learning in cancer prediction and prognosis. Cancer Informat. 2, 59–77.

Cuzick, J., Dowsett, M., Pineda, S., Wale, C., Salter, J., Quinn, E., et al., 2011. Prognostic value of a combined estrogen receptor, progesterone receptor, Ki-67, and human epidermal growth factor receptor 2 immunohistochemical score and comparison with the Genomic Health recurrence score in early breast cancer. J. Clin. Oncol. 29, 4273–4278. https://doi.org/10.1200/JCO.2010.31.2835.

De Abreu, F.B., Schwartz, G.N., Wells, W.A., Tsongalis, G.J., 2014. Personalized therapy for breast cancer. Clin. Genet. 86, 62–67. https://doi.org/10.1111/cge.12381.

De Laurentiis, M., De Placido, S., Bianco, A.R., Clark, G.M., Ravdin, P.M., 1999. A prognostic model that makes quantitative estimates of probability of relapse for breast cancer patients. Clin. Cancer Res. 5, 4133–4139.

Desmedt, C., Haibe-Kains, B., Wirapati, P., Buyse, M., Larsimont, D., Bontempi, G., Mauro, D., et al., 2008. Biological processes associated with breast cancer clinical outcome depend on the molecular subtypes. Clin. Cancer Res. 14 (16), 5158–5165. https://doi.org/10.1158/1078-0432.CCR-07-4756.

Evans, D.G., Howell, A., 2007. Breast cancer risk-assessment models. Breast Cancer Res. 9 (5), 213–220. https://doi.org/10.1186/bcr1750.

Gail, M.H., Brinton, L.A., Byar, D.P., Corle, D.K., Green, S.B., Schairer, C., Mulvihill, J.J., 1989. Projecting individualized probabilities of developing breast cancer for white females who are being examined annually. J. Natl. Cancer Inst. 81 (24), 1879–1886. https://doi.org/10.1093/jnci/81.24.1879.

Haixiang, G., Yijing, L., Shang, J., Mingyun, G., Yuanyue, H., Bing, G., 2017. Learning from class-imbalanced data: review of methods and applications. Expert Syst. Appl. 73, 220–239.

Hippisley-Cox, J., Coupland, C., 2015. Development and validation of risk prediction algorithms to estimate future risk of common cancers in men and women: prospective cohort study. BMJ Open 5(3)e007825. https://doi.org/10.1136/bmjopen-2015-007825.

Jacobi, C.E., de Bock, G.H., Siegerink, B., van Asperen, C.J., 2009. Differences and similarities in breast cancer risk assessment models in clinical practice: which model to choose? Breast Cancer Res. Treat. 115 (2), 381–390. https://doi.org/10.1007/s10549-008-0070-x.

Kim, W., Kim, K.S., Lee, J.E., Noh, D.Y., Kim, S.W., Jung, Y.S., Park, M.Y., Park, R.W., 2012. Development of novel breast cancer recurrence prediction model using support vector machine. J. Breast Cancer 15 (2), 230–238. https://doi.org/10.4048/jbc.2012.15.2.230.

Kira, K., Rendell, L.A., 1992. The feature selection problem: traditional methods and a new algorithm. In: AAAI, pp. 129–134.

Kittaneh, M., Montero, A.J., Glück, S., 2013. Molecular profiling for breast cancer: a comprehensive review. Biomark. Cancer 5, 61–70. https://doi.org/10.4137/BIC.S9455.

Kohavi, R., John, G., 1997. Wrapper for feature subset selection. Artif. Intell. 97 (1-2), 273–324. https://doi.org/10.1016/S0004-3702(97)00043-X.

Kononenko, I., 1993. Inductive and Bayesian learning in medicine diagnostic. Appl. Artif. Intell. 7, 317–337.

Kononenko, I., 1994. Estimating attributes: analysis and extensions of RELIEF. In: European Conference on Machine Learning, pp. 171–182.

Kononenko, I., Kukar, M., 2007. Machine Learning and Data Mining: Introduction to Principles and Algorithms. Horwood Publishing.

Koren, S., Bentires-Alj, M., 2015. Breast tumor heterogeneity: source of fitness, hurdle for therapy. Mol. Cell 60 (4), 537–546. https://doi.org/10.1016/j.molcel.2015.10.031.

Korf, R., 1999. Artificial intelligence search algorithms. In: Atallah, M.J. (Ed.), Handbook of Algorithms and Theory of Computation. CRC Press.

Kourou, K., Exarchos, T.P., Exarchos, K.P., Karamouzis, M.V., Fotiadis, D.I., 2015. Machine learning applications in cancer prognosis and prediction. Comput. Struct. Biotechnol. J. 13, 8–17. https://doi.org/10.1016/j.csbj.2014.11.005.

Kukar, M., Kononenko, I., 2002. Reliable classifications with machine learning. In Proceedings of the 13th European Conference on Machine Learning.

Lee, C.P., Choi, H., Soo, K.C., Tan, M.H., Chay, W.Y., Chia, K.S., Liu, J., Li, J., Hartman, M., 2015. Mammographic breast density and common genetic variants in breast cancer risk prediction. PLoS One 10(9)e0136650. https://doi.org/10.1371/journal.pone.0136650.

Leslie, H.S., Mary, K.G., Christian, W., 2009. TNM Classification of Malignant Tumours. John Wiley & Sons.

Li, H., Qiu, Z., Li, F., Wang, C., 2017. The relationship between MMP-2 and MMP-9 expression levels with breast cancer incidence and prognosis. Oncol. Lett. 14 (5), 5865–5870. https://doi.org/10.3892/ol.2017.6924.

Maglogiannis, I., Zafiropoulos, E., Anagnostopoulos, I., 2009. An intelligent system for automated breast cancer diagnosis and prognosis using SVM based classifiers. Appl. Intell. 30, 24–36. https://doi.org/10.1007/s10489-007-0073-z.

Maimon, O., Rokach, L., 2005. Data Mining and Knowledge Discovery Handbook. Springer, Heidelberg.

Majnik, M., Bosnić, Z., 2013. ROC analysis of classifiers in machine learning: a survey. Intell. Data Anal. 17 (3), 531–558.

Martin, M., Brase, J.C., Ruiz, A., Prat, A., Kronenwett, R., Calvo, L., Petry, C., et al., 2016. Prognostic ability of EndoPredict compared to research-based versions of the PAM50 risk of recurrence (ROR) scores in node-positive, estrogen receptor-positive, and HER2-negative breast cancer. A GEICAM/9906 sub-study. Breast Cancer Res. Treat. 156 (1), 81–89. https://doi.org/10.1007/s10549-016-3725-z.

McGuire, A., Brown, J.A., Malone, C., McLaughlin, R., Kerin, M.J., 2015. Effects of age on the detection and management of breast cancer. Cancers (Basel) 7 (2), 908–929. https://doi.org/10.3390/cancers7020815.

Mealiffe, M.E., Stokowski, R.P., Rhees, B.K., Prentice, R.L., Pettinger, M., Hinds, D.A., 2010. Assessment of clinical validity of a breast cancer risk model combining genetic and clinical information. J. Natl. Cancer Inst. 102 (21), 1618–1627. https://doi.org/10.1093/jnci/djq388.

Newman, C.L., Blake, D.J., Merz, C.J., 1998. UCI Repository of machine learning databases, 1998. University of California, Irvine, Dept. of Information and Computer Sciences.

Owens, W.L., Gallagher, T.J., Kincheloe, M.J., Ruetten, V.L., 2011. Implementation in a large health system of a program to identify women at high risk for breast cancer. J. Oncol. Pract. 7 (2), 85–88. https://doi.org/10.1200/JOP.2010.000107.

Paik, S., Shak, S., Tang, G., Kim, C., Baker, J., Cronin, M., et al., 2004. A multigene assay to predict recurrence of tamoxifen-treated, node-negative breast cancer. N. Engl. J. Med. 351, 2817–2826.

Park, K., Ali, A., Kim, D., An, Y., Kim, M., Shin, H., 2013. Robust predictive model for evaluating breast cancer survivability. Eng. Appl. Artif. Intell. 26, 2194–2205.

Parker, J.S., Mullins, M., Cheang, M.C., Leung, S., Voduc, D., Vickery, T., et al., 2009. Supervised risk predictor of breast cancer based on intrinsic subtypes. J. Clin. Oncol. 27, 1160–1167. https://doi.org/10.1200/JCO.2008.18.1370.

Peng, H., Long, F., Ding, C., 2005. Feature selection based on mutual information: criteria of max-dependency, max-relevance, and min-redundancy. IEEE Trans. Pattern Anal. Mach. Intell. 27 (8), 1226–1238.

Pevec, D., Strumbelj, E., Kononenko, I., 2011. Evaluating reliability of single classifications of neural networks. In: Dobnikar, A., Lotric, U., Ster, B. (Eds.), ICANNGA 2011, Part I, LNCS. In: vol. 6593. Springer, Heidelberg, pp. 22–30.

Prati, R.C., 2012. Combining feature ranking algorithms through rank aggregation. In: The 2012 International Joint Conference on Neural Networks (IJCNN), pp. 1–8.

Rakha, E.A., Reis-Filho, J.S., Baehner, F., Dabbs, D.J., Decker, T., Eusebi, V., Fox, S.B., et al., 2010. Breast cancer prognostic classification in the molecular era: the role of histological grade. Breast Cancer Res. 12 (4), 207. https://doi.org/10.1186/bcr2607.

Reif, M., Shafait, F., Dengel, A., 2012. Meta-learning for evolutionary parameter optimization of classifiers. Mach. Learn. 87 (3), 357–380.

Robnik-Sikonja, M., Kononenko, I., 1997. An adaptation of Relief for attribute estimation in regression. In: Fourteenth International Conference on Machine Learning, pp. 296–304.

Robnik-Sikonja, M., Kononenko, I., 2003. Theoretical and empirical analysis of ReliefF and RReliefF. Mach. Learn. J. 53, 23–69.

Robnik-Sikonja, M., Radovic, M., Djorovic, S., Andjelkovic Cirkovic, B., Filipovic, N., 2018. Modeling ischemia with finite elements and automated machine learning. J. Comput. Sci. 29, 99–106, ISSN 1877-7503. https://doi.org/10.1016/j.jocs.2018.09.017.

Russell, S., Norvig, P., 2003. Artificial Intelligence: A Modern Approach, second ed. Prentice Hall, Upper Saddle River, New Jersey.

SEER, 2010. Surveillance Epidemiology and End Results Program. National Cancer Institute.http://www.seer.cancer.gov. Accessed 11 July 2011.

Steyerberg, E., 2008. Clinical Prediction Models: A Practical Approach to Development, Validation, and Updating. Springer Science & Business Media.

Štrumbelj, E., Bosnić, Z., Kononenko, I., Zakontnik, B., Grašič-Kuhar, C., 2010. Explanation and reliability of prediction models: the case of breast cancer recurrence. Knowl. Inf. Syst. 24 (2), 305–324. https://doi.org/10.1007/s10115-009-0244-9.

Sun, Y., Goodison, S., Li, J., Liu, L., Farmerie, W., 2007. Improved breast cancer prognosis through the combination of clinical and genetic markers. Bioinformatics 23, 30–37. https://doi.org/10.1093/bioinformatics/btl543.

Toss, A., Cristofanilli, M., 2015. Molecular characterization and targeted therapeutic approaches in breast cancer. Breast Cancer Res. 17 (1), 60. https://doi.org/10.1186/s13058-015-0560-9.

Vukicevic, A., Stojanovic, M., Radovic, M., Djordjevic, M., Andjelkovic-Cirkovic, B., Pejovic, T., Jovicic, G., Filipovic, N., 2016. Automated development of artificial neural networks for clinical purposes: application for predicting the outcome of choledocholithiasis surgery. Comput. Biol. Med. 75, 80–89. https://doi.org/10.1016/j.compbiomed.2016.05.016.

Wu, Q., She, H., Liang, J., Huang, Y., Yang, Q., Yang, Q., Zhang, Z., 2012. Expression and clinical significance of extracellular matrix protein 1 and vascular endothelial growth factor-C in lymphatic metastasis of human breast cancer. BMC Cancer 12 (1), 47. https://doi.org/10.1186/1471-2407-12-47.

Yijing, L., Haixiang, G., Xiao, L., Yanan, L., Jinling, L., 2016. Adapted ensemble classification algorithm based on multiple classifier system and feature selection for classifying multi-class imbalanced data. Knowl.-Based Syst. 94, 88–104.

Zhang, G.P., 2000. Neural networks for classification: a survey. IEEE Trans. Human Mach. Syst. 30 (4), 451–462. https://doi.org/10.1109/5326.897072.

CHAPTER 3

Topological and parametric optimization of stent design based on numerical methods

Contents

1 Introduction

Cardiovascular disease is a general term that implies a number of clinical entities that are commonly localized in the heart and/or blood vessels. Cardiovascular and vascular diseases are among the most common diseases in developed countries and are the leading causes of death. According to statistical data related to the structure of mortality, these diseases are among the most common causes of death, accounting for 50% of all deaths (Hagan et al., 2000).

Modern lifestyle, daily stress exposure, irregular nutrition, obesity, and reduced physical activity are just some of the causes of hypertension, which is one of the most significant risk factors for developing cardiovascular diseases. Hypertension is an acute condition or a chronic disease characterized by a

Computational Modeling in Bioengineering and Bioinformatics
https://doi.org/10.1016/B978-0-12-819583-3.00003-5

persistent increase in arterial blood pressure. Often, there are no recognizable symptoms, which is why it is also called a "quiet killer." In most patients, arterial hypertension usually occurs without or with minimal symptoms.

According to the reports from World Health Organizations, cardiovascular diseases represent the greatest health problem for people in the developed world (World Heart Federation, 2015; Go et al., 2014). In addition, as a consequence of infarction, different forms of disability occur (Miljuš et al., 2010). One of the most serious cardiovascular diseases is arteriosclerosis (Mayerl et al., 2006). It is manifested as a process of plaque buildup in the blood vessel wall (Filipovic et al., 2012; Parodi et al., 2012 and Filipovic et al., 2013). Treatment of arteriosclerosis is directed toward the treatment of its complications: myocardial infarction, angina pectoris, arrhythmia, stroke, etc. Arteriosclerotic changes in the arteries range from the lungs to the complete blockage of the blood vessel. There are several ways to treat blood vessel narrowing, and which method will be applied depends on the severity of the disease:

- Pharmacological therapy—the administration of appropriate drugs can increase the flow of blood through the coronary arteries.
- Bypass surgery—surgical intervention under general anesthesia based on the creation of bridging whereby the narrowing and clogging of blood vessels is avoided.
- Percutaneous transluminal coronary angioplasty (PTCA)—the coronary artery is expanded at the place of narrowing to restore a normal blood flow.

About 500,000 of these procedures are being performed in Europe today. Modern methods of treating this disease include the insertion of cardiovascular endoprosthesis—the so-called stents inside diseased blood vessels. The basic tool necessary for performing this procedure is "stent" (Roguin, 2011). A stent is an endoprosthesis of a generally tubular shape made of alloys of biocompatible metals that is introduced through the catheter and placed inside a narrowed part with the aim of expanding or opening the lumen of the hollow tubular structures, usually of the blood vessels and also of other tubular channels (urinary, bile, etc.).

2 Endovascular prosthesis: Stent

The word *stent* (*esténs*, as the word was originally pronounced) (Ring, 2001) comes from the surname of the dentist Charles Thomas Stent (1807–85). Stent became widely known for his improvements and modifications as a dentist gutta-percha, which refers to materials used to fill the treated root of the tooth that he patented. Thanks to his achievements, in 1855, he was awarded a title "Dentist to the Royal Household" (Ring, 2001). He gave his dentistry company the name *Stent* (Ring, 2001). Later, this term in plastic surgery was introduced by a Dutch surgeon Johannes Fredericus Esser (1877–1946). He used Stent materials to reconstruct the injuries of soldiers who were wounded in the trench fights during the First World War. He used these materials for taking dental molds, calling them "stent mold," without the use of capital letters that would associate someone's name or surname (Esser, 1917). Later in the literature, the origin of the word *stent* was explained in more detail, but this term remained to be used among surgeons as a general term for almost any "non-biological material used as a support for biological tissues" (Gillies, 1920). Although very commonly used, in cardiology, the term stent appeared for the first time in 1966 after the implantation of endoprosthesis into aortic homograph (Weldon et al., 1966).

3 History of endovascular prosthesis: *STENT*

Charles Theodore Dotter (1920–85), American clinician and scientist, proposed a new concept of healing the stenosed artery in 1964 (Fig. 3.1A). Today's procedures of stent treatment are functioning according to this principle and the concept of angioplasty. He named this procedure "manually guided dilator" (Dotter and Judkins, 1964). However, angioplasty is widely accepted as a clinical procedure only after a public demonstration and improvement of balloon catheter by Andreas Roland Grüntzig in Zurich 1977 (Short, 2007; Grüntzig et al., 1979).

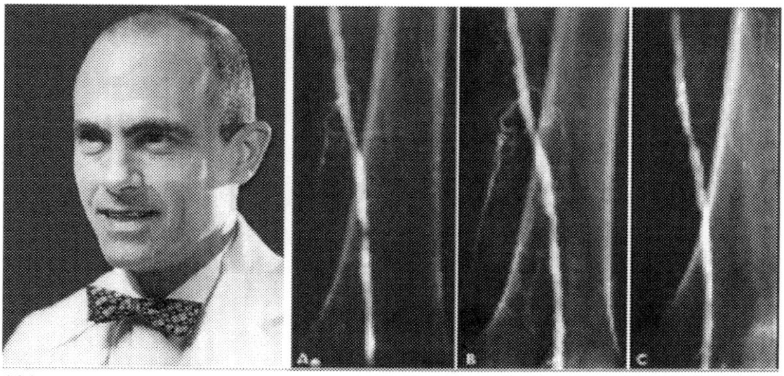

(A) (B) (C) (D)

Fig. 3.1 (A) Charles Theodore Dotter (1920–85) (Payne, 2001), (B–D) angiographic recordings of the first patient who underwent dilated stenosis (Dotter and Judkins, 1964): (B) before dilatation, (C) immediately after dilatation, and (D) a record three weeks after the procedure.

Doter is considered the creator of modern interventional radiology and one of the most deserving people to develop key procedures such as angioplasty, catheterization, and implantation of coronary stents. As in the case of many innovators who preceded him, he did not find the approval of the authorities, so he was credited many years later when many others began to use his methods.

In particular, the first patient treated with angioplasty was an 82-year-old woman who had a left ventricular stenosis (Fig. 3.1B–D). After the patient refused the proposed amputation of legs, Dotter was given permission to test his dilated catheter. The intervention lasted only a few minutes, and the pain and symptoms of vascular disease receded after a few weeks.

That same year, he received a request to perform angiography in a patient, with a reminder by a surgeon: "visualize but do not try to fix" (Fig. 3.2). Despite the warning, Dotter performed dilation, and as he followed the state of his patients after the intervention, he documented and demonstrated the usefulness of their public methods: Dotter (1972) Percutaneous transluminal angioplasty [training video].

A great breakthrough in interventional cardiology occurred on March 28, 1986, in Toulouse, France, by Jacques Puel (1949–2008). Together with a Swiss cardiologist Ulrich Sigwart (1941–) and his collaborators, Jacques Puel implanted the first coronary stent called WALLSTENT (Schneider AG, Bulach, Switzerland) (Puel et al., 1987).

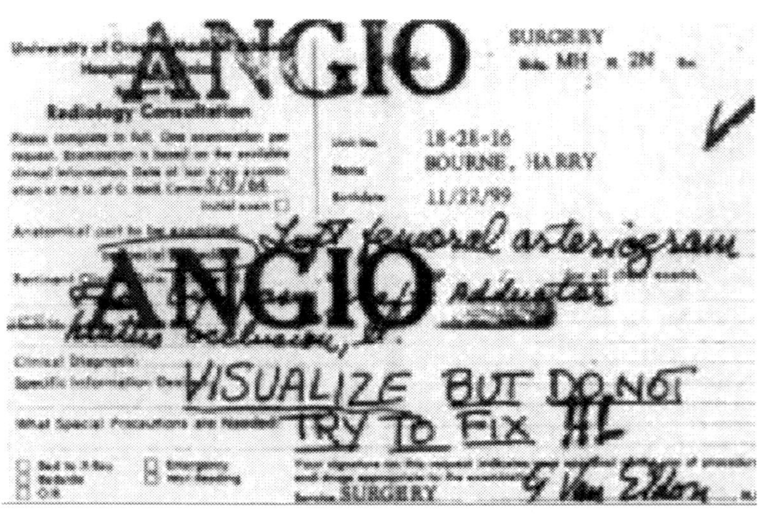

Fig. 3.2 Instructed by the surgeon that Dotter was noticed not to have used his procedure on the patient (Dotter and Judkins, 1964).

To demonstrate the clinical utility of the procedure, the first clinical study was published (Sigwart et al., 1987) that involved the stenting of peripheral and coronary arteries. The study itself describes the authors' experiences in clinics in Lausanne (Switzerland) and Toulouse (France), and it included 10 implanted stents in 6 patients with iliac and femoral arteries and implantation of 24 coronary stents in 19 different patients.

At the same time, in the United States, several groups worked on developing and testing similar procedures and devices. Julio Palmaz (1945–) is known for the procedure for opening stents using balloons for which he obtained the patent in 1985 with Richard Schatz. They made smaller versions of Palmaz-Schatz stent and conducted studies on animals (dogs) in 1987 (Schatz et al., 1987; Roubin et al., 1987).

First Palmaz-Schatz stents (Johnson & Johnson, New Brunswick, NJ, USA) were implanted in a patient in the same year in Sao Paolo, Brazil (Schatz et al., 1991). Because of technical deficiency of initial design and many complications that followed (like in-stent restenosis), Palmaz-Schatz stent was granted a mass-use permit several years later in 1994 when it was proven in extensive clinical studies Belgian Netherlands Stent (BENESTENT) (Serruys et al., 1994) and Stent Restenosis Study (STRESS) (Fischman et al., 1994). These studies present a huge advantage of metal stents (bare-metal stent) in relation to angioplasty.

The rate of occurrence of restenosis for developed designs was in the range of 20%–30%, which triggered a new revolution in the treatment of a diseased coronary artery. Drug-eluting stents represent a new form of stents that release drugs in an artery over time (Sousa et al., 2003). The first drug-eluting stent was implanted by Eduardo Sousa in 1999. However, this new stent form CYPHER (Cordis, Fremont, CA, the United States) was approved in 2004 after extensive clinical studies (Morice et al., 2002; Sousa et al., 2001). Lately, the most attention has been devoted to the biocompatibility of materials from which the stents are made (Dudek et al., 2002; Ramcharitar and Serruys, 2008), as it has been shown that once the stent is consumed, the stent itself becomes a foreign body and causes numerous complications such as malposition, thrombosis, and restenosis. To solve this problem, stents developed after discharge of the drug are disintegrating within the body (currently, there are ongoing global clinical trials designed to test stents and validate their utility).

After obtaining the approval for the use of Palmaz-Schatz stent by the US Food and Drug Administration (FDA), more than 80% of percutaneous coronary interventions have been performed with coronary stents in the period of 4 years. High rate of successful treatment of patients with coronary artery stents (Arjomand et al., 2003) led to the fact that today's global stent claims exceeded the figure of 5 million a year (Laslett et al., 2012).

The development of stenting technology and the development of materials have enabled a wide application of stents in the treatment of many other diseases that cause narrowing or closing of the cavities in the body and not just blood vessels. For example, the so-called airway stents are commonly used in tracheobronchial obstruction (Walser et al., 2004). Stents are as well used for treating problems with narrowing of the urethra and/or obstruction of the prostate (Hussain et al., 2004). There are also stents for support and opening duodenal channels, etc.

The term stent refers to wired devices that are installed at the site of narrowing in the diseased artery or other channel in the body to revitalize the flow of blood or other fluids. The evolution of stent design began with cardiovascular stents, but major modifications were made in coronary stents (Serruys et al., 2006; Nabel and Braunwald, 2012).

The procedure of stenting starts out the same way as coronarography. The puncturing of some of the major arteries is most common in the thigh and on the wrist or upper arm. A guide wire is inserted into the artery through which the catheter comes to the onset of coronary arteries. This procedure belongs to the field of interventional cardiology, and it is professionally referred to as percutaneous coronary intervention (lat. PCI). Then, the stent is positioned by the catheter and coronary angiography at the site of the lesion after which the stent is spread—that is, spreading the diseased segment of the coronary artery and allowing normal blood flow (Fig. 3.3).

The basic parts of a stent, struts, make the ring and bridges of the stent, and the stent itself resembles a series of rings that are longitudinally connected by bridges (Fig. 3.4). It should be noted that mainly modern stent design implies the existence of bridges between the rings. However, there are also versions of stents in which the rings are directly interconnected (there are no bridges) and that can be seen in the design of Palmaz-Schatz stent, which has only one bridge in the middle so it looks like two stents interconnected by a bridge (Fig. 3.5).

By optimizing the geometric shape and number of bridges and struts, the stent design can be adapted to the specific needs of patients depending on the shape of the lesion of the artery. The evolution of stent design is in the direction of reducing the thickness, whereby it is necessary to provide radial stent support forces. The use of various materials and alloys (stainless steel, cobalt, titanium, nitinol, magnesium, etc.) provides us with great potential for stent design.

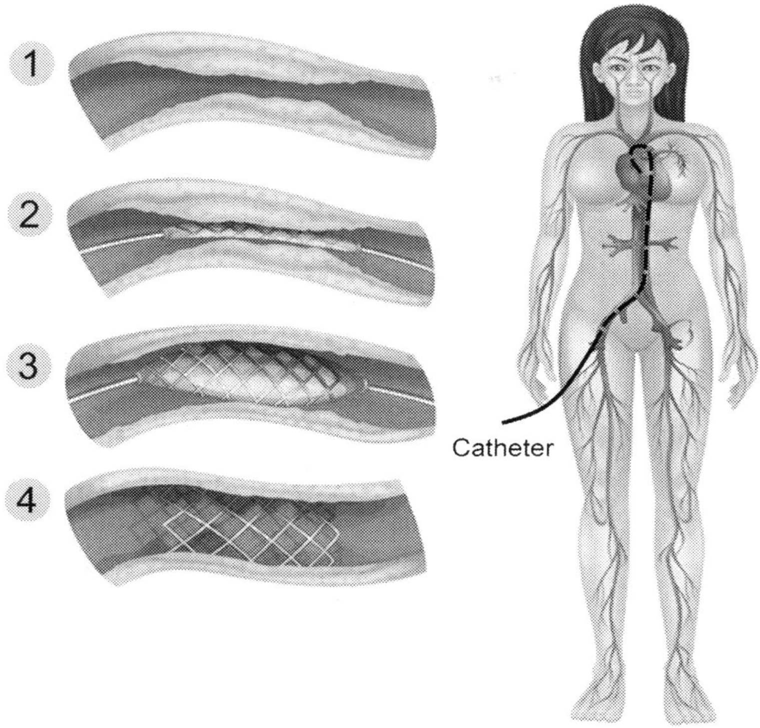

Fig. 3.3 Procedure for implanting coronary stent: (1) place of arterial narrowing, (2) positioning the stent at the point of narrowing, (3) stent dilution procedure with balloon, and (4) extension of catheter and balloon after stenting.

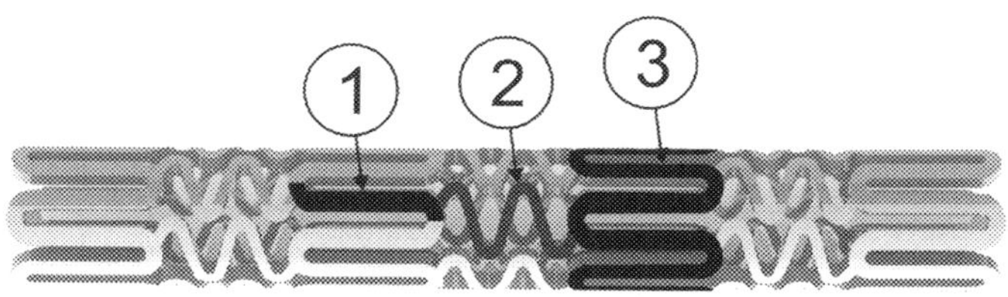

Fig. 3.4 The basic parts of a stent: struts (1) make a stent ring (2) and bridges (3).

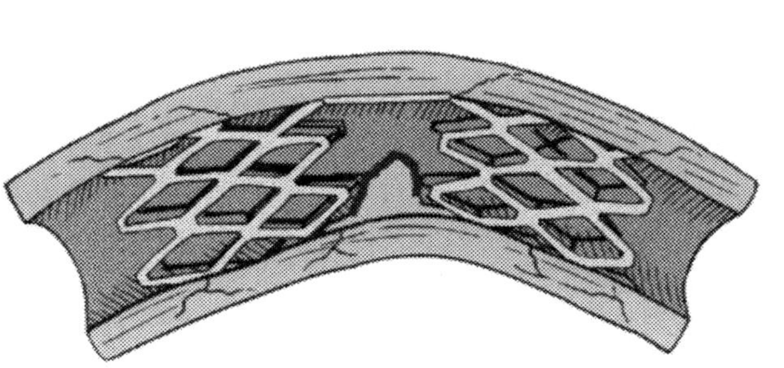

Fig. 3.5 First-generation Palmaz-Schatz stent with a bridge in the middle to increase the flexibility of stents.

4 Classification of stents

Since the first stent was approved for use in patients, there has been an expansion of various patent solutions for stents on the market. In a very short period, engineers developed a very wide range of stent designs with the aim of finding a universal "ideal" solution.

This development process is best described by Stoeckel with associates (Stoeckel et al., 2002). According to their research, there are more than 100 different designs around the world (this number is on the rise), and most of them can be found on the European market. The literature dealing with this matter is mainly focused on cardiovascular stents, and therefore, depending on the literature, this number varies from 43 different designs of coronary stents, according to Serruys and Rensing (2002), to 59, according to Machraoui et al. (2001). However, the attitude of all the authors is that this is not the final number and that this number is constantly increasing. The market value of stents is estimated at over 3 billion US dollars, while the expectation is that this figure is twice as high if the stent market is considered. Most authors divide stents according to their clinical use, for example, vascular and nonvascular and coronary or peripheral arterial stents.

All stents can be roughly divided into three categories:

- According to the method of spread: stents that expand by inflating a balloon (balloon-expanding stents) and stents that expand by themselves (self-expanding stents).
- According to the principle of treatment: noncoated metal stents, bare-metal stent, drug-coated stents, and drug-eluting stents.
- According to persistence in the body: permanent and bioabsorbable stents (Ramcharitar and Serruys, 2008; Venkatraman et al., 2006; Charpentier et al., 2015; Stone et al., 2009).

The focus of this chapter will be on the optimization of self-expanding stents made of NITINOL alloys, that is, the so-called smart material—shape memory alloys (SMA). Other divisions refer to the biochemical properties of stents and/or their effect on the biochemical properties of the tissue. These stent properties do not affect the design itself and, therefore, will not be discussed in this chapter. One of the best divisions of stents is given and illustrated in the form of a pyramid design (Fig. 3.6) by Stoeckel et al. (2002) according to material, shape, model of construction, geometry, and additives.

Among the reasons for the existence of a large number of different stent designs is the very concept of physiological mechanisms within the artery. Stent development evolves in the direction of reducing the stent profile

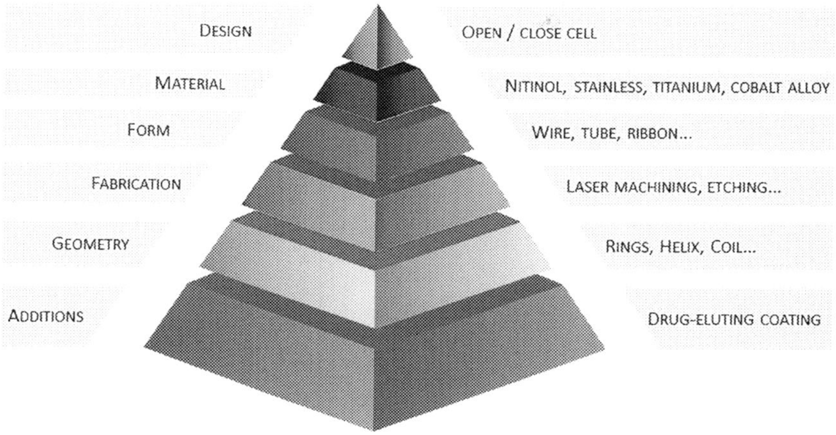

Fig. 3.6 Stent design pyramid.

and increasing the stent flexibility. This produces a new generation of stents that are much safer and easier for delivery in a blood vessel.

The major problem of occurrence of thrombosis in a period of several hours to a few days after insertion of a stent, which can lead to major clinical problems (sudden cardiac death or acute myocardial infarction), is almost completely resolved with oral antiplatelet therapy. However, stent incorporation still carries with it major problems such as lesion coverage to avoid plaque passing and radial support of the artery wall. It is also necessary to provide easier access to the lateral branches through the stent so that the side-branch lesions can be treated. In addition to all of the previously mentioned, it is very important that the stent is radiologically visible to make the positioning easier. For clinicians, it is very important that they can easily select an appropriate stent for the corresponding artery lesion. Some of the general characteristics that an ideal stent must satisfy are the following:

- Great flexibility
- Small profile in the collected state (small diameter in the catheter)
- Easy and reliable expansion
- Biocompatibility
- Thrombosis resistance
- High radial force
- Circular artery coverage
- Small surface in contact with the artery
- Easy monitoring in the artery

- Radiological visibility
- Hydrodynamic compatibility

The stent design pyramid (Fig. 3.6) was created by collecting data about already existing stents and their classification. However, now that it is defined, it can serve as a powerful tool for future design because the correct selection of data from the pyramid can create a stent with desired characteristics very quickly.

5 Stent modeling

5.1 Computer methods for stent design

Today's level of development and availability of technology allows us to develop stent model design in a much simpler and faster way. A large number of CAD[a]/CAM[b] programs from various manufacturers provide us with a wide choice of options and great freedom of work. Although these programs are very similar in their characteristics and capabilities, the way they work in them is very different, which is often an additional problem for engineer's determination. It is difficult to determine which software package is optimal in given technological and industrial conditions, from the aspect of the technical capabilities that it provides to the users, the necessary hardware platforms, license fees, and the required level of IT knowledge, as well as taking into account the way it is accessed and designed in CAD/CAM applications.

5.1.1 Modern CAD/CAM/CAE[c] systems

Globalization of the modern market requires the production of increasingly complex products, with many variants of design solutions, to adapt to specific requirements of the end customer. Competition on the world market, especially in the mechanical engineering industry, the aviation industry, the motor vehicle industry, electronics, and similar industries, requires a constant increase in product quality, lower price, reduced time of market entry, and flexible production. Considered from the technical and technological point of view, the condition for survival on the market is the continuous introduction of new technologies such as programmable and flexible automation and computer-integrated production and new concepts such as agile

[a] Computer-aided design.
[b] Computer-aided manufacturing.
[c] Computer-aided engineering.

production systems and intelligent production systems. In the field of designing new products and technologies, the use of computers by introducing systems such as CAD, CAM, CAE, and CAPP is indispensable. Modern CAD/CAM systems provide a wide choice of different approaches in product design and technology. Possible alternative approaches are explicit, parametric, variant, and modeling based on typical forms—"features." Without explaining in detail all the features of these approaches, it is stated that explicit modeling is standard for all CAD/CAM systems and has a fundamental disadvantage that it is relatively nonflexible, especially if it is a product that, basically, has a geometric and technological similarity and the ability to form based on families of similar parts. Parametric modeling is rooted in CAD systems 3D model to 2D drawing of associativity (change in 3D model leads to a change in 2D drawing). Parametric design has the ability to control model parameters by defining "leading" dimensions and introducing nondimensional geometric and other constraints and relationships (parallel, administrative, etc.). The parameter as the information that fully determines the model extends through all levels of the system and modern CAD/CAM systems have bidirectional activation (change of parameter at some level—the model causes change at all other levels—models). In modern CAD/CAM systems, associations are also introduced between different system components (e.g., CAD and CAM modules) based on a unique data structure that ensures connection of parametric defined geometry with the generated tool path and a very important link between the CAD and CAE modules, which becomes more and more prominent in modern software packages. Such links allow for a simpler creation of a model group based on a reference model representing the group, simply by changing the values of the model parameters. One of the world's leading CAD/CAM systems is SolidWorks Corporation Dassault System, which was used to create a stent model presented in this chapter. Excellent CAE support for this software is Abaqus, which belongs to the very top of the modeling and simulation tool for various machine and building constructions.

The direct connection between the CAD/CAE software packages (Fig. 3.7) is rare but very useful. This direct automatic connection makes manipulation of developing model between CAD-SolidWorks and CAE-Abaqus software easy. Therefore, Dassault software tool used for model design is used in this chapter. Geometry is created in SolidWorks software and then automatically (with one click) switched to Abaqus environment, where the preparation of the model for simulation and presentation of results is performed. If the results do not meet the desired criteria, the change of

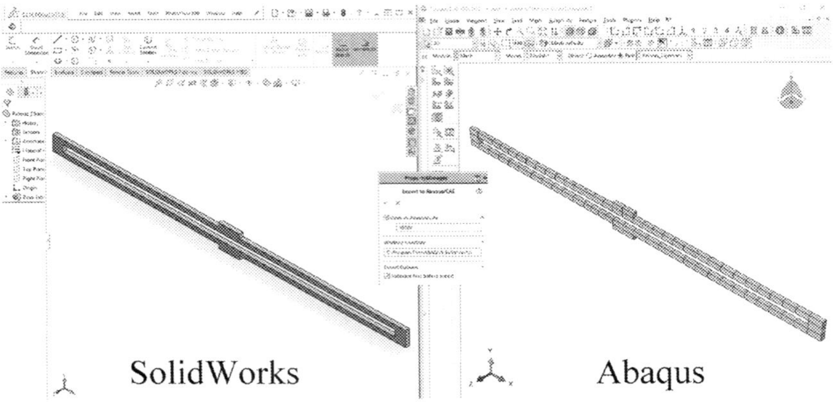

Fig. 3.7 Direct communication of CAD-CAE software (SolidWorks 2016, *left*: Abaqus 2016, *right*).

geometry can easily be done in SolidWorks, and the software will automatically go through the entire procedure of model preparation, setting boundary conditions, solving simulations, and displaying new results. In this way, the time for making a model is shortened a few times, and it is bypassing the so-called communication through files between software that often causes problems due to the way the file is exported when exporting, because the decimal in the file is rounded, and so on.

5.2 Introduction to modeling

Basically, biomechanical stent modeling can be divided into three groups. The first group belongs to modeling related to the mechanical properties of a stent, checking stress states of deformations, stress on the vessel's wall, etc. These tests are mainly based on the finite element method (FEM). It is very important to carry out all the tests at this stage to find out all design defects because any irregularity detected subsequently requires a return to this stage and a retesting, and this can be a very "expensive" step in terms of expenses and time. In the second group, the influence of the stent design itself (e.g., the shape and layout of the struts) on the flow of blood (blood) inside the blood vessel is studied and is based on the computational fluid dynamics (CFD), an area belonging to the FEM. In the third group, the release of the drug from drug-eluting stents (DES), drug spreading through the bloodstream and its penetration into the blood wall, is also investigated and tested. The same simulations of drug release and flow are very important because without them, it is very difficult to understand the way in which the drug spreads through the blood vessel, so in this case, simulations are practically the only tool that can detect all design defects. Moreover, a lot of attention is devoted to merging these groups into a complex simulation that represents fluid-solid interaction (FSI), that is, not only the interaction of the stent with the blood vessel but also the interaction of the fluid/blood with the stent and blood vessel and their interaction. A very interesting review of the fluid-solid method is shown in the paper (Lanoye, 2007).

In this chapter, a focus is placed on mechanical stent behavior or, more precisely, simulation and validation of the stent expansion process and loads during the stent life based on the results obtained by the FEM. Such simulations are very demanding, but they provide insight into the real loads that the stent is exposed to (Mackerle, 2005). Besides many innovations and progress in the development of coronary stents over the last three decades, design validation problems, prediction of the risk of restenosis, injuries to the wall of the coronary artery, and stent fracture and stenosis remain unclear until a prototype is developed. For a projected 10-year life span, the rate of breakdowns of stents varies from 5% to 25% depending on stent design and place of their incorporation (Hernández et al., 2013; Lee et al., 2009; Rits et al., 2008). Pulsating heart rate, with the frequency of about 1–1.5 Hz for an average person (or in the range of 60–90 cycles per minute), causes a pulse, that is, dynamic loadings of the coronary artery and, therefore, of the implanted stent.

For a predetermined 10-year stent, a load of 4×10^8 cycles (Iwasaki et al., 2013) is obtained. During this exploitation period, complex physiological cyclic loads can cause accumulation of structural damage and ultimately lead to the breakdown of the parts or the entire structure of the stent. Based on clinical reports, it is known that stent fracture does not lead directly to the death of the patient but is associated with the process of vascular injuries of the coronary artery wall and in-stent restenosis. It is generally an unwanted event for manufacturers, patients, and clinicians (Edelman and Rogers, 1998). If, for example, it is considered that the typical dimensions of coronary stents are 2–4 mm in diameter and 5–15 mm in length; the mechanical analysis of miniature devices and long-term events in real-life in vivo conditions is practically impossible. On the other hand, experimental in vitro measurements and tests reveal general performance on simplified loads (compared with complex physiological) and require plenty of time and very expensive equipment to perform experiments. For instance, if experimental testing is performed on fracture and fatigue, the result is binary: (1) The stent has not passed the test—that is, a visible fracture appeared on the stent at a given load, or (2) the stent has passed the fatigue test under the given conditions. On the other hand, if the stent has passed the test, it would be useful for engineers in charge of optimizing stent design to know the maximum force this design can withstand before a fracture occurs. Moreover, if there were cracks, it would be useful to know at which speed the cracks would further spread under physiological conditions and how it would affect the coronary artery wall.

In such situations, the finite element method (Kojic et al., 2008) is an ideal tool that can provide acceptable answers to questions that arise during the development of stent design and its analysis, for a reasonable time and price. In addition, FEM is suitable for the use of parametric models and their analysis, experimenting with different materials, and allows visualization of physical quantities (such as stress distribution, deformation, and temperature) throughout the model, which is very complicated to achieve in experimental conditions. Based on the current formulation of standards for testing and validation of vascular devices, the main priority is mechanical testing, followed by animal and clinical studies. At the end, there are numerical simulations, which are recommended for use as a supplement to mechanical tests.

However, due to the high costs and time to be invested in the fabrication of prototype stents for mechanical testing and then for animal and clinical studies, stent development and validation usually begin with computer design and numerical analysis. Only if the results of these analyses prove satisfactory, the next step is the development of prototypes and mechanical endurance tests and, finally, animal studies and clinical studies.

5.3 FEM analysis of stent spreading

5.3.1 Analysis of the initial design

The initial design (initial configuration) was taken from the Palmaz-Schatz stent (Fig. 3.5). The design of the Palmaz-Schatz stent itself is "obsolete" and leaves a lot of space for improvement with modern technologies; therefore, it is taken as a demonstration example in this chapter. Since the Palmaz-Schatz stent consists of several repeated segments for the initial step in modeling, analysis, and optimization, one basic building segment of the stent was used (Fig. 3.8) to make the analysis faster and simpler. Only after obtaining the desired stress and strain on this simple model, the next step began: modeling the entire stent and performing FE analysis on the entire model.

The design starts with the following assumptions:

- The stent will be made of a tube by a laser cutting process.
- The initial diameter of nitinol pipes is Ø1.3 mm and wall thickness 0.07 mm.

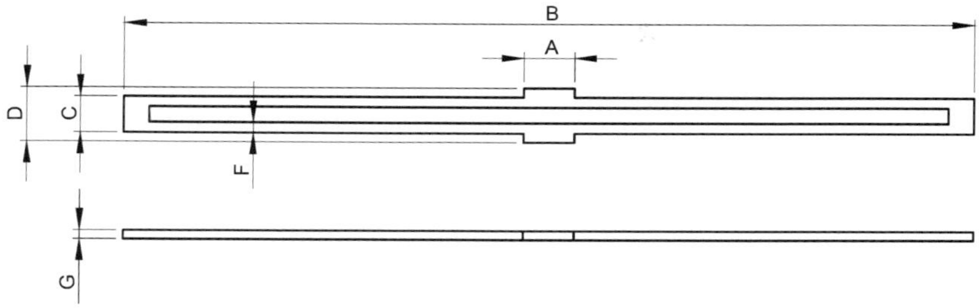

Fig. 3.8 Basic (initial) form of stent design.

- The stent will have 12 closed cells per volume and 6 in length.
- The expected open diameter is around Ø12mm.
- The stent length in closed condition is 33 mm.

Based on these assumptions, the design measures are adopted as the basic material form of the stent: $A = 0.38\text{mm}; B = 5.5\text{mm} = \left(\frac{33\text{mm}}{6kom}\right)$; $C = 0.28\text{mm}; D = 0.34\text{mm} = \frac{Obim}{12kom} = \left(\frac{1.3\text{mm}\cdot\pi}{12kom}\right); F = 0.07\text{mm}$; $G = 0.07\text{mm}$;

To create the initial 3D model, we need to make sure that the model meets the required criteria.

To prevent the problem of the "cartoon[d]" model and obtain a real open stent model, it is necessary to open the model using the FEM. Fig. 3.9 shows the initial model with the given boundary conditions. Since the goal is to open the model as a boundary condition, the displacement of nodes is set; on the contact position (strut/crimping tool), the displacement vectors are in the same direction (collinear) but are of different directions. In this way, the model rises to the desired values, and the simulation mimics the real test performed on the stretching machine (Fig. 3.10). The maximum values for shifts are taken from the cartoon version of the model.

After the analysis, it is necessary to check the results of the model (Fig. 3.11), especially the deformation of the material at this stage, since the displacement of the end points is high. For nitinol (depending on the type), it is expected to tolerate strain in the range of 6%–10% in a slow process of deformation. (For nitinol used in this chapter, it is recommended that the maximum deformation does not exceed 7%–8%.)

[d] Cartoon design—3D CAD model of stent device in open position.

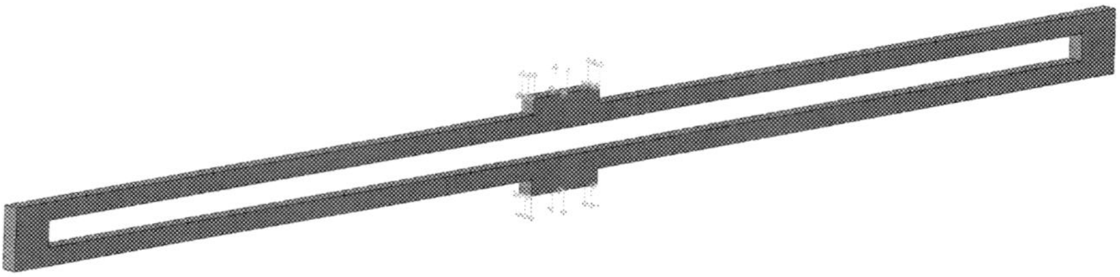

Fig. 3.9 The initial stent model with the given boundary conditions.

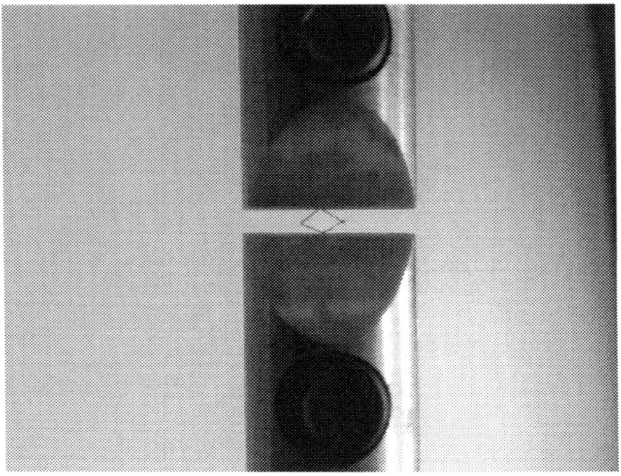

Fig. 3.10 Opening the initial model of the testing machine.

This simple analysis provides insight into most of the shortcomings of the initial design itself, and the most important is the maximum attainable diameter of the stent, in view of the fact that with a maximum permissible deformation of 7.3%, the displacement is ± 1.4 mm. From here, it follows that the maximum stent circumference (suppose that stent has 12 cells per ring) or the maximum possible diameter is approximately Ø12mm.

5.4 UMAT (Abaqus) material model

The theory used in the UMAT material model of nitinol is based on the concept of generalized plasticity and physical principles. In theory, the strain is decomposing into two parts: a purely linear elastic component and a transformation component:

$$\Delta\varepsilon = \Delta\varepsilon^{el} + \Delta\varepsilon^{tr} \tag{3.1}$$

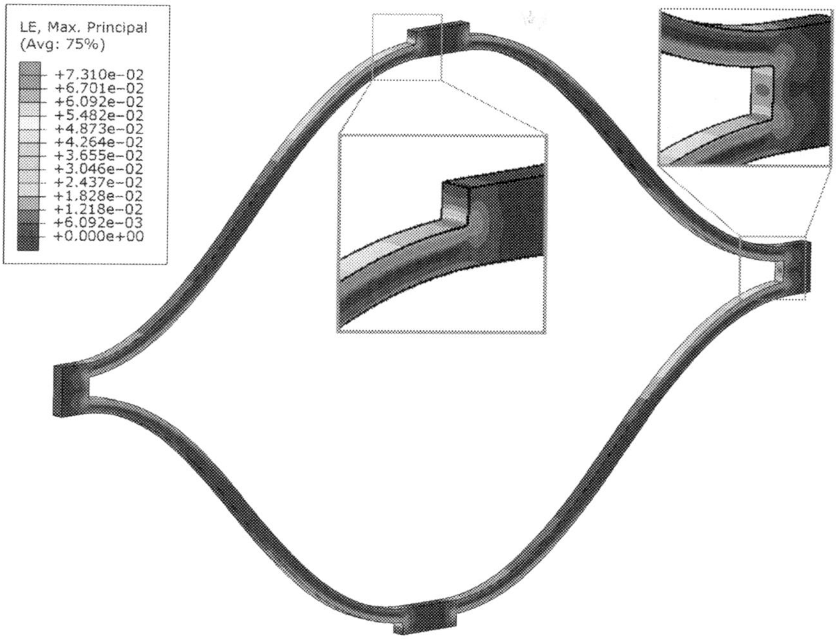

Fig. 3.11 The result of stretching of the initial model; distribution of material deformation.

The martensitic transformation takes place by shifting the atom (by sliding the grid) along a certain plane, called the basic plane or invariant plane, of the crystal that forms the interface between martensite and austenite (Lagoudas, 2010; Grabe, 2007; Panico and Brinson, 2007). There are two models of martensitic appearance in crystal gratings: slip and twinning. As with SMA, twinning is a more common mechanism, and the details are shown in Fig. 3.12.

The transformation of austenite to twinned martensite is conditioned by shaking forces:

$$\Delta\varepsilon^{tr} = a\Delta\zeta\frac{\partial F}{\partial\sigma} \tag{3.2}$$

$$F^S \leq F \leq F^F \tag{3.3}$$

where the ζ share of martensite and F represents the transformational potential. The same goes for reverse transformations, with the stress being different. The intensity of the transformation is defined:

$$\Delta\zeta = f(\sigma,\zeta)\Delta F \tag{3.4}$$

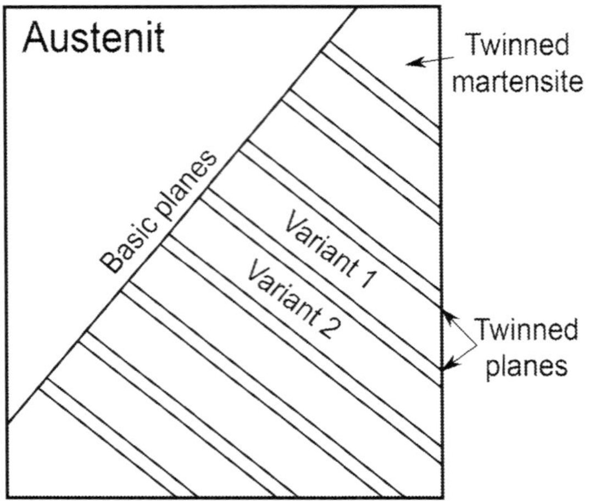

Fig. 3.12 Schematic interface of phase austenite and martensite phase (Dunic, 2015).

Any change in stress direction leads to the reorientation of martensitic grains, which is an additional negligible effort, while temperature change increases stress level. This temperature shift is linear due to the increase in the volume associated with the transformation. It requires less stress in the load on the tensile and more stress at the pressure load. This effect is modeled using Drucker-Prager function for transformation potential:

$$F = \overline{\sigma} - p\tan\beta + CT \tag{3.5}$$

where $\overline{\sigma}$ is the von Mises stress, p basically represents the pressure stress, and T is the temperature. In addition to the usual stress and strain in UMAT routines, the values necessary for the SMA model are monitored. These variables include the distribution of martensite, deformation transformation, equivalent stress, and deformation. Figs. 3.13 and 3.14 show the curve stress—deformation and stress— temperature, and calibration points of the material model, derived from the results of uniaxial tests on Instron device. Key points are obtained by loading, then (returning) unloading the sample, changing the temperature model, etc. Super elastic UMAT material model for SMA defines material based on 15 parameters (Table 3.1).

6 Stent geometry optimization

Although not defined as part of the pyramid design of the stent itself, the thickness and shape of the cross section of the stent strut are very important for the design and durability of the stent and for the effect of the occurrence

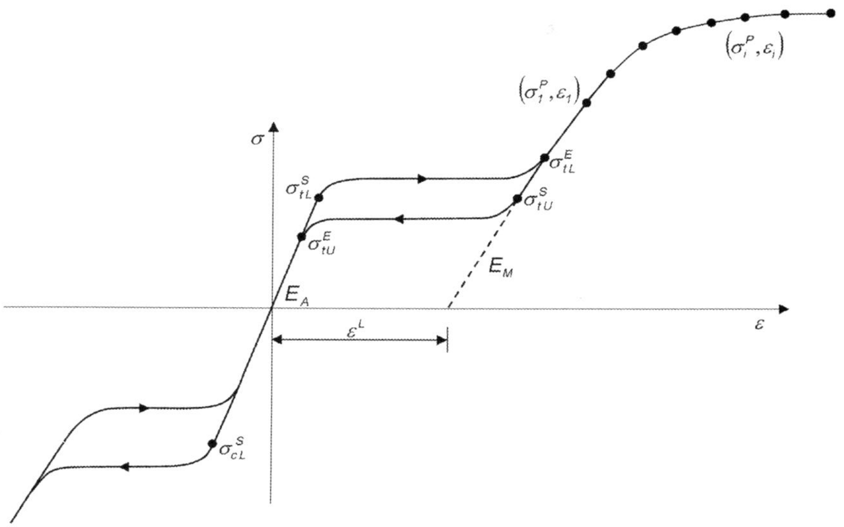

Fig. 3.13 SMA stress-strain curve with calibration points.

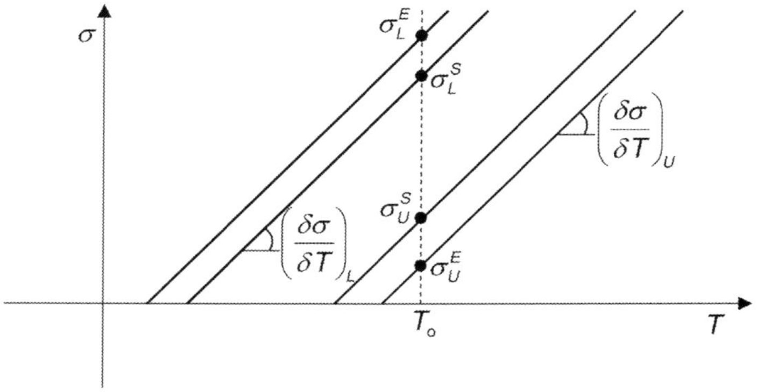

Fig. 3.14 SMA stress-temperature curve with calibration points.

of restenosis. An example of the shape and thickness of the cross-sectional diameter of the strut is given in Fig. 3.15. If the design does not have enough radial force for the desired diameter of the artery, it can be repaired by simply increasing the cross-sectional diameter of the strut. However, studies have shown that the increase in the cross diameter of the strut significantly affects the growth of restenosis.

Clinical studies have shown that the thickness of the strut is a key trigger for restenosis. ISAR-STEREO is the study (Kastrati et al., 2001) in which

Table 3.1 UMAT parameters

Material data		Units	Value
Yang's modulus of elasticity—austenite	E_A	(MPa)	45,000
Poisson's ratio—austenite	(ν_A)	–	0.33
Yang's modulus of elasticity—martensite	E_M	(MPa)	38,000
Poisson's ratio—martensite	(ν_M)	–	0.33
Transformation strain	ε^L	–	0.055
$\delta\sigma/\delta T$ loading	$\left(\frac{\delta\sigma}{\delta T}\right)_L$	(MPaT^{-1})	6.7
Start transformation loading	σ_L^S	(MPa)	590
End transformation loading	σ_L^E	(MPa)	640
Reference temperature	T_0	(°C)	37
$\delta\sigma/\delta T$ unloading	$\left(\frac{\delta\sigma}{\delta T}\right)_U$	(MPaT^{-1})	6.7
Start transformation unloading	σ_U^S	(MPa)	300
End transformation unloading	σ_U^E	(MPa)	270
Start of transformation stress during loading in compression, as a positive value	σ_{CL}^S	(MPa)	–
Volumetric transformation strain		–	0.055
Strain limits		(%)	12
A_f temperature		(°C)	30

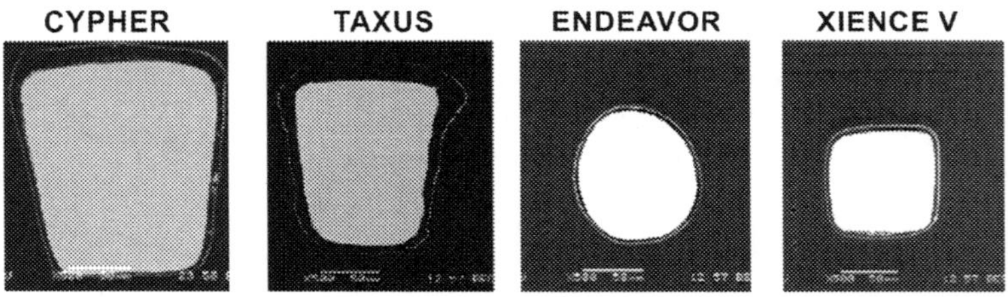

CYPHER	TAXUS	ENDEAVOR	XIENCE V
Strut thickness	**Strut thickness**	**Strut thickness**	**Strut thickness**
140 μm	132 μm	91 μm	81 μm
Polymer coating thickness	**Polymer coating thickness**	**Polymer coating thickness**	**Polymer coating thickness**
12.6 μm	16 μm	5.3 μm	7.6 μm
Total thickness	**Total thickness**	**Total thickness**	**Total thickness**
152.6 μm	148 μm	96.3 μm	88.6 μm

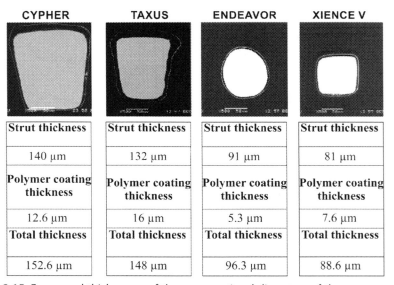

Fig. 3.15 Forms and thicknesses of the cross-sectional diameters of the stent strut.

two similar types of stents with different strut thicknesses were compared. These stents were randomly implanted in 651 patients with lesions in large coronary arteries (>2.8 mm). In the next 6 months, the binary rate of restenosis was higher in the stent with a larger cross-sectional diameter of the struts. Both types of stents were manufactured by the same manufacturer of Guidant, Advanced Cardiovascular Systems (ACS) Multilink RX Duet with a thickness of 0.14 mm and stents ACS RX Multilink with a 0.05-mm thickness of the sticks.

Similar results were also made by the subsequent study ISAR-STEREO-2 (Pache et al., 2003). In this study, two stents of different design and thickness of the struts were implanted randomly in 611 patients, where the stent Guidant, Advanced Cardiovascular Systems (ACS) RX Multilink with a 0.05-mm-thin strut had fewer angiographic and clinical restenosis compared with the stent with a 0.14-mm-thick strut. In addition, subsequent analyses carried out by Briguori and associates (Briguori et al., 2002) showed that the thickness of the struts plays an important role in the development of restenosis (Fig. 3.16). Based on this information, the conclusion is that a detailed optimization is necessary in designing stent devices to obtain a stent with as thin as possible struts.

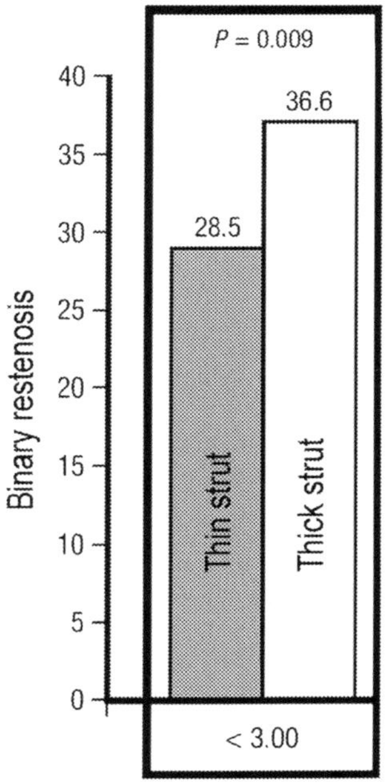

Fig. 3.16 Influence of the thickness of the rods on restenosis (Briguori et al., 2002).

7 Nonparametric optimization of the model

The Palmaz-Schatz stent design itself is very simple, so the initial configuration from which it begins is very simplified and full of flaws. If analyzed in more detail in Fig. 3.11, more positions with a concentration of strains or stress can be noticed. Since only the change in geometric parameters cannot make a great improvement to such a simple design, it is necessary to introduce certain geometric improvements. These improvements are mainly introduced based on the experience of constructors and assumptions.

Nonparametric optimization is a good tool for understanding the model and extending stress and deformation through the model, as well as precisely defining zones in which there is a potential excess of matter and areas where additional materials need to be added.

Simulia TOSCA (TOSCA, 2016) is a software tool used in this chapter for the so-called topological optimization of design. The topology represents the connection of the elements of the system. In the topological optimization, the structure of domains is changed based on two coupled fields: deformation of the model and shape of the model volume; if it is a steel stent, it is advisable to follow the stress field and the shape of the stent volume. Topological optimization itself is not an ideal tool, and a special problem occurs when it is applied to structures such as stents where the building blocks of strut have one dimension (length) much larger than the other two (width and thickness). However, if the results of the analysis are interpreted in the right way, it can greatly reduce the process of "wandering" in finding the ideal solution for stent design.

Start-point model that will be used for future optimization in this case is the Palmaz-Schatz stent. This stent has already been analyzed, and it is well known by its characteristics.

The basic parameters in optimization are as follows:

- Strain—remains unchanged (fixed)
- Model volume—changes (decreases or increases as needed)

For the initial optimization model, the Palmaz-Schatz stent was loaded with tensile. The boundary conditions are simple: the displacement on the contact nodes is given as shown in Fig. 3.10. Input parameters for starting the optimization are the strain obtained from the tensile analysis shown in Fig. 3.11.

During the optimization process in each step of model simulation, the software analyzes the field of deformation propagation and the current volume of the model. Based on this analysis, the software "concludes" not only which elements of the model are not burdened but also which elements of the model are very burdened. In this way, the sites of critical strain and places without strain on the model are defined. At the end of the simulation, the software adds material to the critical distortion points and in this way increases the volume of the model, while in the places where there are no strains, it takes away material and reduces the volume of the model. In Fig. 3.17, a model of the Palmaz-Schatz stent is presented after

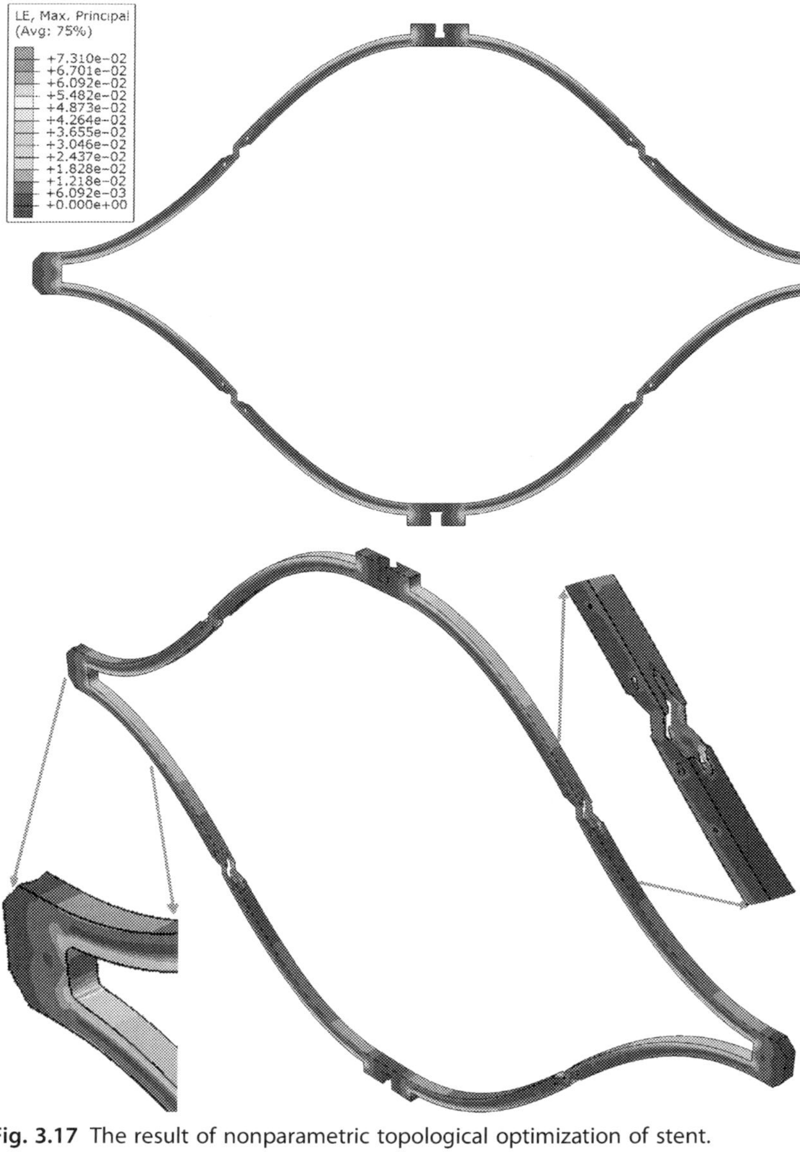

Fig. 3.17 The result of nonparametric topological optimization of stent.

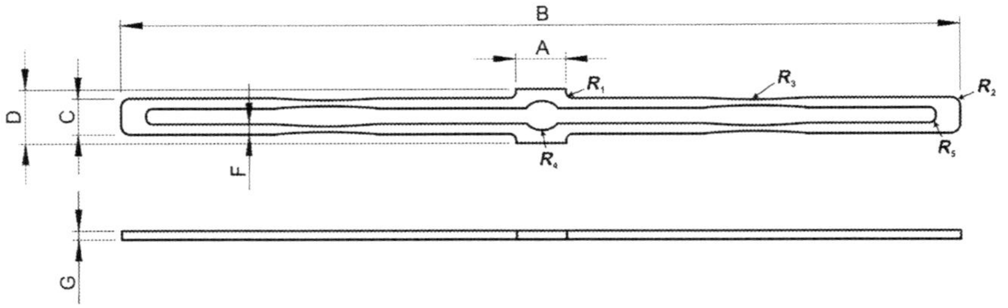

Fig. 3.18 Parametric model sketch based on topological optimization results (new radius introduced: R_1, R_5, adding material, and R_2, R_3, R_4, removing of materials).

completion of the topological optimization. Thanks to the result obtained from the topological analysis, it is possible to predict in detail the places on the original starting model that needs to be changed or to introduce new geometric elements.

Based on the results of the topological optimization shown in Fig. 3.17, new parameters are introduced in the geometry, which are shown in Fig. 3.18. In places where stress concentration has occurred, the material is added by introducing the external radii R_1 and R_5. In all places where the analysis showed that it is necessary to remove the excess material, internal radii R_2, R_3, and R_4 are introduced. The radius R_4 was introduced on the opposite side in relation to the result obtained from topological optimization to align with the radius R_1, while the resulting optimization effect remained similar. According to the results of the analysis, the largest removal of the material should be carried out on half of the stent strut, since half of the strut is the least loaded in this type of load. However, due to the hardly feasible technological process, because at this point the cross section has very small dimensions, the result of the analysis is not fully followed, and large radius R_3 has been introduced, whose function is to reduce the cross-sectional diameter of the strut in half and evenly spread the stress and deformation along the strut. The dimensions of these radii are assumed and certainly far from ideal, so another set of tests and different parameters needs to be carried out to determine the optimal dimensions. This procedure can be performed manually but requires a lot of work and time; thus the so-called parametric optimization is used.

8 Parametric optimization

In a complex process of product development and production environments, designers and engineers have to use a wide range of software tools to design and simulate their products. During the process of exiting one software, often the next software enters, especially when it comes to simulations. In the complex optimization process, several thousand simulations need to be performed over some work. This process is very extensive and requires a lot of manual work and time. The parametric analysis itself is performed in a closed loop that needs to be done many times (Fig. 3.21). Different types of tools (software) are used to facilitate this procedure, but one of the most famous is Isight v2016 (product of Dassault Systèmes Simulia Corp., Johnston, RI, USA) (Isight and The Simulia Execution Engine, 2016).

The geometry of the model was created based on the proposal obtained from the topological optimization given in Fig. 3.17 in software package SolidWorks 2016. An example of the new model is given in Fig. 3.19.

Because of the complex optimization process, that is, due to the large number of necessary simulations to optimize the model, the model itself was created in such a way that the change of a very small number of parameters can change the entire geometry.

A list of parameters that define geometry is shown in Fig. 3.20:

k—Width of the contact part

L—Length of the strut

D_0—Thickness of the strut at the end

D_1—Thickness of the strut in the middle

R_0—Central radius of indentation

R_1—External radius

The optimization process is a closed loop created in the Isight software package. Isight uses algorithms and techniques to run various external applications, performs the processes based on the required parameters, and then passes the results to the next application without interfering users.

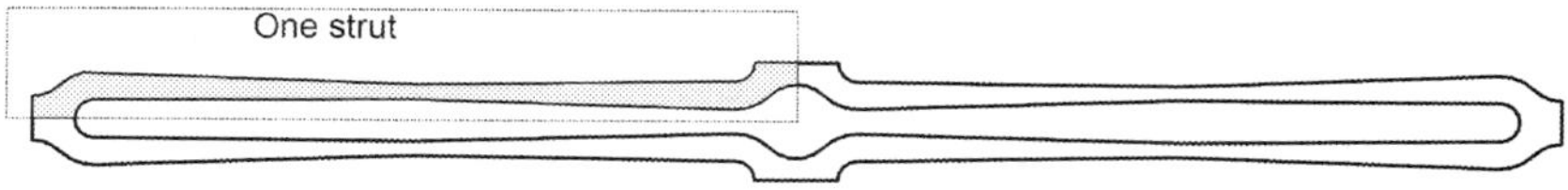

Fig. 3.19 Parametrically complex stent model based on assumptions obtained from nonparametric optimization.

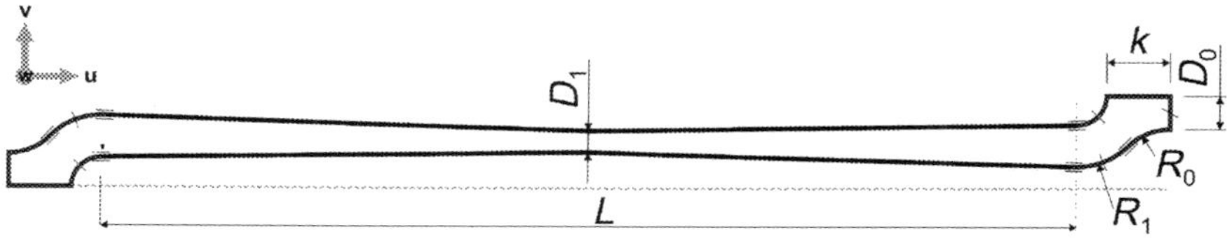

Fig. 3.20 Segment of the stent model shown in Fig. 3.19 with geometric parameters that define the entire model.

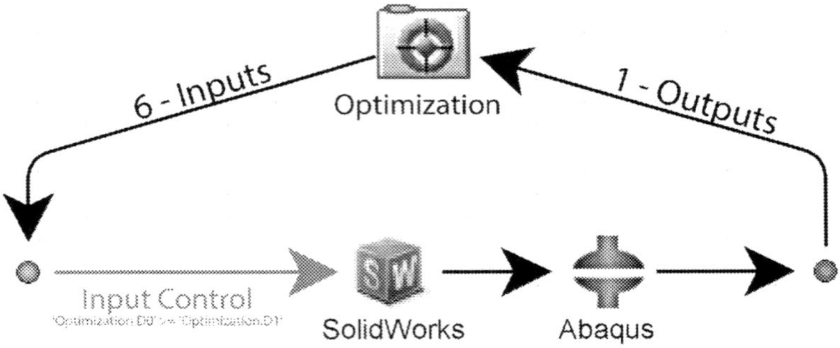

Fig. 3.21 Graphical representation of the closed parameter optimization circuit.

Name	Lower	Upper	Objective	Target
D0	0.02	0.12		
D1	0.04	0.12		
k	0.09	0.2		
L	2.8	3.2		
R0	0.07	0.14		
R1	0.1	0.2		
Max. Strain	0.055	0.09	target	0.07

Fig. 3.22 List of input/output parameters in Isight 2016 software package.

The input data (in this case, six geometry parameters) are forwarded to the 3D model creation software (SolidWorks). Then, the geometry obtained is directly forwarded to the simulation tool on the 3D model (Abaqus) The result obtained from the simulation (on this case, the maximum strain) passes the optimization algorithm on the basis of which the optimization algorithm creates new input parameters. This process is repeated when the conditions of convergence are not fulfilled or the closest possible solution is not found.

For the entire process to be performed, the user enters the input geometric parameters and their possible ranges like the required maximum output value, that is, strain and permitted range of deviations (Fig. 3.22).

The technique that was used for optimization is Pointer Automatic Optimizer, or short Pointer, integrated in Isight software. This technique is used for automatic optimization, and it contains a set of several standard optimization algorithms. Now, four optimization algorithms (method) are being used:

Genetic algorithm (Holland, 1975)
Nelder and Mead downhill simplex (Nelder and Mead, 1965)
NLPQL—Nonlinear Programming by Quadratic Lagrangian (Falk, 1967)
Linear Solver

This hybrid combination of algorithms enables this technique to have a high robustness in operation and puts it first in the selection of optimization methods. Different approaches in the implemented optimization algorithms make it possible for Pointer to be applied to any problem that greatly facilitates work. At the start point, the "Pointer" optimization technique can control only one or all four implemented algorithms, and as optimization progresses, the technique determines which algorithm is most successful and continues to optimize using this method based on internal control parameters (step size, number of repeats, number of repeat startups, etc.). This control procedure is hidden from the user.

Based on the range of input parameters and the step of changing geometric parameters, the $\Delta = 0.005$ algorithm provides 4750 possible combinations (required simulations). To reduce the number of combinations, the boundary conditions at the SolidWorks input are introduced. The input control function contains some of the geometric constraints to prevent the creation of bad geometric models, for example

$$\text{'Optimization.}D_0\text{'} >= \text{'Optimization.}D_1\text{'}$$

By design requirements, it is possible to conclude that the thickness of the strut in the middle must be equal to or less than the thickness of the strut at the ends (Fig. 3.20). By introducing this boundary condition, the problem is reduced to 3600 possible geometric models.

Manual analysis of such a large number of models would require too much time. By introducing the optimization algorithms, the problem is solved in 117 steps, and the result is shown in Fig. 3.23. A number of approximate solutions were obtained (marked in gray), while the most optimal solution is indicated in green in Fig. 3.23.

Analyzing the new model under the same conditions as the Palmaz-Schatz stent model, it can be concluded that the field of strain and maximal value of strain are very much improved. The results are presented in Fig. 3.24 and can be easily compared with the results obtained from the old model presented in Fig. 3.11.

9 Creation of a 3D model of the whole stent

After completed analyses on a small model, that is, on one part of the stent, it is necessary to create a model of the whole stent, as the results would not only verify but also simulate other more complex loads. A large number of simulations cannot be done on a small model, for example, three-point stent bending, four-point stent bending, pressure load, concentrated

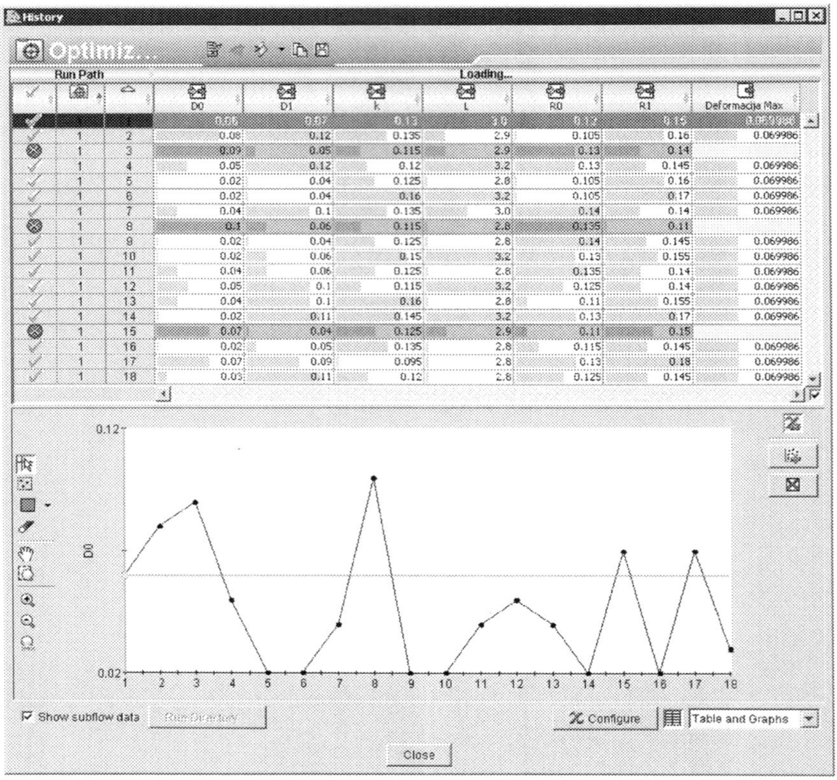

Fig. 3.23 Parameter optimization solutions.

pressure load between two plates, and stent wedge opening. Therefore, it is necessary to create a whole stent model.

Based on the parameters obtained from optimization, the geometry of the stent model was created (Fig. 3.25) consisting of a series of repeated stent cells (6 in the longitudinal direction and 12 cells in the radial direction) and meeting the required criteria.

The whole stent model was created based on the following parameters:

- $k = 0.13$mm—width of the contact part.
- $L_{strut} = L = 3.0$mm—length of the strut.
- $D_0 = 0.06$mm—thickness of the strut at the ends.
- $D_1 = 0.07$mm—thickness of the rod in the middle.
- $R_0 = 0.12$mm—central radius of indentation.
- $R_1 = 0.15$mm—external radius.
- $H_{cell} = 2 \cdot (R_0 + D_0) = 0.36$mm—width of the basic building cell.

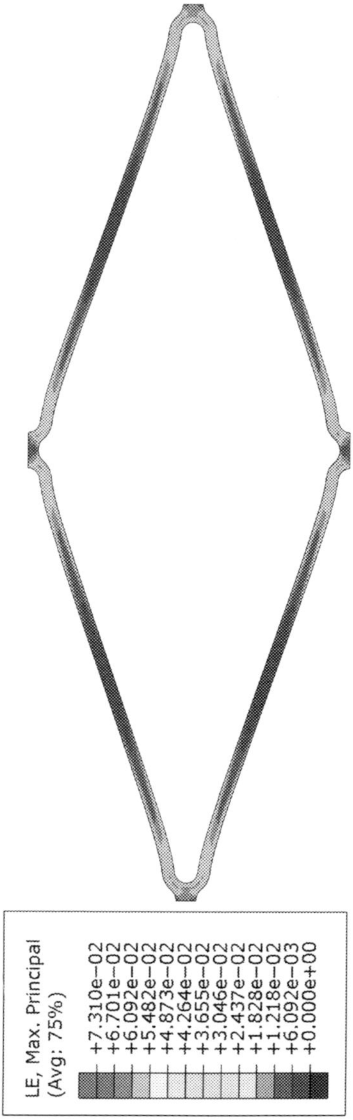

Fig. 3.24 Result of stretching of the new model; strain distribution.

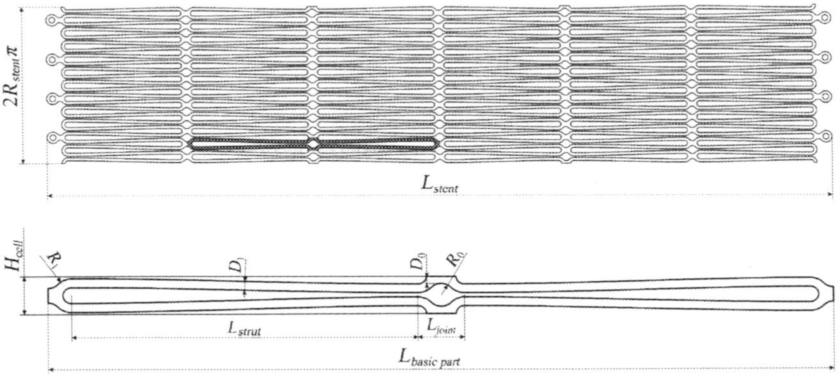

Fig. 3.25 The entire stent geometry.

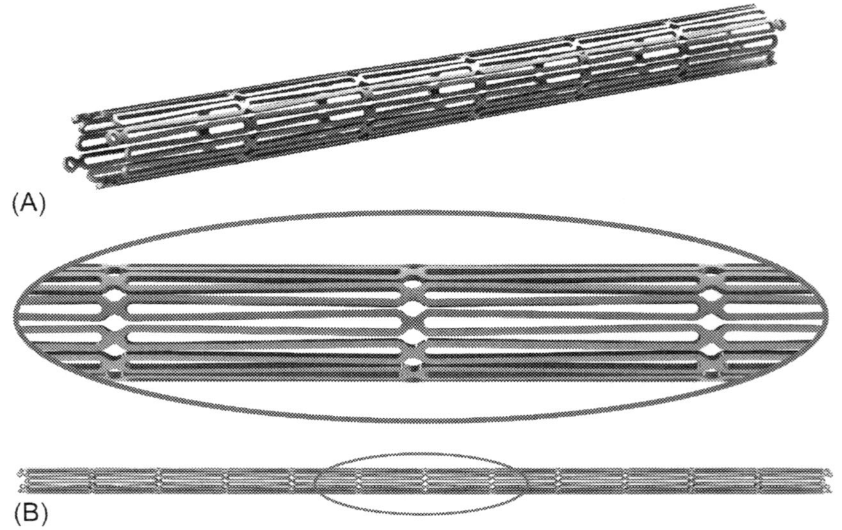

Fig. 3.26 Three-dimensional model of stent obtained by parametric optimization.

- $L_{joint} = 2 \cdot ((R_0 - D_0) + k) = 0.38\text{mm}$—length of joint.
- $L_{basic\ part} = 2 \cdot (L_{joint} + L_{strut}) = 6.72\text{mm}$—length of the basic building cell.
- $2R_{stent}\pi = 12 \cdot H_{cell} = 4.32\text{mm}$—circumference of stent.
- $L_{stent} = 6 \cdot L_{basic\ part} = 40.32\text{mm}$—length of stent.

A detailed model of the whole 3D stent was created in the SolidWorks software package and is shown in Fig. 3.26. This type of model allows engineers to carry out counted tests that can largely predict the behavior of the real prototype before it is made. This shortens the time of producing a prototype and reduces the possibility of producing a wrong prototype with critical mistakes.

10 Conclusion

Cardiovascular diseases are the leading cause of death in all developed countries. Modern lifestyle, reduced physical activity, daily stress exposure, irregular nutrition, and obesity are some of the causes of hypertension, which is one of the most significant risk factors for developing cardiovascular diseases.

Stent restenosis is one of the key problems that doctors are facing with when stents are used in treating coronary arteries. As presented in this chapter, several studies show that stent strut thickness has significant influence on this phenomenon.

In this chapter, it is shown that three-dimensional simulations were conducted to determine the mechanical behavior of the initial design Palmaz-Schatz stent and calculate stress and strain inside this device during stretch loading.

Based on these results, nonparametric and parametric optimization of stent geometry was performed. The aim of the optimization was to decrease stent volume and strut thickness and to keep the same values of stress and strain inside the device and, therefore, to keep the same structural integrity.

New stent design was created using the results obtained in the optimization process. This design was analyzed in the same way as the initial Palma-Schatz stent in the previous section of this chapter.

From the analysis of the old and new design, it is concluded that the new design has many advantages comparing with the old one. The distribution of stress and strain is more uniform, and compared with the old design, there are no places of concentration in the new design.

The aim of this chapter is to present new methods for fast and easy optimization process in designing stents. These methods can greatly facilitate the process of stent design in the future and prevent possible "expensive" mistakes that are detected at a later stage of prototype testing.

References

Arjomand, H., Turi, Z.G., McCormick, D., Goldberg, S., 2003. Percutaneous coronary intervention: historical perspectives, current status, and future direction. Am. Heart J. 146, 787–796.

Briguori, C., Sarais, C., Pagnotta, P., Liistro, F., Montorfano, M., Chieffo, A., Sgura, F., Corvaja, N., Albiero, R., Stankovic, G., Toutoutzas, C., Bonizzoni, E., Di Mario, C., Colombo, A., 2002. Instent restenosis in small coronary arteries: impact of strut thickness. J. Am. Coll. Cardiol. 40, 403–409.

Charpentier, E., Barna, A., Guillevin, L., Juliard, J.M., 2015. Fully bioresorbable drug-eluting coronary scaffolds. Arch. Cardiovasc. Dis. 108 (6–7), 385–397.

Dotter, C.T., 1972. Percutaneous transluminal angioplasty [training video]. ca. 1972.

Dotter, C.T., Judkins, M.P., 1964. Transluminal treatment of arteriosclerotic obstruction. Description of a new technic and a preliminary. Report of its application. Circulation 30, 654–670.

Dudek, D., Onuma, Y., Ormiston, J.A., Thuesen, L., Miquel-Hebert, K., Serruys, P.W., 2002. Four-year clinical follow-up of the ABSORB everolimus-eluting bioresorbable vascular scaffold in patients with de novo coronary artery disease: the ABSORB trial. EuroIntervention 7 (9), 1060–1061.

Dunic V, 2015. Development and Implementation of Thermo-Mechanical Constitutive Model for Numerical Analysis of Shape Memory Alloys, Doktorska disertacija, Univerzitet u Kragujevcu.

Edelman, E.R., Rogers, C., 1998. Pathobiologic responses to stenting. Am. J. Cardiol. 81, 4E–6E.

Esser, J.F., 1917. Studies in plastic surgery of the face. Ann. Surg. 65, 297–315.

Falk, E.J., 1967. Lagrange multipliers and nonlinear programming. J. Math. Anal. Appl. 19 (1), 141–159.

Filipovic, N., Nikolic, D., Saveljic, I., Milosevic, Z., Exarchos, T., Pelosi, G., Parodi, O., 2013. Computer simulation of three dimensional plaque formation and progression in the coronary artery. Comput. Fluids 88, 826–833.

Filipovic, N., Rosic, M., Tanaskovic, I., Milosevic, Z., Nikolic, D., Zdravkovic, N., Peulic, A., Kojic, M., Fotiadis, D., Parodi, O., 2012. ARTreat project: three-dimensional numerical simulation of plaque formation and development in the arteries. IEEE Trans. Inform. Technol. Biomed. 16 (2), 272–278.

Fischman, D.L., Leon, M.B., Baim, D.S., Schatz, R.A., Savage, M.P., Penn, I., Detre, K., Veltri, L., Ricci, D., Nobuyoshi, M., Cleman, M., Heuser, R., Almond, D., Teirstein, P.S., Fish, R.D., Colombo, A., Brinker, J., Moses, J., Shaknovich, A., Hirshfeld, J., Bailey, S., Ellis, S., Rake, R., Goldberg S for the Stent Restenosis Study Investigator, 1994. A randomized comparison of coronary-stent placement and balloon angioplasty in the treatment of coronary artery disease. N. Engl. J. Med. 331, 496–501.

Gillies, H.D., 1920. Plastic Surgery of the Face. Oxford University Press, London.

Go, A.S., Mozaffarian, D., Roger, V.L., Benjamin, E.J., Berry, J.D., Blaha, M.J., Dai, S., Ford, E.S., Fox, C.S., Franco, S., Fullerton, H.J., Gillespie, C., Hailpern, S.M., Heit, J.A., Howard, V.J., Huffman, M.D., Judd, S.E., Kissela, B.M., Kittner, S.J., Lackland, D.T., Lichtman, J.H., Lisabeth, L.D., Mackey, R.H., Magid, D.J., Marcus, G.M., Marelli, A., Matchar, D.B., McGuire, D.K., Mohler, I.I.I.E.R., Moy, C.S., Mussolino, M.E., Neumar, R.W., Nichol, G., Pandey, D.K., Paynter, N.P., Reeves, M.J., Sorlie, P.D., Stein, J., Towfighi, A., Turan, T.N., Virani, S.S., Wong, N.D., Woo, D., Turner, M.B., On behalf of the American Heart Association Statistics Committee and Stroke Statistics Subcommittee, 2014. AHA statistical update: heart disease and stroke statistics—2014 update: a report from the American Heart Association. Circulation 129, e28–e292.

Grabe C, 2007. Experimental Testing and Parameter Identification on the Multidimensional Material Behavior of Shape Memory Alloys. PhD Thesis, Institut für Mechanik, Ruhr-Universität Bochum, Germany.

Grüntzig, A.R., Senning, A., Siegenthaler, W.E., 1979. Nonoperative dilation of coronaryartery stenosis: percutaneous transluminal coronary angioplasty. N. Engl. J. Med. 301, 61–68.

Hagan, P.G., Nienaber, C.A., Isselbacher, E.M., Bruckman, D., Karavite, D., Russman, P., 2000. The International Registry of Acute Aortic Dissection (IRAD): new insights into an old disease. JAMA 283, 897–903.

Hernández, F.H., Román, A.J., Tejada, J.G., Martín, M.V., González-Trevilla, A.A., Pérez Juan, C.T., 2013. Intravascular diagnosis of stent fractures: beyond X-ray imaging. Rev. Esp. Cardiol. 66 (9), 751–753.

Holland, J.H., 1975. Adaptation in Natural and Artifcial Systems. University of Michigan Press, Ann Arbor.
Hussain, M., Greenwell, T.J., Shah, J., Mundy, A., 2004. Long-term results of a self-expanding wallstent in the treatment of urethral stricture. BJU Int. 94 (7), 1037–1039.
Isight & The Simulia Execution Engine, 2016. Isight Documentation. Dassault Systèmes, Providence, RI, USA.
Iwasaki, K., Kishigami, S., Arai, J., Ohba, T., Zhu, X., Yamamoto, T., Hikichi, Y., Umezu, M., 2013. Flexibility and stent fracture potentials against cyclically bending coronary artery motions: comparison between 2-link and 3-link DESs. Am. J. Cardiol. 111 (7), 26B.
Kastrati, A., Mehilli, J., Dirschinger, J., Dotzer, F., Schuhlen, H., Neumann, F.J., Fleckenstein, M., Pfafferott, C., Seyfarth, M., Schomig, A., 2001. Intracoronary stenting and angiographic results: strut thickness effect on restenosis outcome (ISARSTEREO) trial. Circulation 103, 2816–2821.
Kojic, M., Filipovic, N., StojanovicB, K.N., 2008. Computer Modeling in Bioengineering—Theoretical Background, Examples and Software. John Wiley and Sons, ISBN: 978-0-470-06035-3.
Lagoudas, D., 2010. Shape Memory Alloys: Modeling and Engineering Applications. Springer.
Lanoye L, 2007. Fluid-Structure Interaction of Blood Vessels. PhD thesis, Ghent University.
Laslett, L.J., Alagona, P., Clark, B.A., Drozda, J.P., Saldivar, F., Wilson, S.R., Poe, C., Hart, M., 2012. The worldwide environment of cardiovascular disease: prevalence, diagnosis, therapy, and policy issues: a report from the American College of Cardiology. J. Am. Coll. Cardiol. 60 (25), 1–49.
Lee, S.E., Jeong, M.H., Kim, I.S., Ko, J.S., Lee, M.G., Kang, W.Y., Kim, S.H., Sim, D.S., Park, K.H., Yoon, N.S., Yoon, H.J., Kim, K.H., Hong, Y.J., Park, H.W., Kim, J.H., Ahn, Y.K., Cho, J.G., Park, J.C., Kang, J.C., 2009. Clinical outcomes and optimal treatment for stent fracture after drug-eluting stent implantation. J. Cardiol. 53 (3), 422–428.
Machraoui, A., Grewe, P., Fischer, A., 2001. Koronarstenting. Steinkopff Verlag, Darmstadt.
Mackerle, J., 2005. Finite element modelling and simulations in cardiovascular mechanics and cardiology: a bibliography 1993-2004. Comput. Methods Biomech. Biomed. Eng. 8 (2), 59–81.
Mayerl, C., Lukasser, M., Sedivy, R., Niederegger, H., Seiler, R., Wick, G., 2006. Atherosclerosis research from past to present—on the track of two pathologists with opposing views, Carl von Rokitansky and Rudolf Virchow. Virchows Arch. 449, 96–103.
Miljuš D, Mickovski–Katalina N, Plavšić S, Živković S, Rakočević I, Janković J, Savković S, Božić Z, Institut za javno zdravlje Srbije "Milan Jovanović Batut", 2010. Incidencija i mortalitet od akutnog koronarnog sindroma u Srbiji. Registar za akutni koronarni sindrom u Srbiji, Izveštaj br. 5, ISBN: 978-86-7358-045-6,
Morice, M.C., Serruys, P.W., Sousa, J.E., Fajadet, J., Ban Hayashi, E., Perin, M., Colombo, A., Schuler, G., Barragan, P., Guagliumi, G., Molnàr, F., Falotico, R., 2002. A randomized comparison of a sirolimus-eluting stent with a standard stent for coronary revascularization. N. Engl. J. Med. 346 (23), 1773–1780.
Nabel, E.G., Braunwald, E., 2012. A tale of coronary artery disease and myocardial infarction. N. Engl. J. Med. 366 (1), 54–63.
Nelder, J.A., Mead, R., 1965. A simplex method for function minimization. Comput. J. 7 (4), 308–313.
Pache, J., Kastrati, A., Mehilli, J., Schuhlen, H., Dotzer, F., Hausleiter, J., Fleckenstein, M., Neumann, F.J., Sattelberger, U., Schmitt, C., Muller, M., Dirschinger, J., Schomig, A., 2003. Intracoronary stenting and angiographic results: strut thickness effect on restenosis outcome (ISAR-STEREO-2) trial. J. Am. Coll. Cardiol. 41, 1283–1288.
Panico, M., Brinson, L., 2007. A three-dimensional phenomenological model for martensite reorientation in shape memory alloys. J. Mech. Phys. Solids 55 (11), 2491–2511.

Parodi, O., Exarchos, T., Marraccini, P., Vozzi, F., Milosevic, Z., Nikolic, D., Sakellarios, A., Siogkas, P., Fotiadis, D., Filipovic, N., 2012. Patient-specific prediction of coronary plaque growth from CTA angiography: a multiscale model for plaque formation and progression. IEEE Trans. Inform. Technol. Biomed. 16 (5), 952–965.

Payne, M.M., 2001. Charles Theodore Dotter The Father of Intervention. Tex. Heart Inst. J. 28, 28–38.

Puel, J., Joffre, F., Rousseau, H., Guermonprez, B., Lancelin, B., Valeix, B., Imbert, G., Bounhoure, J., 1987. Endo-protheses coronariennes autoexpansives dans la prevention des restenoses apres angioplastie transluminale. Arch. Mal. Coeur Vaiss. 8, 1311–1312.

Ramcharitar, S., Serruys, P.W., 2008. Fully biodegradable coronary stents: progress to date. Am. J. Cardiovasc. Drugs 8 (5), 305–314.

Ring, M.E., 2001. How a dentist's name became a synonym for a life-saving device: the story of Dr Charles Stent. J. Hist. Dent. 49, 77–80.

Rits, J., van Herwaarden, J.A., Jahrome, A.K., Krievins, D., Moll, F.L., 2008. The incidence of arterial stent fractures with exclusion of coronary, aortic, and non-arterial settings. Eur. J. Vasc. Endovasc. Surg. 36 (3), 339–345.

Roguin, A., 2011. Stent: the man and word behind the coronary metal prosthesis. Circ. Cardiovasc. Interv. 4 (2), 206–209.

Roubin, G.S., Robinson, K.A., King III, S.B., 1987. Early and late results of intracoronary arterial stenting after coronary angioplasty in dogs. Circulation 76, 891–897.

Schatz, R.A., Baim, D.S., Leon, M., Ellis, S.G., Goldberg, S., Hirshfeld, J.W., Cleman, M.W., Cabin, H.S., Walker, C., 1991. Clinical experience with the Palmaz-Schatz coronary stent. Initial results of a multicenter study. Circulation 83 (1), 148–161.

Schatz, R.A., Palmaz, J.C., Tio, F.O., Garcia, F., Garcia, O., Reuter, S.R., 1987. Balloonexpandable intracoronary stents in the adult dog. Circulation 76, 450–457.

Serruys, P.W., de Jaegere, P., Kiemeneij, F., Macaya, C., Rutsch, W., Heyndrickx, G., Emanuelsson, H., Marco, J., Legrand, V., Materne, P., Belardi, J., Sigwart, C., Colombo, A., Goy, J.J., van den Heuvel, P., Delcan, J., Morel M-A for the Benestent Study Group, 1994. A comparison of balloon-expandable-stent implantation with balloon angioplasty in patients with coronary artery disease. N. Engl. J. Med. 331, 489–495.

Serruys, P.W., Kutryk, M.J.B., Ong, A.T.L., 2006. Coronary-artery stents. N. Engl. J. Med. 354 (6), 483–495.

Serruys, P.W., Rensing, B.J., 2002. Handbook of Coronary Stents, fourth ed. Martin Dunitz Ltd, London.

Short, R., 2007. In search of Andreas Roland Grüntzig, MD (1939-1985). Circulation 116 (9), 49–53.

Sigwart, U., Puel, J., Mirkovitch, V., Joffre, F., Kappenberger, L., 1987. Intravascular stents to prevent occlusion and restenosis after transluminal angioplasty. N. Engl. J. Med. 316, 701–706.

Sousa, J.E., Costa, M.A., Abizaid, A.C., Rensing, B.J., Abizaid, A.S., Tanajura, L.F., Kozuma, K., Van Langenhove, G., Sousa, A.G., Falotico, R., Jaeger, J., Popma, J.J., Serruys, P.W., 2001. Sustained suppression of neointimal proliferation by sirolimus-eluting stents: one-year angiographic and intravascular ultrasound follow-up. Circulation 104 (17), 2007–2011.

Sousa, J.E., Serruys, P.W., Costa, M.A., 2003. New frontiers in cardiology: drug-eluting stents: part I. Circulation 107, 2274–2279.

Stoeckel, D., Bonsignore, C., Duda, S., 2002. A survey of stent designs. Minim. Invasive Ther. Allied Technol. 11 (4), 137–147.

Stone, G.W., Lansky, A.J., Pocock, S.J., Gersh, B.J., Dangas, G., Wong, S.C., Witzenbichler, B., Guagliumi, G., Peruga, J.Z., Brodie, B.R., Dudek, D., Möckel, M., Ochala, A., Kellock, A., Parise, H., Mehran, R., HORIZONS-AMI

Trial Investigators, 2009. Paclitaxel-eluting stents versus bare-metal stents in acute myocardial infarction. N. Engl. J. Med. 360 (19), 1946–1959.
TOSCA, 2016. TOSCA Documentation. Dassault Systèmes, Providence, RI, USA.
Venkatraman, S.S., Tan, L.P., Joso, J.F.D., Boey, Y.C.F., Wang, X., 2006. Biodegradable stents with elastic memory. Biomaterials 27 (8), 573–1578.
Walser, E.M., Robinson, B., Raza, S.A., Ozkan, O.S., Ustuner, E., Zwischenberger, J., 2004. Clinical outcomes with airway stents for proximal versus distal malignant tracheobronchial obstructions. J. Vasc. Interv. Radiol. 15 (5), 471–477.
Weldon, C.S., Ameli, M.M., Morovati, S.S., Shaker, I.J., 1966. A prosthetic stented aortic homograft for mitral valve replacement. J. Surg. Res. 6, 548–552.
World Heart Federation, 2015. State of the Heart—Cardiovascular Disease Report. Available at the website: http://www.world-heart-federation.org/publications/reports/state-of-the-heart-cvd-report/ (Accessed 10 January 2017),

CHAPTER 4

Lung on a chip and epithelial lung cells modeling

Contents

1 Introduction

Organ on a chip is a new concept in the field of bioengineering, which provides better insight in certain organ functions and behavior of the building cells. Observed cells are seeded into microfluidic chip, which simulates bioactivities and mechanical and physiological behavior of the organ. Microchips are small in size; usually have range of computer memory stick. Each individual organ on a chip contains hollow microfluidic channels. These microfluidic channels are lined by living human cells of developing artificial organ and these human cells interface with human endothelial cells, creating on that way artificial vasculature. Artificial vasculature creation is important because each cell in human body lives thanks to nutrients that come from blood supplies to the tissues. With generated artificial

Computational Modeling in Bioengineering and Bioinformatics
https://doi.org/10.1016/B978-0-12-819583-3.00004-7

vasculature, communication between living cells of the observed organ and endothelial vascular cells can be achieved, thus providing more detailed model in comparison with the currently used in vitro testing. To specific organ on a chip can be applied external mechanical forces in order to mimic the physical microenvironment of living organs. Building artificial organs, which can be validated, requires knowledge in cellular manipulation, together with knowledge in physiological behavior of tested organ and response of it to any event.

In vitro experiments are good starting point in understanding the biology of the cells and processes that occur within them. For in vitro testing, cells are extracted from the living body and examined in the laboratory conditions. Cells are placed on the glass surface and examined. Thanks to in vitro testing knowledge of cells, physiological behavior is enriched, and correlation with medical research is established. It is a less invasive technique in comparison with the in vivo testing, where cells are observed in the living organism. In vitro testing can represent many processes that occur in the living body but in validation with in vivo testing is observed that isolation from the system leads to the loss of some cell functions due to disabled interaction with surrounding cells, in the first row the one originated from the vascular tissue. That's why there raised a need for improvement of experimental environment when it comes to cell testing. One of the proposed solutions to this problem is the development of microfluidic chips.

As already mentioned before, in comparison with the in vitro experiments, where cells are isolated from the organ for testing in laboratory conditions, organ on a chip is a more complex system and better represents cell environment. Many researchers and institutions are currently employed on the development of the different organ on a chip—the heart (Kitsara et al., 2019), lungs (Huh, 2015), liver (Knowlton and Tasoglu, 2016), kidney (Lee and Kim, 2018), brain (Wheeler, 2008), bone (Hao et al., 2018), etc. Kitsara et al. (2019) gave a review of the state of the art when it comes to heart on a chip. In the paper are reported main approaches of microfluidics for cardiac culture systems, and their assessment is provided. Huh (2015) described microphysiological system that replicates the functional unit of the human lung alveolar-capillary interface. The model is able to show dynamic mechanical activity and physiological function of the breathing lung on a chip. The model can be used for modeling complex human disease processes of the lungs. He is one of the founders of lung-on-a-chip model. Knowlton and Tasoglu (2016) focused on the development of liver-on-a-chip model. The liver is the organ in charged in detoxification of the metabolic products. Because of that, the liver is subjected to the high frequency of drug-induced injuries. Liver-on-a-chip tissue model is needed to be developed for drug screening and hepatotoxicity testing. Lee and Kim (2018) reviewed the state of the art for kidney-on-a-chip models. Various drugs applied to the human body can result in kidney dysfunction, and because of that, drug development process needs to include this issue. The advantage of using kidney-on-a-chip model in comparison with in vivo animal experiments is primary in reduced discrepancies in drug pharmacokinetics and pharmacodynamics between humans and animals. More realistic pathophysiological responses of the cells can be achieved in microfluidic, 3D culture systems, and then in conventional 2D culture systems. Several kidney-on-a-chip models show ability to reproduce microenvironment of the kidney tubule and drug nephrotoxicity. With these microfluidic chips, drug-induced biological responses can be measured. Kidney-on-a-chip models can be used in kidney disease modeling and development of novel therapies according to identification of drug effects, interactions, and nephrotoxicity. Wheeler (2008) researched on brain-on-a-chip systems. These models are very complicated for building, because working of the brain is not quite determined. In spite of the great knowledge about brain functions, a great amount of brain processes are not still revealed. Brain-on-a-chip models include nerve cells growing in the culture with ability of signal conduction and thus stimulation of nerve cells. Hao et al. (2018) investigated bone-on-a-chip model in relation of bone metastasis to breast cancer cells. About 70% of metastatic breast cancer leads to metastasis of bone tissue. The mechanism of bone tissue "infection" by breast cancer in metastasis is not completely revealed. For purpose of bone-on-a-chip model development, mature osteoblastic tissue is naturally formed. Osteoblastic tissue contains heavily mineralized collagen fibers. Coculture of metastatic breast cancer is examined with osteoblastic tissue. The model shows opportunity to investigate cancer cell interaction with bone matrix. Unique hallmarks of breast cancer bone colonization confirmed with in vivo experiments can be observed with this model.

This chapter focuses on the development of the lung-on-a-chip and lung models. Several cell lines can be used for the development of artificial lung models. So far, mostly used cell lines with this purpose are nonprimary cell lines A549 (adenocarcinomic human alveolar basal epithelial cells) and Calu-3 (epithelial lung cancer cell line) and primary cells NHBE (normal human bronchial epithelial cells). A549 cells were first determined from cancerous lung tissue of 58-year-old Caucasian male in 1972 by Giard et al.[a]

[a] http://www.a549.com/.

Calu-3 cell line is also cell line derived from cancerous tissue, first observed in pleural effusion of a 25-year-old Caucasian male with a lung adenocarcinoma in 1975 by Trempe and Fogh.

First lung-on-a-chip system is developed at Wyss Institute for Biologically Inspired Engineering at Harvard University[b]. This system contains two chambers separated by the membrane. Through the upper chamber circulates the air and through the lower chamber circulates the blood. Endothelial cells are seeded on the upper side of the membrane and endothelial cells on the bottom side of the membrane. On that way, upper canal permits airflow and represents the airways, while the lower canal permits blood flow and represents the blood vessel, supplying the lung tissue. To the chambers is applied vacuum, which produces cyclic mechanical stretching and mimicking the breathing.

This and similar systems not only can simulate behavior of the healthy lungs but also can serve for testing several lung diseases and malfunctioning of the lungs. Lung-on-a-chip device can be used for the analysis of interaction of nanoparticles, derived from the air and water pollution, causing the inflammatory process, as well as for testing of new drugs and therapies for specific lung diseases.

In the Wyss laboratory, alveolus-on-a-chip model is developed by Huh for the simulation of pulmonary thrombosis. The upper chamber contains alveolar epithelial cells, and the lower chamber contains pulmonary microvascular endothelial cells. For simulation of thrombosis, tumor necrosis factor (TNF-α) is added to the upper chamber with aim to provoke inflammation. Inflammation involves cytokines, which alert leukocytes and platelet blood cells, causing formation of thrombi. The drug parmodulins is tested for the reduction of inflammation, and the obtained result is reduced vascular inflammation without endothelial layer damage. Benam et al. (2015) analyzed physiological reactions during acute asthmatic crisis. Several researchers concluded that airway on a chip can be used as a representation of the healthy human lungs and for analysis of asthma and chronic obstructive pulmonary disease (COPD). Huh (2015) tested pulmonary edema, condition where blood clots and fluid are filling the lung and occur in case of heart failure or as a side effect of cancer drugs. He injected cancer

[b] Wyss Institute. (November 07, 2012). Available online at: https://wyss.harvard.edu/wyss-institute-models-a-human-disease-in-an-organ-on-a-chip.

drug into the lower chamber, chamber representing the blood vessel, to achieve pulmonary edema as a side effect of cancer drug. Fluid and protein migration into the upper chamber was noticed. Model showed increase in leakage due to breathing, not due to reaction of the immune system. Yang et al. (2018) developed tumors on a chip with PLGA electrospinning nanofiber membrane. They investigated Gefitinib antitumor drug and resistance of A549 cells in the coculture with HFL1 cells. Secretion of HFL1 cells can be cause of low response of the tumor cells to the chemotherapeutic drugs. Besides that, A549 cells cause apoptosis and death of the endothelial cells, which will permit spreading of tumor cells. Application of this model in personalized, patient-specific tumor case can significantly improve clinical research. Detailed overview of the currently developed lung-on-a-chip models can be found in Nikolic et al. (2018b).

Besides in vitro testing and development of the organ-on-a-chip devices, there is a need for the development of in silico models, in order to reduce experiments on animals and clinical trials. In silico models represent computer models that can simulate the behavior of realistic systems. Improved technology yields to reduce time of the system responses. Results from the calculations are collected faster. Different types of the models can be developed at different scales, depending on observed phenomena. Models are presented usually at three scales—macro, meso, and micro (or nano). If behavior of whole macrosystem is needed, model will be presented at macroscale (e.g., system is solved with finite element method, FEM). Behavior of the systems at the level of molecular clusters can be presented with models at mesoscale (system can be solved with mesomethod; for instance dissipative particle dynamics, DPD). If behavior of the system is required at atomistic, molecular level, then model will be presented at microlevel (system can be solved with molecular dynamics). When models are set, calculations are run and results collect for observation and analysis. Results of in silico model simulations need to be confirmed with accompanying in vitro or in vivo testing. The final goal is to develop adequate in silico models, reduce costs and time of experimental measurements and testing on animals and humans in clinical trial, and provide satisfactory results.

When it comes to the lungs, there are not many in silico lung models. Most of the computer developed lung models include imaging methods for cell biology. Konar et al. (2016) analyzed the application of lung organoid with several different imaging techniques—microscopic analysis. They explain currently mostly used imaging technique—confocal laser scanning microscopy, two-photon and multiphoton microscopy, transmission electron microscopy, scanning electron microscopy and time-lapse imaging. In integration of organoid models with microfluidics, an organ on a chip can be produced. Developed model mimic the microenvironment of the lung tissue and replicate air-liquid interface in the lung. The paper is quite detailed but lacks in mathematical description of the observed system. Results are obtained upon image reconstruction and analysis. Numerical solution to the model is not proposed.

Hancock and Elabbasi (2018) created lung model in COMSOL commercial software based on Huh (2015) model. COMSOL software is used for the simulation of the fluid-structure interaction, laminar fluid flow, nonlinear materials, and dilute species transport, and there is capability of particle tracing (e.g., nanoparticles inhaled into the lungs). They model the polydimethylsiloxane (PDMS) membrane between the upper and lower chamber as neo-Hookean, Mooney-Rivlin, or Ogden nonlinear material models with thickness of 10 μm. Vacuum pressure waveform is applied to deliver the desired membrane strain waveform. Obtained simulation is time dependent and includes membrane and channels deformations. Through the upper chamber flows the air, and through the lower chamber the blood. Particle tracing can be used for the observation of nanoparticles uptake, modeling of bacteria, drug, or nutrient transport within the culture media across the porous membrane. COMSOL model well simulates lung model developed by Huh (2015). Still, mathematical model is not explained and given in the paper.

Airway epithelial model with wound and its closure is mathematically modeled by Savla et al. (2004). The repair of airway epithelium caused by injury is necessary in order to restore epithelial barrier function. Repair of the wound is difficult because of cyclic elongation and compression by which lungs are subjected in the breathing process. Developed mathematical model takes into account spreading, migration and proliferation of the epithelial cells. Proliferation of the cells doesn't occur immediately after cells are injected at the wound borders. Usually, it takes 2 days for proliferation to start. Starting time point of proliferation differs from cells to cells. Mathematically described model represents extended diffusion equation. During the initial phase of the reepithelialization, there is influx of cells into the wound. Simulation results of the proposed model were compared with the experimental measurements of cell density and the rate of wound closure. Parameters for the model are selected based on experimental measurement or taken from the literature. This developed mathematical model is used as reference point in the development of A549 cell model.

Mathematical and computer models of lung-on-a-chip device should be improved in the future. Two models that will be presented and mathematically and numerically explained in this chapter are related to the lung-on-a-chip device, and these models will be further developed. In the literature, many mathematically developed and numerically solved models cannot be found—in silico models of existing organ-on-a-chip devices. Therefore, it is important to work in this area, to harmonize computer models with experimental setups and to achieve faster responses of the systems, with precisely enough results of the in silico models. This will improve organ-on-a-chip research field and will lead to reduced experiments on the animals and reduced clinical trials on the humans.

2 Model of the bioreactor for organ-on-a-chip usage

This section presents model of the bioreactor through which fluid flows containing monocyte cells. Model is developed with purpose to test immune response to biomaterials (Moriarty et al., 2013) and level of biomaterial compatibility with immune cells. The goal is to develop a granuloma-on-a-chip system and to simulate the behavior of the monocytes for optimal physiological attachment to the biomaterial surfaces (Nikolic et al., 2019). Model of the bioreactor can be used as vascular component in lung-on-a-chip model. Currently developed bioreactor model provides insights into fluid flow through the bioreactor and monocyte trajectories. Description of the model with geometrical and physical parameters used for simulation is written in the succeeding text. Section continues with description of mathematical model and applied numerical analysis and finalizes with obtained results. Mathematical model is developed according to observation of the problem—fluid flows through the bioreactor with the constant inlet velocity for which Navier-Stokes equation can be used and monocytes travel together with the fluid, driven by fluid velocity, which can be explained as mass transfer in fluid domain. Model is numerically solved with finite element method, and detailed derivation of the finite element matrices can be found in the following text. Calculations are performed in open source, in-house PAK solver (Kojic et al., 1998a, 2001), which works with finite element method. For purpose of the visualization of the results, CAD application is used.

2.1 Modeling of monocytes distribution inside the bioreactor

The first model is the model of bioreactor presented in Fig. 4.1. Through the bioreactor flows fluid with constant inlet velocity, which is provided with

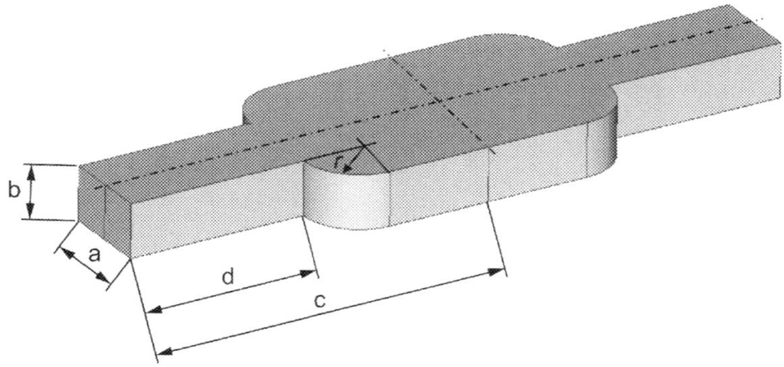

Fig. 4.1 The bioreactor model with marked geometrical parameters.

Table 1 List of the bioreactor geometrical parameters—values and units.

Geometrical parameter	Value	Unit
a	4	[mm]
b	2.5	[mm]
c	15	[mm]
d	8	[mm]
r	2.5	[mm]

peristaltic pump, which pumps flow from the reservoir. Fluid contains the monocyte cells, and their distribution through the bioreactor is observed.

Values of the bioreactor geometrical parameters are similar to the parameters proposed by Sharifi et al. (2019) and they are given in Table 1.

Mesh of the bioreactor model is created with 3D eight-node finite elements. Total number of nodes is 47,841, and total number of elements is 40,356. The upgraded model of the bioreactor should include porous membrane at which monocytes will be bounded during the time and circulation process through the bioreactor. On that way, model can be used to represent immune response of the system and selected cells.

2.2 Mathematical derivation of the bioreactor model

Fluid inside the bioreactor is considered as incompressible, and behavior of the fluid is presented with Navier-Stokes equation (4.1):

$$\rho \frac{\partial V}{\partial t} - \mu \nabla^2 V + \rho (V \cdot \nabla) V + \nabla p = 0 \tag{4.1}$$

Here, in Eq. (4.1), V stands for fluid velocity vector, μ is dynamic viscosity of the fluid, ρ is density of the fluid, p is fluid pressure, t is time, and ∇ is operator nabla. For the stationary flow, the first member in Eq. (4.1) disappears, since there is no change of velocity in time.

Beside Navier-Stokes equation, fluid should fulfil continuity equation. Fluid is considered as incompressible (fluid density is constant), so continuity equation has form written with Eq. (4.2):

$$\nabla \cdot V = 0 \tag{4.2}$$

Fluid flow through the bioreactor is laminar, and expected fluid velocity profile is parabolic. Boundary conditions of the model are constant fluid velocity at inlet boundary (0.01667 $\frac{\text{mm}}{\text{s}}$) and zero pressure at the outlet boundary (open end).

Monocytes inside the fluid are presented as a concentration and modeled as mass transfer in fluid (Eq. 4.3). They are traveling together with fluid, and they are driven by fluid flow:

$$\frac{\partial N}{\partial t} + \nabla(-D\nabla N + NV) = 0 \tag{4.3}$$

In Eq. (4.3), D represents diffusion coefficient of the monocytes and N is concentration (number of cells).

Physical parameters used for simulation of the bioreactor model are listed in Table 2.

In order to simulate presented model of the bioreactor, Eqs. (4.1)–(4.3) ought to be solved. These equations make coupled system, because fluid velocity impacts monocyte distribution. Derivation of the coupled system is given in the following. Coupled system is adjusted to be solved with finite element methods. Monocyte cells are small in size but fall for the physics of the macroscopic level, because movement of the cells occurs due to the motion of the fluid. Because of that, finite element method can be applied, and model can be solved on the macroscopic level. Development of the

Table 2 List of the bioreactor physical parameters—values and units.

Parameter	Name	Value	Unit
ρ	Density of the fluid	$0.99 \cdot 10^{-3}$	[g/mm^3]
μ	Dynamic viscosity	$0.69 \cdot 10^{-3}$	[g/(mm s)]
V_0	Inlet velocity	0.01667	[mm/s]
D	Diffusion coefficient	$4.583 \cdot 10^{-4}$	[mm^2/s]
N	Number of monocytes cells	$6 \cdot 10^3$	[cells/mm^3]

model in more details including chemical reaction of the monocytes may require change of the model scale and transition to lower scale-meso- or microscale.

2.3 Application of the finite element method to model of the bioreactor

Finite element method is well-known numerical method in the field of engineering. It can be applied to many different problems related to fluid and solid behavior, heat transfer, electrical potential, etc. Usually, finite element method is applied when mathematical model consists of higher order partial differential equations. Domain of interest, geometry of the model, which is analyzed, is divided into many finite elements, and equations are solved for each element and value of each node in the element is determined. Equations to be solved are simplified in comparison with the differential partial equations, but number of these equations is very large, and it is almost impossible to solve them analytically. With development of the computers, finite element method becomes often used, since computer enables calculation of large number of the algebraic equations in a short time.

On the model of bioreactor, presented previously, finite element method is applied. Eqs. (4.1)–(4.3) are derived in order to obtain finite element matrices.

Model of bioreactor is analyzed in Cartesian coordinate system. Eqs. (4.4)–(4.8) represent Eqs. (4.1)–(4.3) projected in this coordinate system:

$$\rho\frac{\partial V_x}{\partial t}-\mu\left(\frac{\partial^2 V_x}{\partial x^2}+\frac{\partial^2 V_x}{\partial y^2}+\frac{\partial^2 V_x}{\partial z^2}\right)+\rho\left(V_x\frac{\partial V_x}{\partial x}+V_y\frac{\partial V_x}{\partial y}+V_z\frac{\partial V_x}{\partial z}\right)+\frac{\partial p}{\partial x}=0 \tag{4.4}$$

$$\rho\frac{\partial V_y}{\partial t}-\mu\left(\frac{\partial^2 V_y}{\partial x^2}+\frac{\partial^2 V_y}{\partial y^2}+\frac{\partial^2 V_y}{\partial z^2}\right)+\rho\left(V_x\frac{\partial V_y}{\partial x}+V_y\frac{\partial V_y}{\partial y}+V_z\frac{\partial V_y}{\partial z}\right)+\frac{\partial p}{\partial y}=0 \tag{4.5}$$

$$\rho\frac{\partial V_z}{\partial t}-\mu\left(\frac{\partial^2 V_z}{\partial x^2}+\frac{\partial^2 V_z}{\partial y^2}+\frac{\partial^2 V_z}{\partial z^2}\right)+\rho\left(V_x\frac{\partial V_z}{\partial x}+V_y\frac{\partial V_z}{\partial y}+V_z\frac{\partial V_z}{\partial z}\right)+\frac{\partial p}{\partial z}=0 \tag{4.6}$$

$$\frac{\partial V_x}{\partial x}+\frac{\partial V_y}{\partial y}+\frac{\partial V_z}{\partial z}=0 \tag{4.7}$$

$$\frac{\partial N}{\partial t}-D\left(\frac{\partial^2 N}{\partial x^2}+\frac{\partial^2 N}{\partial y^2}+\frac{\partial^2 N}{\partial z^2}\right)+\left(\frac{\partial N}{\partial x}V_x+\frac{\partial N}{\partial y}V_y+\frac{\partial N}{\partial z}V_z\right)=0 \tag{4.8}$$

In order to obtain finite element matrices, Eqs. (4.4)–(4.8) are multiplied with interpolation functions h_I and integrated over the volume of finite element (Kojic et al., 1998b).

Calculated quantities are presented as the product of interpolation functions and values of the quantities at the appropriate nodes (Eq. 4.9). In this model, quantities of interest are fluid velocity (V), pressure of fluid (p), and monocyte concentration (N):

$$V_i = h_J V_i^J, i = 1,2,3; p = \hat{h}_J P^J; N = h_J N^J \tag{4.9}$$

In order to simplify the form of Eqs. (4.4)–(4.8), index notation is used. Indices $i, j = 1, 2, 3$ stand for x, y, z directions, respectively. First, Navier-Stokes equation is written in index notation:

$$\int_V h_I \rho \frac{\partial V_i}{\partial t} dV - \int_V h_I \mu V_{i,jj} dV + \int_V h_I \rho V_j V_{i,j} dV + \int_V h_I p_{,i} dV = 0 \tag{4.10}$$

$$\begin{aligned} &\int_V h_I \rho \frac{\partial (h_J V_i^J)}{\partial t} dV - \int_V h_I \mu (h_J V_i^J)_{,jj} dV \\ &+ \int_V h_I \rho \left(h_J V_j^J\right) \left(h_J V_i^J\right)_{,j} dV + \int_V h_I \left(\hat{h}_J P^J\right)_{,i} dV = 0 \end{aligned} \tag{4.11}$$

Interpolation functions are function of coordinates, and they are not depending on time. That is why interpolation functions have derivative by coordinates and values of quantities in nodes have derivative by time.

In Eq. (4.11), the second member contains the second partial derivative of velocities by coordinates. On that, member applied Gauss's theorem in order to avoid the second derivative and to obtain surface integral instead (Eq. 4.12). The divergence theorem states that the outward flux of a tensor field through a closed surface is equal to the volume integral of the divergence over the region inside the surface:

$$\begin{aligned} \int_V h_I \mu (h_J V_i^J)_{,jj} dV &= \int_V \mu \left(h_I \cdot \left(h_J V_i^J\right)_{,j}\right)\Big)_{,j} dV - \int_V \mu h_{I,j} \left(h_J V_i^J\right)_{,j} dV \\ &= \int_S \mu h_I \cdot \left(h_J V_i^J\right)_{,j} \cdot n_j dS - \int_V \mu h_{I,j} h_{J,j} V_i^J dV \\ &= \int_S \mu q_n n_j dS - \int_V \mu h_{I,j} h_{J,j} V_i^J dV \end{aligned} \tag{4.12}$$

Gauss's theorem is also applied on integral from Eq. (4.11), which contains pressure:

$$\int_V h_I\left(\hat{h}_J P^J\right)_{,i} dV = \int_S h_I \hat{h}_J P^J n_i dS - \int_V h_{I,i} \hat{h}_J P^J dV \tag{4.13}$$

Integrals from Eqs. (4.12) and (4.13) are substituted into Eq. (4.11):

$$\begin{aligned}\int_V \rho h_I h_J dV \dot{V}_i^J &+ \int_V \mu h_{I,j} h_{J,j} dV V_i^J + \int_V \rho h_I\left(h_J V_j^J\right) h_{J,j} dV V_i^J \\ &- \int_V h_{I,i} \hat{h}_J dV P^J = \int_S \mu q_n n_j dS - \int_S h_I \hat{h}_J P^J n_i dS\end{aligned} \tag{4.14}$$

$$M_t \dot{V}_i^J + \left(K_{V\mu} + K_{V\rho}\right) V_i^J + K_{VP} P^J = F_S = 0 \tag{4.15}$$

Matrices from Eq. (4.14) are written in the succeeding text:

$$M_t = \rho \int_V h_I h_J dV \tag{4.16}$$

$$K_V = K_{V\mu} + K_{V\rho} \tag{4.17}$$

$$K_{V\mu} = \mu \int_V h_{I,j} h_{J,j} dV \tag{4.18}$$

$$K_{V\rho} = \rho \int_V h_I\left(h_J V_j^J\right) h_{J,j} dV \tag{4.19}$$

$$K_{VP} = -\int_V h_{I,i} \hat{h}_J dV \tag{4.20}$$

In Eq. (4.15) on the right-hand side, there is a force, F_S, representing total surface force, which is equal to zero due to natural boundary condition.

With Eq. (4.21), continuity equation is written in index notation, and further derivation is performed to form finite element matrix:

$$V_{i,i} = 0 \tag{4.21}$$

$$\int_V \hat{h}_I V_{i,i} dV = 0 \tag{4.22}$$

$$\int_V \hat{h}_I\left(h_J V_i^J\right)_{,i} dV = 0 \tag{4.23}$$

$$\int_V \hat{h}_I h_{J,i} dV V_i^J = K_{PV} V_i^J = 0 \tag{4.24}$$

$$K_{PV} = \int_V \hat{h}_I h_{J,i} dV = K_{VP}{}^T \tag{4.25}$$

Mass transfer in index notation and derivation is presented in the succeeding text:

$$\frac{\partial N}{\partial t} - DN_{,ii} + N_{,i} V_i = 0 \tag{4.26}$$

$$\int_V h_I \frac{\partial (h_J N^J)}{\partial t} dV - \int_V h_I D (h_J N^J)_{,ii} dV + \int_V h_I (h_J N^J)_{,i} (h_J V_i^J) dV = 0 \tag{4.27}$$

In Eq. (4.27) occurs second derivative of monocyte concentration by coordinates, and on that, member should apply Gauss's theorem:

$$\begin{aligned} \int_V h_I D (h_J N^J)_{,ii} dV &= \int_V D\left(h_I \cdot (h_J N^J)_{,i}\right)\big)_{,i} dV - \int_V D h_{I,i} \cdot (h_J N^J)_{,i} dV \\ &= \int_S D h_I \cdot (h_J N^J)_{,i} \cdot n_i dS - \int_V D h_{I,i} h_{J,i} N^J dV \\ &= \int_S D q_n n_i dS - \int_V D h_{I,j} h_{J,j} N^J dV \end{aligned} \tag{4.28}$$

Transformed second derivative of monocyte concentration by Cartesian coordinates from Eq. (4.28) is substituted in Eq. (4.27):

$$\int_V h_I h_J \dot{N}^J dV + \int_V D h_{I,j} h_{J,j} N^J dV + \int_V h_I h_{J,i} N^J h_J V_i^J dV = \int_S D q_n n_i dS = 0 \tag{4.29}$$

Surface force from Eq. (4.29) is equal to zero, since there is no leakage, that is, mass transfer through the walls of the bioreactor is disabled.

Matrices from Eq. (4.29) are listed in the succeeding text:

$$N_t = \int_V h_I h_J dV \tag{4.30}$$

$$K_{CN} = \int_V D h_{I,i} h_{J,i} dV \tag{4.31}$$

$$K_{CV} = \int_V h_I h_{J,j} N^J h_J dV \tag{4.32}$$

Finally, obtained system of the equations in matrix form is presented with Eq. (4.33):

$$\begin{bmatrix} M_t & 0 & 0 \\ 0 & 0 & 0 \\ 0 & 0 & N_t \end{bmatrix} \begin{Bmatrix} \dot{V} \\ \dot{P} \\ \dot{N} \end{Bmatrix} + \begin{bmatrix} K_V & K_{VP} & 0 \\ K_{PV} & 0 & 0 \\ K_{CV} & 0 & K_{CN} \end{bmatrix} \begin{Bmatrix} V \\ P \\ N \end{Bmatrix} = \begin{Bmatrix} F \\ 0 \\ 0 \end{Bmatrix} \tag{4.33}$$

Vector V contains three component of velocity (V_x, V_y, and V_z) for each node.

Matrix equation is solved with implicit scheme, by writing balance equations for the $t + \Delta t$ moment. Solving equations by iterations requires correction of the right-hand side of the equations (Kojic et al., 1998b).

2.4 Results of the bioreactor model

Finite element bioreactor model, explained previously, is solved in open source in-house PAK solver, developed at the University of Kragujevac and BioIRC (Kojic et al., 1998a, 2001, 2008). PAK solver has been used previously for solving many different engineering problems, and solutions are validated in comparison with the solutions from the other commercial finite element solvers as well as to the experimental measurements. Great amount of PAK applications relates to the bioengineering problems. PAK solver has been successfully used for modeling of vascular problems—the coronary arteries (Filipovic et al., 2012; Parodi et al., 2012), aorta (Filipovic et al., 2013) and stent implementation (Nikolic et al., 2014), modeling of inner (Vulovic et al., 2019) and middle ear (Isailovic et al., 2014), etc.

Results of the bioreactor model obtained in PAK solver are compared with the results obtained in commercial software ANSYS (ANSYS Fluent [version 6.3, ANSYS, the United States]). In Fig. 4.2, the obtained fluid velocities in PAK and ANSYS software are presented. Results showed good matching in velocity profile and in values of fluid flow velocity for inlet velocity of 0.01667 mm/s (Nikolic et al., 2018a).

Beside velocities, pressure distribution through the bioreactor obtained in the PAK is compared with the one obtained in ANSYS. Results are presented in Fig. 4.3.

One of the results from the bioreactor model solved in PAK solver can be shear stress. Obtained shear stress distribution of this model is shown in Fig. 4.4.

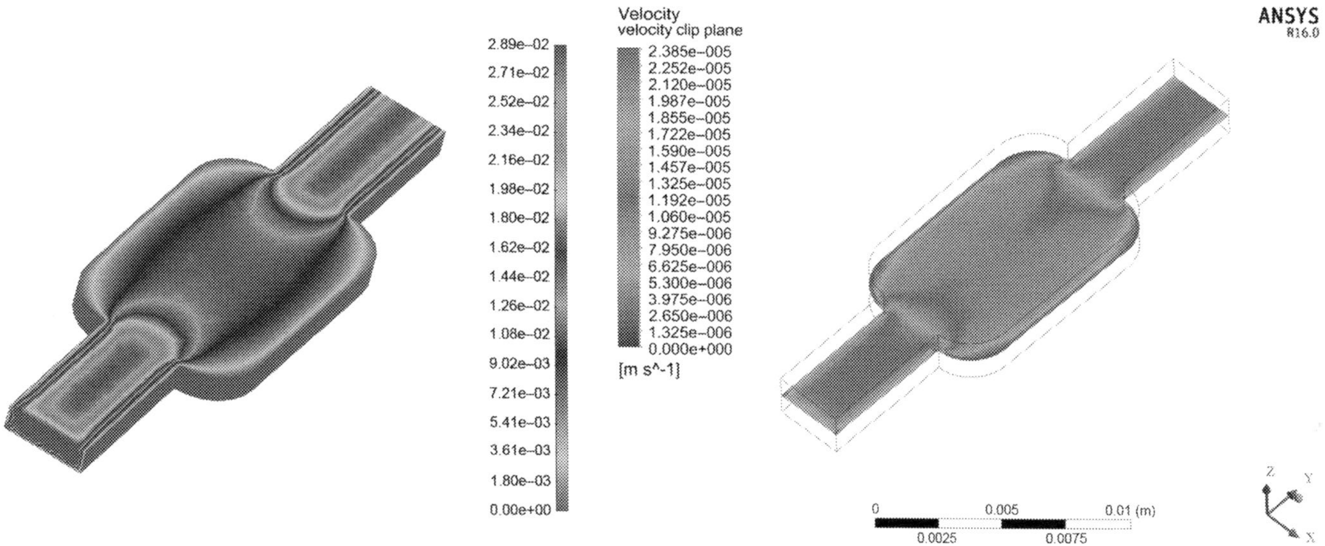

Fig. 4.2 Fluid velocity—PAK (*left*: half of the model, mm/s), ANSYS (*right*: half of the model, m/s).

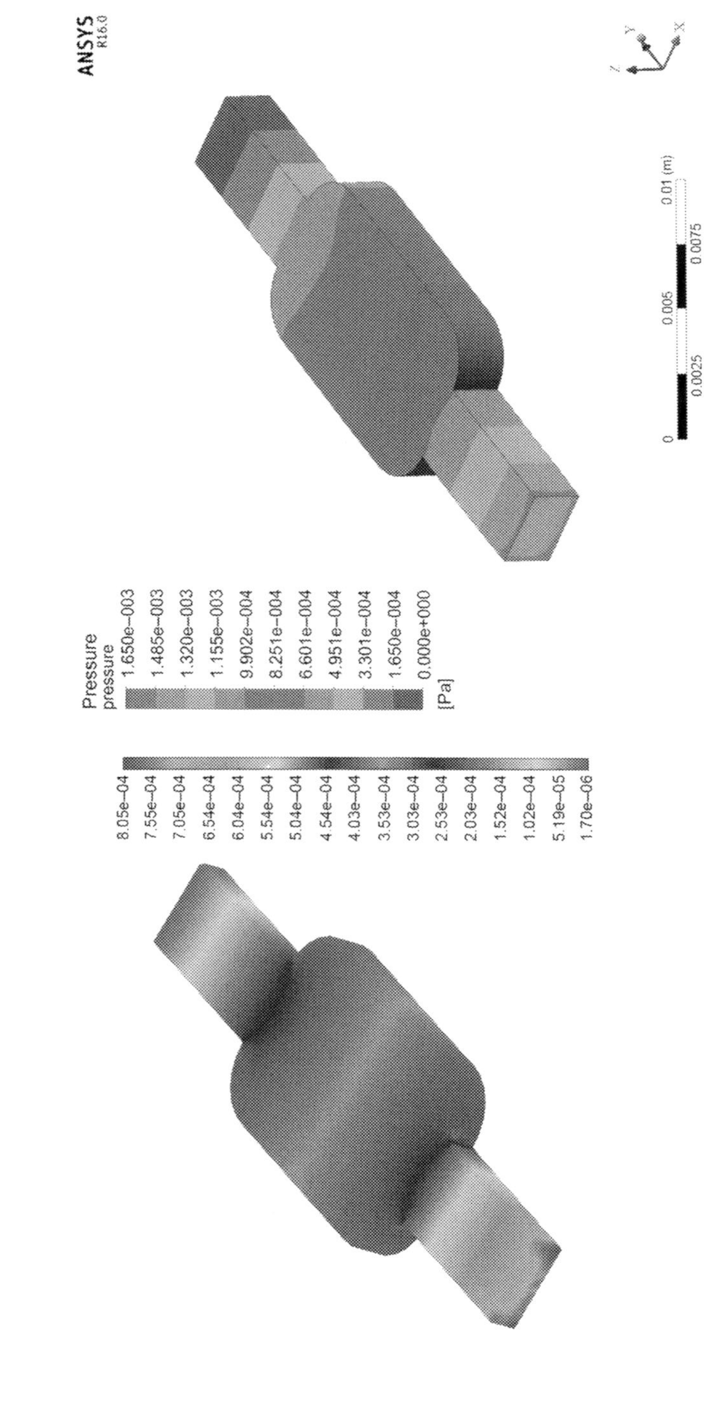

Fig. 4.3 Pressure distribution—PAK (*left*: full model, Pa), ANSYS (*right*: full model, Pa).

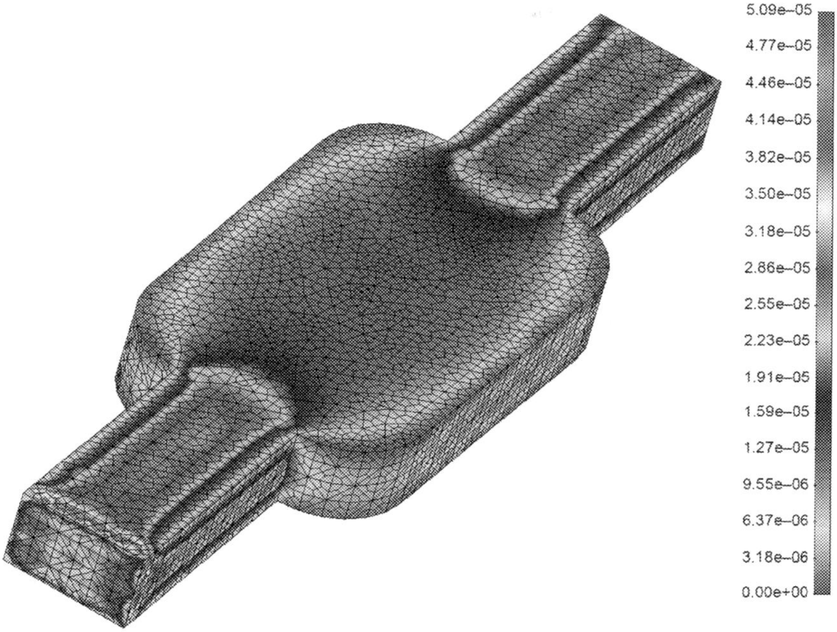

Fig. 4.4 Shear stress distribution—PAK (Pa).

As far as monocytes are concerned, their distribution along the bioreactor is monitored. Monocytes movement in the model is presented as mass transfer in fluid. Monocytes are driven by fluid velocity—drag force and Stoke's law:

$$\vec{F_D} = 6\pi\mu R\,\vec{V} \tag{4.34}$$

In Eq. (4.34), F_D stands for drag force; μ is dynamic viscosity of the fluid; R is radius of the spherical particle, in this case radius of the monocytes; and V is flow velocity. Velocity of the monocytes is calculated according to the second Newtonian law, where force on the right-hand side of the equation is drag force.

Concentration, as a number of monocyte cells, is prescribed at the inlet side, and further distribution is observed. Monocyte trajectories obtained as a result from PAK solver and ANSYS software are given in Fig. 4.5.

The presented model of the bioreactor with fluid that contains monocytes and flows through it is a good starting point for further analysis of monocytes and their bounding to the substrate, to the bottom of the bioreactor. Monocytes bounding to the substrate will serve for simulation of

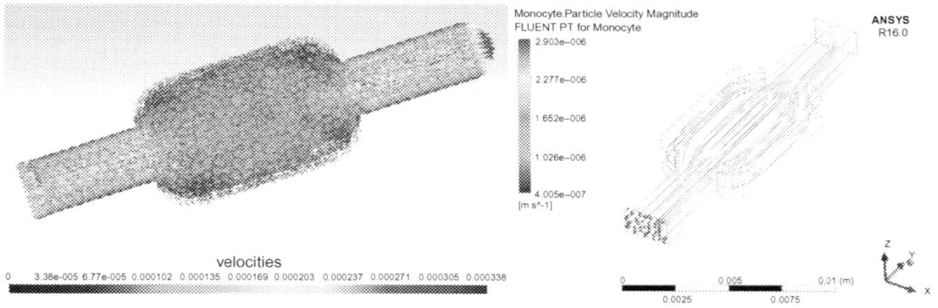

Fig. 4.5 Monocyte trajectories—PAK (*left*, mm/s), ANSYS (*right*, m/s).

inflammatory process. Inflammatory process assumes attraction of the monocytes to the certain place and formation of the macrophage. As a concept of monocyte, bounding can be used as a model of atherosclerosis and LDL bounding to the blood vessel, found in Filipovic et al. (2014).

3 Model of the A549 lung epithelial cell line

A549 cell line is a cancerous cell line, first extracted from cancerous lung tissue of 58-year-old Caucasian male in 1972. A549 cells can be used as alveolar pulmonary epithelium (Foster et al., 1998) in lung models and thus used for examination of the metabolic processes of lung tissue. A549 cell line share many characteristics with human primary alveolar epithelial cells (Mazzarella et al., 2007). A549 cells can also serve well in lung models for drug testing. Lee et al. (2014) treated the human lung adenocarcinoma cell line A549 with variety of drugs and different drug concentrations and observed the cell viability with standardized MTT assay. Their research showed that drug acriflavine (ACF) can inhibit cell growth in A549 cells and provide a new antitumor strategy against nonsmall cell lung cancer (NSCLC). Yang et al. (2018) tested gefitinib on lung-on-a-chip model, containing A549 cell line. Many in vitro testing were performed on A549 cell line either in examination of the metabolic processes that occur in lungs or in examination of existing and novel drugs. A549 cells are also used in the development of several lung-on-a-chip models. In silico models of A549 cell line are mostly based on the application of artificial network (Taghipour et al., 2015; Zendehdela and Shirazi, 2015). Peng et al. (2013) include finite element analysis in their paper using commercial software COMSOL (COMSOL Multiphysics [Ver. 4.3, COMSOL Inc., Burlington, MA]) but without derivation of the mathematical model. Due to wide application range of A549 cell line and lack of finite element models, we start developing in silico model of A549 cell line using finite element method.

3.1 Modeling of A549 cell behavior and barrier formation

With the second model, behavior of lung epithelial cell line A549 was investigated. A549 cells, virally immortalized epithelial cells from lung carcinoma, are seeded on the glass slide (substrate). Diameter of the glass slide used for simulation is $Dg = 1$ mm, and number of seeded cells is 104 cells per glass slide (equal to 33 cells/mm^2). Between seeded cells and substrate, thin layer of fluid is located. Thickness of fluid layer used for simulation is 10 μm. Model with geometrical parameters is presented in Fig. 4.6.

Cells initially seeded at the top of the model move toward the bottom of the model. This movement cannot be explained only by diffusion process. Realistic experimental setup shows that cells go downward, which can be explained by sedimentation process. Because of that, sedimentation coefficient of the A549 cells should be determined from experimental measurement. Beside sedimentation process, cells show spreading behavior. At certain point of time, cells will start to proliferate. To include spreading and proliferation occurrence in the model, corresponding coefficients should be calculated. For calculation of the mentioned coefficient, imaging technique should be applied on the real experiments. Cell behavior can be observed at certain time points; images of the cells on the slide should be taken and used for analysis and calculation of the spreading, diffusion, and proliferation coefficients. Based on these parameters, computer simulations can be conducted. The information of interest is how fast the substrate surface can be covered by a specific cell type. The simulated response time is time for functional epithelial barrier to be formed.

Mesh of the epithelial lung cell model is created with 3D eight-node finite elements. Total number of nodes is 17430, and total number of

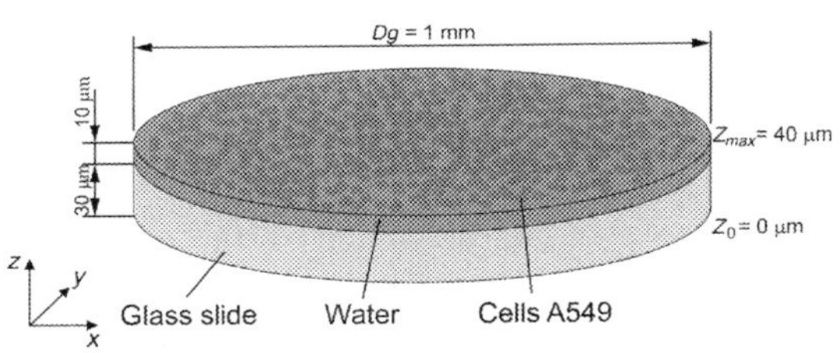

Fig. 4.6 Glass slide where A549 cells are seeded.

elements is 14466. After validation of the proposed model, the upgraded second model can include cell death into calculation. The model can become more complex if there raise a need for modeling of cell interaction with surroundings or with endothelial cells. Mathematical model of the currently developed A549 cell model is given in the following.

3.2 Mathematical derivation of the A549 cell model

Cells are seeded on the glass surface. Initially, between seeded cells and substrate, there is a layer of fluid with thickness of 10 μm. Movement of the cells toward the substrate can be described with diffusion equation (Eq. 4.35):

$$\frac{\partial n}{\partial t} = D\left(\frac{\partial^2 n}{\partial x^2} + \frac{\partial^2 n}{\partial y^2} + \frac{\partial^2 n}{\partial z^2}\right) \tag{4.35}$$

Here, in Eq. (4.35), n represents cell density [cells/mm^2], t means time [h], and D is diffusion coefficient [mm^2/h]). In case of the stationary flow, the first member of Eq. (4.35), the member on the left-hand side of the equation, is neglected (equal to zero).

Beside diffusion process, cells show spreading, and at certain time after seeding, they start to proliferate. These characteristics should be included in model development. Savla et al. (2004) modeled wound closure of epithelial barrier and took into account the spreading and proliferation of the cells. Based on this work, diffusion equation has been extended for model of A549 cell line barrier formation. Proliferation of the cells usually starts a day or two after the seeding and for simulations within this time only spreading of the cells is taken into account (Eq. 4.36):

$$\frac{\partial n}{\partial t} = D\left(\frac{\partial^2 n}{\partial x^2} + \frac{\partial^2 n}{\partial y^2} + \frac{\partial^2 n}{\partial z^2}\right) + k_s\left(1 - \frac{n}{n_0}\right)n \tag{4.36}$$

For simulations of the A549 cell behavior when they start to proliferate, Eq. (4.37) is used:

$$\frac{\partial n}{\partial t} = D\left(\frac{\partial^2 n}{\partial x^2} + \frac{\partial^2 n}{\partial y^2} + \frac{\partial^2 n}{\partial z^2}\right) + \left(k_s + k_p\right)\left(1 - \frac{n}{n_0}\right)n \tag{4.37}$$

Newly introduced parameters in Eqs. (4.36) and (4.37) are spreading coefficient of A549, k_s, [1/h], proliferation coefficient of A549, k_p, [1/h], and initial number of cells, n_0, [cells/mm^2].

Model of epithelial cell line should show sedimentation of the cells on the glass substrate, so Mason-Weaver equation is used.

Finally, equation to be solved for simulation of the A549 cell line behavior on the substrate has form given with Eq. (4.38):

$$\frac{\partial n}{\partial t} = D\left(\frac{\partial^2 n}{\partial x^2} + \frac{\partial^2 n}{\partial y^2} + \frac{\partial^2 n}{\partial z^2}\right) + \left(k_s + k_p\right)\left(1 - \frac{n}{n_0}\right)n + sg\frac{\partial n}{\partial z} \tag{4.38}$$

In Eq. (4.38), beside already defined properties, s stands for coefficient of sedimentation [h] and g is gravitational acceleration ($g = 9.81\frac{m}{s^2} = 127,14 \cdot 10^9\ \frac{mm}{h^2}$).

Product of sedimentation coefficient and corresponding acceleration, in this case gravitational, defines terminal velocity, V_t (Eq. 4.39):

$$V_t = s \cdot g = \sqrt{\frac{2mg}{\rho A C_d}} \tag{4.39}$$

Parameters affecting sedimentation coefficient are as follows: mass of the falling objects, m (seeded cells); density of the fluid through which objects falling, ρ (water is fluid between cells and substrate); projected area from the objects, A (cross-sectional area of the cells); and drug coefficient, C_d.

Boundary conditions of the extended diffusion equation are initial cell density at the top of the model, where the cells are seeded:

$$n(z_{max}, 0) = n_0 = 33\ \frac{\text{cells}}{\text{mm}^2} \tag{4.40}$$

Boundary condition for sedimentation process is that there is no leaking at the top and bottom of the system. Presented in Eq. (4.41), boundary condition has the following form:

$$D\frac{\partial n}{\partial z} + \text{sgn} = 0 \tag{4.41}$$

In the following section, application of the finite element method to the model of epithelial lung cells A549 is presented.

3.3 Application of the finite element method to the A549 cells model

Once again, finite element method is employed to solve A549 cell model. Radius of the epithelial cells is about 10 μm, but settling of the A549 cells on the glass slide can be presented on macroscale, due to sedimentation and diffusion processes; thus finite element method can be appropriate for this purpose.

In order to formulate finite element matrices, Eq. (4.38) is multiplied with interpolation functions (h_I) and integrated over the volume of the finite elements. Model is projected in Cartesian coordinate system:

$$\begin{aligned}\int_V h_I \frac{\partial n}{\partial t} dV &= \int_V h_I D\left(\frac{\partial^2 n}{\partial x^2} + \frac{\partial^2 n}{\partial y^2} + \frac{\partial^2 n}{\partial z^2}\right) dV \\ &+ \int_V h_I (k_s + k_p)\left(1 - \frac{n}{n_0}\right) n dV + \int_V h_I s g \frac{\partial n}{\partial z} dV\end{aligned} \tag{4.42}$$

Density of the A549 cells can be presented as product of interpolation functions and values of the concentration in the appropriate nodes:

$$n = h_J N^J \tag{4.43}$$

Interpolation functions are depending on coordinates and not depending on time:

$$\begin{aligned}\int_V h_I \frac{\partial (h_J N^J)}{\partial t} dV &= \int_V h_I D\left(\frac{\partial^2}{\partial x^2} + \frac{\partial^2}{\partial y^2} + \frac{\partial^2}{\partial z^2}\right)(h_J N^J) dV \\ &+ \int_V h_I (k_s + k_p)\left(1 - \frac{h_J N^J}{n_0}\right) h_J N^J dV + \int_V h_I s g \frac{\partial (h_J N^J)}{\partial z} dV\end{aligned} \tag{4.44}$$

Eq. (4.44) is written in index notation and presented in the succeeding text:

$$\begin{aligned}\int_V h_I \frac{\partial (h_J N^J)}{\partial t} dV &= \int_V D h_I (h_J N^J)_{,ii} dV \\ &+ \int_V (k_s + k_p) h_I \left(1 - \frac{h_J N^J}{n_0}\right) h_J N^J dV + \int_V s g h_I (h_J N^J)_{,3} dV\end{aligned} \tag{4.45}$$

On the first member on the right-hand side of Eq. (4.45), Gauss's theorem is applied (divergence theorem), so the second partial derivative of concentration (cell density) is replaced with one surface integral and one volume integral, which includes only first partial derivative:

$$\begin{aligned}\int_V D h_I (h_J N^J)_{,ii} dV &= \int_V D\left(h_I (h_J N^J)_{,i}\right)_{,i} dV - \int_V D h_{I,i} (h_J N^J)_{,i} dV \\ &= \int_S D h_I (h_J N^J)_{,i} n_i dS - \int_V D h_{I,i} h_{J,i} N^J dV \\ &= \int_S D q_n n_i dS - \int_V D h_{I,i} h_{J,i} N^J dV\end{aligned} \tag{4.46}$$

Presented integral transformation (Eq. 4.46) is included in Eq. (4.45). After arranging, Eq. (4.45) has the following form:

$$\int_V h_I h_J \dot{N}^J dV + \int_V D h_{I,i} h_{J,i} N^J dV - \int_V (k_s + k_p) h_I \left(1 - \frac{h_J N^J}{n_0}\right) h_J N^J dV$$
$$- \int_V sgh_I h_{J,3} N^J dV = \int_S D q_n n_i dS = F_S \tag{4.47}$$

Eq. (4.47) is then written in matrix form:

$$C\dot{N}^J + (K_D + K_{SP} + K_S) N^J = F_S = \int_S D q_n n_i dS \tag{4.48}$$

Matrices from Eq. (4.48) are defined in the succeeding text:

$$C = \int_V h_I h_J dV \tag{4.49}$$

$$K_D = D \int_V h_{I,i} h_{J,i} dV \tag{4.50}$$

$$K_{SP} = -(k_s + k_p) \int_V h_I \left(1 - \frac{h_J N^J}{n_0}\right) h_J dV \tag{4.51}$$

$$K_S = -sg \int_V h_I h_{J,3} dV \tag{4.52}$$

Eq. (4.41) defining boundary condition for sedimentation process is integrated over the surface:

$$\int_S h_I D (h_J N^J)_{,3} n dS + \int_S h_I sgh_J N^J n dS = 0 \tag{4.53}$$

$$\int_S D h_I h_{J,3} N^J n dS + \int_S sgh_I h_J N^J n dS = 0 \tag{4.54}$$

From Eq. (4.54) comes out surface force, F_S, figuring in Eq. (4.48) if diffusion is considered only in z direction:

$$F_S = \int_S D h_I h_{J,3} N^J n dS = -\int_S sgh_I h_J N^J n dS \tag{4.55}$$

Matrix equation is solved with implicit scheme, by writing balance equations for the $t + \Delta t$ moment. Solving equations by iterations requires correction of the right-hand side of the equations (Kojic et al., 1998b).

3.4 Results of the A549 cell model

At the beginning of the simulation, there is an initial concentration at the top of the model, representing the number of A549 cells. Half of the model in the first time step is presented in Fig. 4.7.

Time step used in simulation is 10^{-4}h, and values of the needed coefficients are given in Table 3.

In the Figs. 4.8 and 4.9 are presented values of the cell concentration at different time points.

As it can be seen from the Figs. 4.8 and 4.9, concentration of the cells is moving toward the bottom surface. After certain time period, all the cells should be placed at the bottom of the model. Quantity of interest is time for cells to reach the bottom surface and to form monolayer.

The model showed strong dependence on the values of the parameters, resulting in the issue with convergence. This model needs to be validated with experimental measurements. It means that values of the coefficients should be determined from the experimental observation. The easiest way to define the values of the coefficients is to use certain imaging technique to make images of the cells in different time points and then to analyze images in order to obtain values of the parameters for simulation. With precisely defined parameters, simulation will be run, and time and concentration of the A549 cells at the bottom surface will be compared in order to validate the A549 developed model.

4 Discussion and conclusions

Organ-on-a-chip devices represent new epoch in bioengineering and biotechnology. It is a very interesting area for research with tremendous contribution to medicine. These devices can be used for simulation of healthy organs with their building cells observed in detailed physiological environment. That is a major advantage of organ-on-a-chip devices over *in vitro* testing where parts of the tissues and cells are isolated and observed with reduced physiological conditions. Final goal of organ-on-a-chip technique is to simulate functioning of the certain organ in state of disease and observe related processes that occurs and reaction to the applied drugs, whether they are already approved by health sector or in a stage of clinical trial.

This chapter provides detailed introduction section with thoroughly explained organ-on-a-chip device concept and major advantages of this new approach in organ behavior modeling in comparison with currently

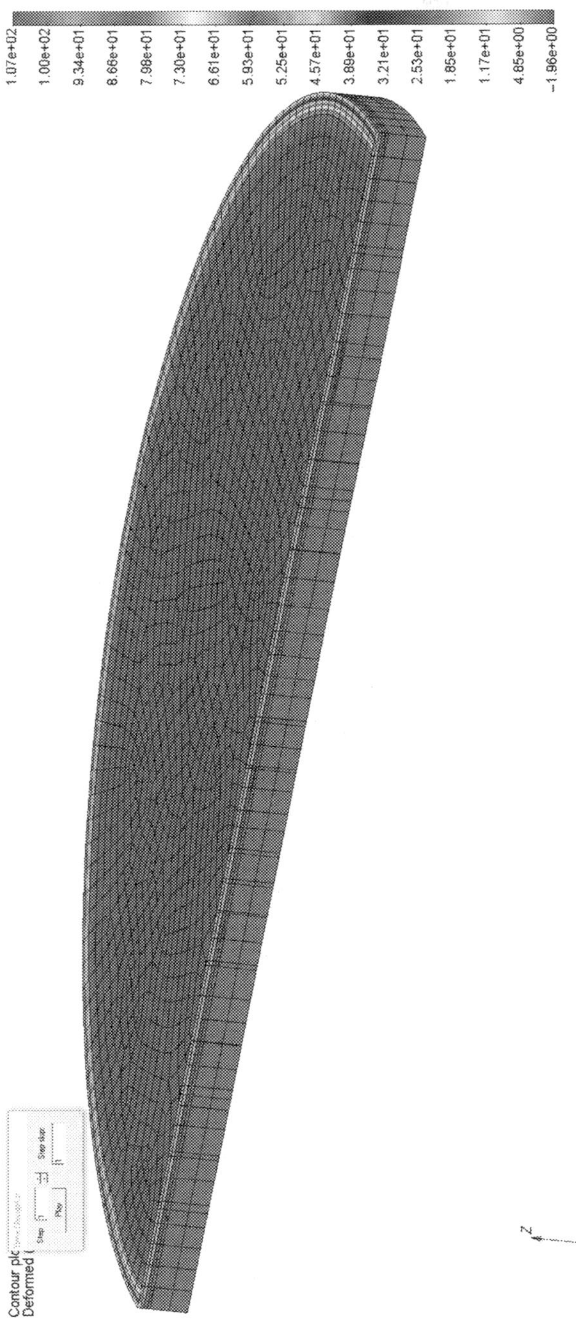

Fig. 4.7 Half of the A549 cell model at the first time step.

Table 3 Parameters used for simulation of the A549 cell model.

Parameter	Name	Value	Unit
D	Diffusion coefficient	0.0031	[mm^2/h]
k_s	Spreading coefficient	0.6	[1/h]
k_p	Proliferation coefficient	0.03	[1/h]
V_t	Terminal velocity	2	[mm/h]
Δt	Time step	$1 \cdot 10^{-4}$	[h]

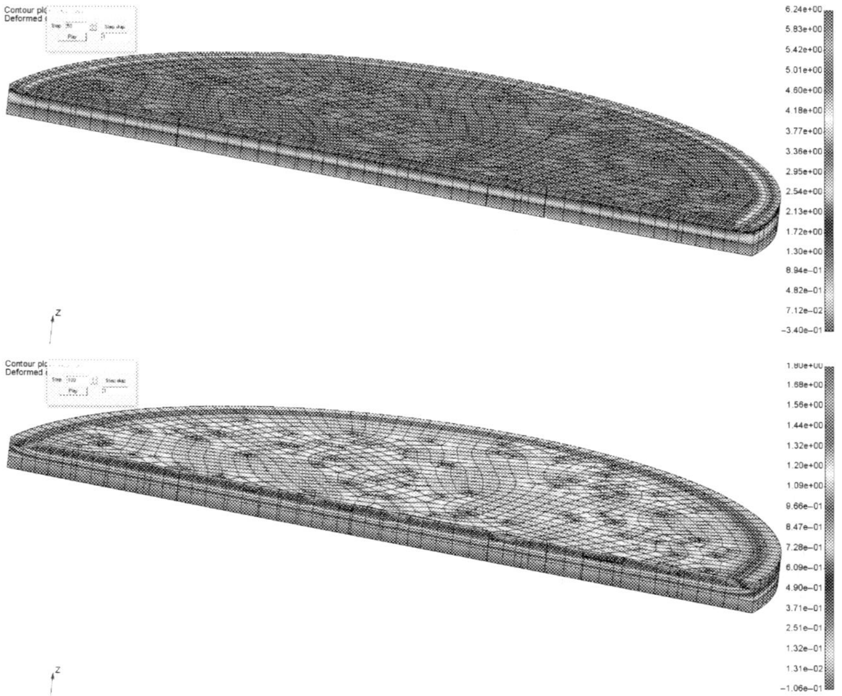

Fig. 4.8 A549 cell concentration in the time steps—$t=50$ steps (*upper*) and $t=100$ steps (*lower*).

available in vitro and in vivo testing. In the continuation of the introductory part, which organs are made on chips are briefly mentioned. This chapter focuses on the lung-on-a-chip model. That is why the progress of the developed lung-on-a-chip model is given in introduction section. Introduction section further includes description of the in silico models, specifically few lung-on-a-chip in silico models, that is found in literature. The first model presented in this chapter is the model of the bioreactor, which will

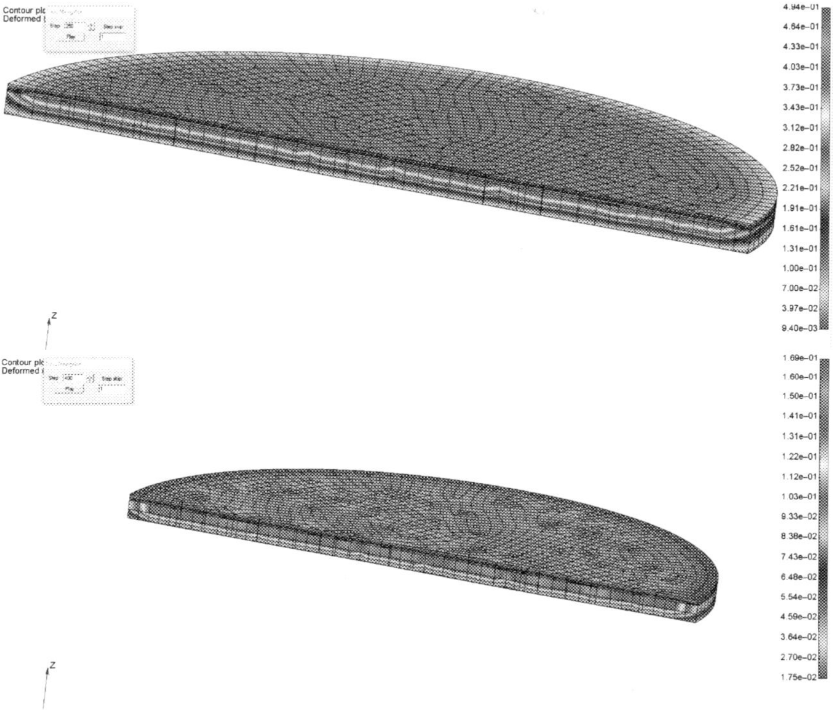

Fig. 4.9 A549 cell concentration in the time steps—$t=250$ steps (*upper*) and $t=400$ steps (*lower*).

further be extended in order to present granuloma-on-a-chip model. Currently developed bioreactor model includes fluid flow through the bioreactor together with monocytes inside of it. The model is mathematically derived and numerically solved using finite element method. Calculations are performed with PAK solver. Obtained results are compared with the one obtained in commercial finite element software ANSYS. Compared results show good matching in fluid velocity and fluid pressure, as well as in monocyte velocities. This model will be further extended to include monocytes attachment to the bottom of the bioreactor in order to simulate immune response of the system. More detailed insight into monocytes behavior can be achieved by the development of the model on the lower scale. The second model presented in this chapter relates to the model of A549 epithelial cells. Initially, model is developed to simulate settling of the seeded cells from top of the model to the bottom of the model, on the glass substrate. Presented model includes diffusion, sedimentation, spreading, and proliferation of the cells. Cell death is omitted from this model and can be included in the next version of the A549 cell model. The next item in the development of the A549 cell model is validation of the model with experimental measurements. The plan is to determine appropriate coefficients of the model via experiments and later to compare obtained results between the in silico model and experimental observations. This model with epithelial barrier formed on the substrate will be extended to application in the lung-on-a-chip model and subjected to several different environmental conditions.

Presented models in this chapter are not complex in physical manner. The bioreactor model is confirmed with ANSYS model, and A549 cell model needs to be validated with experimental measurements. Nevertheless, they are good starting point for further upgrading. The bioreactor model will be upgraded to include monocytes bounding to the substrate and to simulate immune response of the system with final goal to form granuloma-on-a-chip model that will be validated with experimental measurements during European PANBioRA project. A549 cell model first needs to be set in tune with experimental measurements and then to be upgraded to produce more physiological properties of the cells with final goal to be applied for lung-on-a-chip model. There are not many in silico models developed to accompany created organ-on-a-chip devices. Few in silico models that can be found in the literature are mainly based on cell biology and imaging technique. Research papers including mathematical model of the organ on a chip are rare to find as well. The goal of the organ-on-a-chip devices is to reduce and eliminate animal testing and clinical trials and usage of computer simulations has important role in that aspect.

Organ-on-a-chip systems are new area of research, but still there are many scientists who are skeptical when it comes to replacement of the animal testing and reduction of clinical trials. Their belief is that human organism is a very complex system and that further improvement of the organ-on-a-chip devices still will not include all the parameters from the environment affecting the tissue and the cells. Of course, that is true, but organ-on-a-chip devices can give better insight into functioning of the healthy organs and in finding more details of specific organ diseases. Specifically, they can be useful for testing of existing and novel drugs.

Lung-on-a-chip devices, which are in focus of this chapter, include lung airway epithelium cells and microvascular endothelium cells and interaction between these cells. Microvascular endothelium cells are in charged for activation of the immune system—activation of the immune cells, such as neutrophils and monocytes—and due to that fact, inflammation process can be simulated. Besides regular behavior of the lung, these microfluidic devices can be used for collecting information and new knowledge about lung diseases, as well as for testing novel drugs approaches and treatments.

The research process should go this way: *in vitro* testing is the starting point in formation of organ-on-a-chip devices. Upon organ on a chip, detailed in silico models should be developed, which should provide fast response resulting in reduced animal testing and clinical trial. Finally, organ-on-a-chip devices together with in silico models need to be verified with in vivo testing. Currently, there are good bases for several organs developed on-chip technology, but there are not many accompanying in silico models. Our future investigation will go in that direction—development of the in silico models.

Acknowledgments

This work was supported by the Ministry of Education, Science and Technological Development of the Republic of Serbia, project number III41007 and by the European Project H2020 PANBioRA (Personalised and generalised integrated biomaterial risk assessment), grant number 760921. This chapter reflects only the author's view. The commission is not responsible for any use that may be made of the information it contains.

References

Benam, K.H., Villenave, R., Lucchesi, C., Varone, A., Hubeau, C., Lee, H.H., et al., 2015. Small airway-on-a-chip enables analysis of human lung inflammatory and drug responses in vitro. Nat. Methods 13 (2), 151–157. ISSN 1548-7105. https://doi.org/10.1038/nmeth.3697.

Filipovic, N., Nikolic, D., Saveljic, I., Djukic, T., Adjic, O., Kovacevic, P., Cemerlic-Adjic, N., Velicki, L., 2013. Computer simulation of thromboexclusion of the complete aorta in the treatment of chronic type B aneurysm. Comput. Aided Surg. 18 (1-2), 1–9. ISSN 1097-0150. https://doi.org/10.3109/10929088.2012.741145.

Filipovic, N., Rosic, M., Tanaskovic, I., Milosevic, Z., Nikolic, D., Zdravkovic, N., Peulic, A., Fotiadis, D., Parodi, O., 2012. ARTreat project: three-dimensional numerical simulation of plaque formation and development in the arteries. IEEE Trans. Inf. Technol. Biomed. 16 (2), 272–278. ISSN 1089-7771, PMID: 21937352. https://doi.org/10.1109/TITB.2011.2168418.

Filipovic, N., Zivic, M., Obradovic, M., Djukic, T., Markovic, Z., Rosic, M., 2014. Numerical and experimental LDL transport through arterial wall. Microfluid. Nanofluid. 16 (3), 455–464. ISSN 1613-4982. https://doi.org/10.1007/s10404-013-1238-1.

Foster, K.A., Oster, C.G., Mayer, M.M., Avery, M.L., Audus, K.L., 1998. Characterization of the A549 cell line as a type II pulmonary epithelial cell model for drug metabolism. Exp. Cell Res. 243 (2), 359–366. ISSN 0014-4827. https://doi.org/10.1006/excr.1998.4172.

Hancock M. J., Elabbasi N., 2018. Modeling a Lung-On-A-Chip Microdevice, COMSOL Software. Available online at: https://www.comsol.com/paper/download/257421/hancock_abstract.pdf.

Hao, S., Ha, L., Cheng, G., Wan, Y., Xia, Y., Sosnoski, D.M., Mastro, A.M., Zheng, S.Y., 2018. A spontaneous 3D bone-on-a-chip for bone metastasis study of breast cancer cells. Small 14 (12). https://doi.org/10.1002/smll.201702787. ISSN 1613-6810.

Huh, D.D., 2015. A human breathing lung-on-a-chip. Ann. Am. Thorac. Soc. 12 (1), S42–S44. ISSN 2325-6621. https://doi.org/10.1513/AnnalsATS.201410-442MG.

Isailovic V., Nikolic M., Milosevic Z., Saveljic I., Nikolic D., Radovic M. and Filipovic N., (2014), Finite element coiled cochlea model, 12th International Workshop on the Mechanics of Hearing, June 23–29, 2014, Cape Sounio, Greece, AIP Conference Proceedings, vol. 1703, No. 070015-1–070015-4, ISBN 978-0-7354-1350-4, ISSN 0094-243x, doi https://doi.org/10.1063/1.4939389.

Kitsara, M., Kontziampasis, D., Agbulut, O., Chen, Y., 2019. Heart on a chip: micro-nanofabrication and microfluidics steering the future of cardiac tissue engineering. Microelectron. Eng. 203 (204), 44–62. ISSN 0167-9317. https://doi.org/10.1016/j.mee.2018.11.001.

Knowlton, S., Tasoglu, S., 2016. A bioprinted liver-on-a-chip for drug screening applications. Trends Biotechnol. 34 (9), 681–682. ISSN 0167-7799. https://doi.org/10.1016/j.tibtech.2016.05.014.

Kojic, M., Filipovic, N., Stojanovic, B., Kojic, N., 2008. Computer Modeling in Bioengineering—Theoretical Background, Examples and Software. John Wiley and Sons, Chichester, England. ISBN: 978-0-470-06035-3.

Kojic, M., Filipovic, N., Živkovic, M., Slavkovic, R., Grujovic, N., 1998a. PAK-F Finite Element Program for Laminar Flow of Incompressible Fluid and Heat Transfer. Kragujevac, Serbia.

Kojic, M., Filipovic, N., Živkovic, M., Slavkovic, R., Grujovic, N., 2001. PAK-FS Finite Element Program for Fluid-Structure Interaction. Kragujevac, Serbia.

Kojic, M., Slavkovic, R., Zivkovic, M., Grujovic, N., 1998b. Metod konacnih elemenata I—Linearna analiza (en Finite element method I—Linear analysis). Masinski fakultet u Kragujevcu (en Faculty of Mechanical Engineering Kragujevac), ISBN 86-80581-27-5.

Konar, D., Devarasetty, M., Yildiz, D.V., Atala, A., Murphy, S.V., 2016. Lung-on-a-chip technologies for disease modeling and drug development. Biomed. Eng. Comput. Biol. 7 (1), 17–27. ISSN 1179-5972. https://doi.org/10.4137/BECB.S34252.

Lee, J., Kim, S., 2018. Kidney-on-a-chip: a new technology for predicting drug efficacy, interactions, and drug-induced nephrotoxicity. Curr. Drug Metab. 19 (7), 577–583. ISSNe 1875-5453, ISSNp 1389-2002. https://doi.org/10.2174/1389200219666180309101844.

Lee, C.J., Yue, C.H., Lin, Y.J., Lin, Y.Y., Kao, S.H., Liu, J.Y., Chen, Y.H., 2014. Antitumor activity of acriflavine in lung adenocarcinoma cell line A549. Anticancer Res. 34 (11), 6467–6472. ISSN 0250-7005.

Mazzarella, G., Ferraraccio, F., Prati, M.V., Annunziata, S., Bianco, A., Mezzogiorno, A., Liguori, G., Angelillo, I.F., Cazzola, M., 2007. Effects of diesel exhaust particles on human lung epithelial cells: an in vitro study. Respir. Med. 101 (6), 1155–1162. ISSN 0954-6111. https://doi.org/10.1016/j.rmed.2006.11.011.

Moriarty, T.F., Zaat, S.A., Busscher, H.J. (Eds.), 2013. Biomaterials Associated Infection—Immunological Aspects and Antimicrobial Strategies. Springer Science & Business Media. ISBN 978-1-4614-1031-7.

Nikolic, D., Radovic, M., Aleksandric, S., Tomasevic, M., Filipovic, N., 2014. Prediction of coronary plaque location on arteries having myocardial bridge, using finite element models. Comput. Methods Prog. Biomed. 117 (2), 137–144. ISSN 0169-2607. https://doi.org/10.1016/j.cmpb.2014.07.012.

Nikolic, M., Sustersic, T., Filipovic, N., 2018b. In vitro models and on-chip systems: biomaterial interaction studies with tissues generated using lung epithelial and liver metabolic cell lines. Front. Bioeng. Biotechnol. 6 (120), 1–13. eISSN: 2296-4185. https://doi.org/10.3389/fbioe.2018.00120.

Nikolic, M., Sustersic, T., Muller, C.B., Zhang, Y.S., Vrana, N.E., Filipovic, N., 2019. Monocyte behaviour under perfusion conditions for development of granuloma on-a-chip. In: TERMIS EU 2019, 27th to 31st of May 2019, Rhodes, Greece.

Nikolic, M., Sustersic, T., Saveljic, I., Vrana, N.E., Filipovic, N., 2018a. Modelling of Monocytes Behaviour inside the Bioreactor. In: Belgrade BioInformatics Conference, BelBi 2018, June 18-22, 2018, Belgrade, Serbia, Book of Abstracts, Biologia Serbica-vol. 40, p. 123. No. 1 (Special Edition). ISSN 2334-6590, UDK 57 (051).

Parodi, O., Exarchos, T., Marraccini, P., Vozzi, F., Milosevic, Z., Nikolic, D., Sakellarios, A., Siogkas, P., Fotiadis, D., Filipovic, N., 2012. Patient-specific prediction of coronary plaque growth from CTA angiography: a multiscale model for plaque formation and progression. IEEE Trans. Inf. Technol. Biomed. 16 (5), 952–956. ISSN 1089-7771. https://doi.org/10.1109/TITB.2012.2201732.

Peng, C.C., Liao, W.H., Chen, Y.H., Wu, C.Y., Tung, Y.C., 2013. A microfluidic cell culture array with various oxygen tensions. Lab Chip 13 (16), 3239–3245. ISSN 1473-0197. https://doi.org/10.1039/c3lc50388g.

Savla, U., Olson, L.E., Waters, C.M., 2004. Mathematical modeling of airway epithelial wound closure during cyclic mechanical strain. J. Appl. Physiol. 96 (2), 566–574. pISSN 8750-7587, eISSN 1522-1601. https://doi.org/10.1152/japplphysiol.00510.2003.

Sharifi, F., Htwe, S.S., Righi, M., Liu, H., Pietralunga, A., Yesil-Celiktas, O., Maharjan, S., Cha, B.H., Shin, S.R., Dokmeci, M.R., Vrana, N.E., Ghaemmaghami, A.M., Khademhosseini, A., Zhang, Y.S., 2019. A foreign body response-on-a-chip platform. Adv. Healthcare Mater. 8 (4), 1801425. https://doi.org/10.1002/adhm.201801425.

Taghipour, M., Vand, A.A., Rezaei, A., Karim, G.R., 2015. Application of artificial neural network for modeling and prediction of MTT assay on human lung epithelial cancer cell lines. J. Biosens. Bioelectron. 6 (2), 1–6. ISSN 2155-6210. https://doi.org/10.4172/2155-6210.1000170.

Vulovic, R., Nikolic, M., Filipovic, N., 2019. Smart Platform for the Analysis of Cupula Deformation caused by Otoconia Presence within SCCs. Comput. Methods Biomech. Biomed. Eng. 22 (2), 130–138. ISSN 1476-8259. https://doi.org/10.1080/10255842.2018.1539166. Published online: 24 Dec 2018.

Wheeler, B.C., 2008. Building a brain on a chip. Conf. Proc. IEEE Eng. Med. Biol. Soc. 2008, 1604–1606. https://doi.org/10.1109/IEMBS.2008.4649479.

Yang, X., Li, K., Zang, X., Liu, C., Guo, B., Wen, W., et al., 2018. Nanofiber membrane supported lung-on-a-chip microdevice for anti-cancer drug testing. Lab Chip 18 (3), 486–495. ISSN 1473-0197. https://doi.org/10.1039/c7lc01224a.

Zendehdela, R., Shirazi, F.H., 2015. Discrimination of human cell lines by infrared spectroscopy and mathematical modeling. Iran. J. Pharm. Res. 14 (3), 803–810. ISSN 1735-0328.

CHAPTER 5

Aortic dissection: Numerical modeling and virtual surgery

Contents

Computational Modeling in Bioengineering and Bioinformatics
https://doi.org/10.1016/B978-0-12-819583-3.00005-9

1 Introduction

Cardiovascular disease is a general term that implies a number of clinical entities that are commonly localized in the heart and/or blood vessels. Cardiovascular and vascular diseases are among the most common diseases in developed countries and are the leading causes of death. According to statistical data relating to the structure of mortality, these diseases are among the most common causes of death, accounting for 50% of all deaths (Hagan et al., 2000).

Modern lifestyle, daily stress exposure, irregular nutrition, obesity, and reduced physical activity are just some of the causes of hypertension, which is one of the most significant risk factors for the development of cardiovascular diseases. Hypertension is an acute condition or a chronic disease in which arterial blood pressure is increased. Often, there are no recognizable symptoms, which is why it is also called a "quiet killer." In most patients, arterial hypertension usually occurs without or with minimal symptoms, and in most cases, it is only aneurysm and aortic dissection that reveal it; thus, early detection and treatment are necessary.

Dissection of the aorta is a very serious and life-threatening condition in which the inner layer of the aortic wall breaks down and stretches. The mortality rate is 1% per hour after the onset of dissection, and about 98% of patients die in the first few days. It begins with a small crack in the intimal layer, but as the blood flowing through the aorta has high pressure (over 120 mmHg during the systole), a small crack becomes rapidly wider. The blood penetrates the aortic wall through the process, continues to split and spread the layer of the medium, and creates a new channel in the aortic wall, creating another, the so-called false lumen.

Aortic dissection is a life-threatening event because it can lead to the development of a stroke, a sudden heart failure, a break in blood flow to the main organs, or a complete split of the aorta—a burst of aorta and deadly bleeding. There are two systems in relation to the extensiveness of the dissection process according to which the aortic dissection classification is performed: Stanford and DeBakey classification. The Stanford classification divides the aortic dissection into type A (proximal) and type B (distal). Type A dissection always involves the ascending aorta and can spread to the arch of the aorta and possibly along the descending aorta, while type B always involves a downward aorta and can retrograde the aortic arch. DeBakey classification divides aortic dissection into subgroups. Type I affects the entire aorta, type II involves the ascending aorta, and type III involves the descending aorta. Basically, Stanford type A implies DeBakey type I and II, while Stanford type B implies DeBakey type III (Reul et al., 1975). All dissection groups can be seen in both acute and chronic phases. Chronic dissection is characterized by more than 14 days of acute anxiety from the onset of first symptoms.

The basic hypotheses that underlie this chapter are based on the analysis of the current state in scientific fields of this scientific research. From the standpoint of fluid mechanics, the cardiovascular system is considered as a system in which the flow occurs with the domination of viscous or inertial forces. For the flow of blood in the aorta, the dominant inertial forces are viscous, while the ratio of inertial (nonstationary) and viscous forces is defined by Womersley number. Reynolds numbers at the maximum systole are of the order of $Re = 4000$. Calculating the wall shear stress is obtained over the friction velocity. The wall shear stress and pressure in the fluid domain, as well as the von Mises stress in the solids domain, indicate the potential locations of the false lumen rupture and the reverse process of the false lumen into the true lumen.

Numerical simulations of dynamic behavior of fluids, that is, the blood in the aorta with and without dissection, and simulation of flows and pressures in the obstructed aortic branches due to dissection are of great importance in medicine. Simulations of this type can help physicians a lot because they can provide insight into further development of the disease. In support of this, according to the International Registry of Acute Aortic Dissection (IRAD), mortality after surgery is about 25%–30%, in the case of type I and II dissection according to DeBakey, so these serious and urgent conditions in cardiac surgery require full attention to better understand this phenomenon. Also, it is very important to perform virtual surgery of the ascending aorta and replacement with the tubus graft after which it is easy to determine the flows and pressures in the surgically treated aorta, while the operation is not physically performed.

2 Aortic dissection

2.1 History of aortic dissection

English doctor Frank Nicholls (1699–1778), studying the arteries, more precisely aorta and aortic aneurysm, observed in 1728 that aneurysm rupture may occur or only the rupture of the inner layers allows the extension of the external lumen. As the personal doctor of the King of England, George II, he had the opportunity to record the first official case of aortic dissection.

The King of England died suddenly in 1760 at his palace. It was discovered by autopsy that the King died of acute aortic dissection of type A that affects the ascending part of the aorta. In 1819, Rene Laënnec (1791–1826), a French physician and inventor of the stethoscope, first introduced the term of aortic dissection with the aim of describing splitting of the intima distally from the aortic valve and in the longitudinal space in aortic wall with a 67-year-old man who suddenly died. Much later, in 1863, Thomas Bevill Peacock (1812–82), a famous English cardiologist, published his research on 80 patients describing the three stages of the disease. The first phase is splitting the intima layer, the second is extending of the dissected hematoma with possible cracking, and the third is the lumen recanalization. Until the onset of the aortography in 1939, the aortic dissection was only postmortemally diagnosed. The first successful dissection operation on the descending part of the aorta was carried out in 1955 by the famous American surgeon Michael Ellis DeBakey (1908–2008). The next successful operation on the ascending aortic part was carried out in 1963 by the American cardiac surgeon Morris G.C. Jr. In the early 1980s, echocardiography, computerized tomography, and magnetic resonance techniques were included in the diagnostic process. Since aortic dissection is a rare diagnosis and the number of patients in hospitals is relatively small, the International Registry of Acute Aortic Dissection was established in 1996 with the aim of collecting data on etiological factors, clinical characteristics, and treatment and patient outcomes of patients worldwide. The organization consists of 30 large centers in 11 countries around the world and all with the same goal—to better understand this phenomenon and contribute to reducing the percentage of mortality.

2.2 Classification of aortic dissection

Dissection of the aorta or dissecting aortic aneurysm is a disorder of the aorta characterized by the longitudinal splitting of its wall, whereby the blood enters the newly formed duct and forms another, false aortic lumen. It starts with a tear in the intima, allowing the blood to enter the inner layer of the blood vessel. Usually, tears occur transversely and do not include the full extent of the aorta (Khan and Nair, 2002). The so-called false lumen represents the filled space between the separated layers of the aorta. Depending on the volume that the tear occupies, the dilation of a false lumen can quite compress the true lumen (Fig. 5.1).

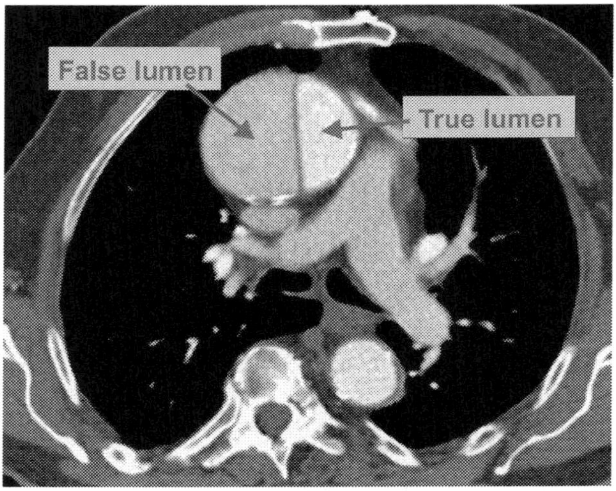

Fig. 5.1 Two-dimensional image of true and false lumens.

During the systole, the true lumen expands and collapses in the diastole due to pressure change. The blood in it flows forward and is usually closer to the inner curvature of the aortic arch. Unlike the true lumen, the false lumen is narrowed down during the systole and is located in the outer part of the aorta. Blood flows slowly through it, and thrombus can also be present. Based on the thickness of the wall and intima cleft, it is possible to determine the starting and ending points of the dissection. Depending on the flow in the false lumen, there are communicating and noncommunicating aortic dissections. In communicating dissections, the flow in the false lumen can be forward, backward, or in the form of a delayed flow. In the case of non-communicating dissection, there is a thrombus in the false lumen while the flow is present. It occurs as a result of rupture of the media layer leading to intramural bleeding.

Aortic dissection is classified as acute and chronic depending on the duration of symptoms. Aortic dissection diagnosed within 2 weeks after the onset of symptoms, when most complicated life-threatening situations occur, is considered acute, and if it is present for 2 weeks, it is considered chronic. Anatomically, there are three diagrams of the aortic dissection classification that can be found in the literature. They are classified according to the site of injection of tears and based on proximal or distal involvement. Sixty-five percent of aortic dissection occurs in the ascending part of the aorta, 20% in the downstream part, 10% in the aortic artery, and 5% in the abdominal

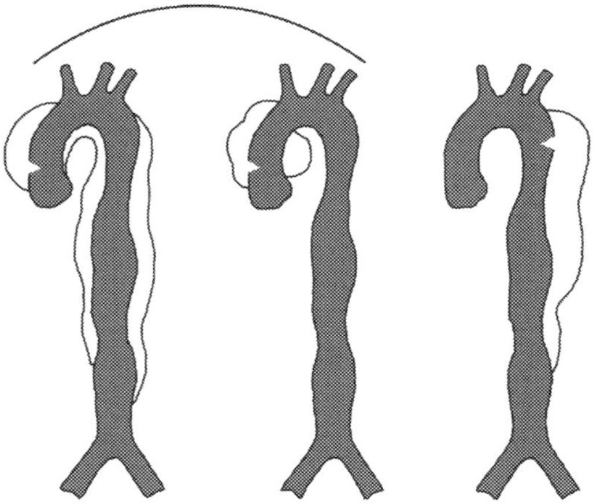

Fig. 5.2 Classification of aortic dissection.

aorta (Hagan et al., 2000). In 1965, American surgeon DeBakey proposed three classification schemes (Fig. 5.2).

DeBakey I begins in the ascending part of the aorta, propagates through the aortic arch, and includes the entire aorta. This type of dissection is most commonly seen in patients less than 65 years of age and represents the deadliest form of the disease. DeBakey II only affects the ascending aorta and is limited to that part, while the DeBakey III classification begins in the descending part of the aorta, rarely includes proximal aorta but extends distally and captures the rest of the aorta. The Stanford classification divides the aorta dissection into type A, which only affects the ascending part of the aorta, and type B, which only affects the descending part of the aorta.

Acute dissection of the type A aorta is very deadly, with mortality between 1% and 2% per hour after the first symptoms occur. The characteristic symptom of this type of dissection is surprisingly strong, punctuating pain in the chest (in about 85% of cases) or behind the chest (46%) and also stomach pain (22%), sudden loss of consciousness (13%), impaired pulse (20%), and rarely a stroke (6%) (Hagan et al., 2000; Mehta et al., 2002). Acute dissection of type A always requires surgical treatment. Treatment using only drugs is associated with a mortality rate of almost 24% to 24 h, 30% to 48 h, 40% to 7 days, and 50% to 1 month. Even after surgery, the mortality rate is 10% to 24 h, 13% to 7 days, and almost 20% to 30 days (Suzuki et al., 2003).

Aortic rupture, stroke, visceral ischemia, and cardiac tamponade are the most common causes of death. Current surgical techniques on the ascending aorta have an aim to replace or repair aortic root and aortic valve (if necessary). Elimination of the remaining false lumen and repair of aortic dissection on the descending aorta are not in the foreground. Replacement or repair of the ascending aorta does not eliminate the full flow and pressure in the distal part of the false lumen. It happens that in less than 10% of surgical operations of aortic dissection of this type, there is a development of a postoperative false lumen that is destroyed later. Acute dissection of the type B aortic is less lethal than type A but does not differ too much in the clinical picture. The most common symptoms of this disorder are back pain (64%) and chest pain (63%), as well as sudden abdominal pain (43%) and hypertension. Stroke occurs slightly less (21%), but ischemia can also occur (Suzuki et al., 2003). Patients with minor complications of type B disorder have a mortality rate of 10% within 30 days. In contrast, in the case of developing leg ischemia, renal insufficiency, or visceral ischaemia, urgent surgical repair of the aorta is required. The mortality rate in these cases is 20% per day and 25% per month. The two most common sites in which the entry tear is formed are the aortic root, in type I and II dissection, and several centimeters below the left subclavian artery, for the third type of dissection, according to DeBakey.

3 Diagnostic techniques

In diagnosing aortic dissection, various direct and indirect and invasive and noninvasive imagine techniques are used. The choice of diagnostic techniques depends on the patient's stability, local expertise, and availability of the technique itself. The main objectives of the diagnostic methods are the rapid confirmation of the presence (or absence) of acute aortic dissection, classification of the type of dissection, identification and location of intimal tears, confirmation of the presence of true/false lumen and thrombus (if present), etc. Within this chapter, various imaging techniques will be presented, taking into account technical capabilities, specific diagnostic methods, and precision of the method itself.

3.1 Transthoracic and transesophageal echocardiography

Echocardiography, or ultrasound of the heart, is a noninvasive diagnostic method that uses ultrasonic waves to show the heart and blood vessels.

It has been used in cardiology since 1953. There is a one-dimensional (M-mod), two-dimensional, and three-dimensional echocardiography. In addition to the previously mentioned, there is also Doppler echocardiography that measures the flow of blood in the heart, as well as a colored echocardiography that shows the flows of blood in the heart and blood vessels in blue and red with the goal of determining the direction of flow.

Transthoracic echocardiography (TTE) is a method that determines the morphological appearance and function of the heart muscle, the valve, and the pericardium, as well as the ascending aorta. The transducer works with a 1–5 MHz band probe. The patient needs to lie down on the left side. Then, using a probe to which the gel has previously been placed, look for a heart image in the foreseen position intersections. Transcranial echocardiography is performed in standard cross sections: parasternal, apical, and subcostal. The ascending aorta can be best look in the left parasternal projections, while the aortic arch is best seen from suprasternal cross section. Information on the extension of dissection to adjacent arteries, a brachiocephalic tree, a left carotid, or subclavian, may give orthogonal and longitudinal fields of recording.

Transesophageal echocardiography (TEE) uses an ultrasound probe inserted into the esophagus via the endoscope. As only the esophagus wall divides the heart and blood vessels from the heart, blood vessels are much better seen than in transthoracic echocardiography. TEE works with a 3.5- and 7-MHz probe. More modern TEE probes allow one or more cross sections to be recorded, allowing viewing of the aorta at any angle, and therefore, images obtained at different angles can be used for 3D reconstruction.

The diagnosis of aortic dissection by using these echocardiographic methods has the task of detecting the intima cleft in the aortic wall. The sensitivity of TTE echocardiography is from 77% to 80%, while the specificity ranges from 93% to 96% for the ascending aorta (Mintz et al., 1979). The diagnostic value of transthoracic echocardiography is limited in patients with altered chest cone contraction and narrow intercostal spaces, in patients with permanent enlargement of respiratory tract, and in obese people. Fortunately, all of these restrictions have been overcome by using transesophageal echocardiography. The European cooperative study group, composed of highly experienced echocardiographists, showed that TEE sensitivity is as high as 99%, and the specificity is 89%. A positive predictive value of 89% was also determined, while the negative predictive value was 99% (Erbel et al., 1989).

3.2 Computerized tomography

Computerized tomography (CT) is a radiological method of recording, which, in addition to X-rays, uses tomography, a method based on the mathematical procedure of recording images using modern computers. Conventional CT devices (first-generation devices, rough images, and scanning lengths of about 30 min) provide for each cycle of X-ray tubing data used to reconstruct axial shots, while newer versions, such as spin CT (third-generation devices; shorter recording procedure), have the ability to simultaneously move the table on which the patient lies and exposure to X-rays. With the use of newer versions of CT devices, the length of the scan is considerably shortened, as it is possible to get the image during only one air retention. This diagnostic method is the most used one in patients who are suspected of having aortic dissection. Transaxial images can be displayed at various planes—sagittal, coronal, and axial. In the part of the body being shot, it is injected at a rate of 1 to 3 mL/s of 120 mL of low-molecular-weight contrast. The first part of the examination covers the chest aorta, and the second part, also by the same technique, includes the abdominal aorta and the iliac branches. The task of computerized tomography is to display a split of intima layer that separates the true lumen from the false and the expansion of the aorta. As with the previous diagnostics technique, in this case, there are certain limitations of interpretation—artifacts due to the presence of veins and the displacement of the aorta. Artifacts due to veins are identified on the basis of their direction, which varies from cross section to cross sections and because they cross the boundaries of the aortic wall, while artifacts due to aortic motion are reduced by using linear interpolation algorithms of 180 degree. In the early 1990s, large studies were done to analyze aortic dissection using CT, and it was found that the sensitivity of this method was 83%–94%, and the specificity of 87%–100% (Erbel et al., 1989). By comparing the spiral CT with a conventional one, it should be said that the spiral CT has an advantage over the length of the examination, and it also enables a better analysis of the images. The sensitivity of the spiral CT apparatus is 93% and the specificity of 98% (Sommer et al., 1996).

3.3 Magnetic resonance

The magnetic resonance (MR) is a noninvasive and precise diagnostic method that gives a picture of the health of certain organs. It is completely painless and harmless to the patient, so the examination can be repeated several times. MR devices record signals from the hydrogen nuclei found in molecules of the human body, which were previously placed in a strongly homogeneous magnetic field. The magnetic field unit is tesla (T). Depending on the strength of the magnetic field, MR devices are divided into devices with low field strengths, up to 0.5T; medium field strengths, 0.5–1T; and high field strengths, 1T and more. Conventional MR techniques require a lot of time to record the entire aorta. A typical representative is a spin-echo technique. A three-dimensional contrast MR aortography is one of the newer MR techniques in the diagnosis of aortic dissection. The obtained images are similar to invasive aortography, but the field of view is much larger. Much of the literature (Prince et al., 1996; Krinsky et al., 1997) shows that this method can be successfully used in the diagnosis of aortic dissection. Therefore, the combination of MR aortography with cross sections of MR images provides a good technique for displaying the anatomy of the entire aorta, which can be significant when planning the surgical intervention and details related to the dissecting aortic membrane. The magnetic resonance, unlike TEE, gives much more information about the expansion of aortic dissection, distal part of the ascending aorta and aortic arch. It shows in a much better manner the place of the beginning and the end of the dissection, while the sensitivity in these cases is 90% (Deutsch et al., 1994). A significant advantage of MR is that this technique can be easily and rapidly dissected in the aorta to proximal and distal part.

3.4 Aortography

The first precise method for evaluating patients with suspected dissection of the aorta was a retrograde aortography. The first published work containing the diagnosis of dissection setup on the basis of aortography was in 1939 by Robb and Steinberg. Twenty years later, aortography has become a routine method in estimating the aortic dissection. The diagnosis of aortic dissection is based on direct angiographic signs, such as splitting intima and the existence of two separate lumens, and indirect signs, such as an irregular contour of lumen aorta, wall thickening, aortic abnormality, and aortic regurgitation (Iliceto et al., 1984). For the presentation of breast aortic dissection, an adequate contrast agent is used, which is injected at a rate of 20–25 mL/s in a volume of 40–50 mL. (Sanders, 1990). The accuracy of the aortography is slightly greater than 95%, while the sensitivity is lower than in other techniques in the case of atypical forms of aortic dissection (Khandheria, 1993). A European cooperative study conducted in 1989 showed that the sensitivity and accuracy of the aortography for the diagnosis of dissection was 88% (Erbel et al., 1989). The percentage was not greater due to the inability of the technique to distinguish two lumens in the aorta in the case of a completely thrombosis of false lumen (Erbel et al., 1989; Erbel et al., 1993). The imperfection of aortography is because it is an invasive technique, which requires careful handling of the catheter, which can only be provided by an experienced physician. It can be very dangerous to push the catheter because it can lead to injury if it is pushed into a false lumen. Because of this, on the basis of the difference in pressure and use of the contrast medium, it is necessary to determine the difference between the false and the true lumen. Also, the intestinal split, as well as the distal end of aortic dissection can be difficult to define clearly if there is a small flow. The greatest problem is in the case of a patient with an unstable hemodynamic condition. The preparation of the aortography teams, as well as the necessary equipment, can last for too long, which certainly affects the increase in mortality (Deeb et al., 1997). In some patients with aortic dissection, before the surgical procedure, it is necessary to display the anatomy of the coronary arteries. Despite new and precise techniques, coronarography is still widely used when assessing the condition of the entire coronary tree (Creswell et al., 1995). This has to be done, because although chronic coronary atherosclerotic disease does not have to affect the dissection of the aorta, it may have an effect on the outcome of surgical treatment.

3.5 Intravascular ultrasound

Intravascular ultrasound is a medical imaging technique using a small ultrasound probe located at the top of the catheter. In this way, it is possible to display the inside of the patient's blood vessel. The recording process lasts up to 10 min. Using a 10-MHz probe, it is possible to reach a 3.3-mm caliber. In this case, the probe is positioned over a catheter with a diameter of 8.9 mm. The advantage of this technique is that it can complement the information that doctors receive by angiography (Yamada et al., 1995). As the IVUS technique shows the structure of the aorta lumen blood vessel wall, it is very accurate to see the characteristics of the aortic wall. Intravascular ultrasound is a great method in the case of classical dissections, because it shows the split of intima or media layer, as well as its size, and also the degree of lumen narrowing. This technique has an excellent result in determining the most distant part of the chest aorta disorder. In the literature, data on sensitivity and accuracy of almost 100% can be found. This technique reveals thrombosis in a false lumen with much greater precision and sensitivity than TEE (Yamada et al., 1995).

Intravascular ultrasound can show a change in the aortic wall due to bleeding in the media layer due to the increase in wall thickness (Alfonso et al., 1995). This technique is very precise in describing the dimensions of the hematoma, and as hematomas are semicircular or circular, they can easily be seen as a thickening of the aorta. Also, IVUS takes advantage in the case when other techniques do not give results, especially when there is a case of exclusion of the existence of intima pulsing tear, as in the case of deep penetrating atherosclerotic ulceration. Disadvantages of intravascular ultrasound are seen in the fact that it is difficult to precisely localize the initial splitting that can be seen more often in abdominal than in the thoracic aorta and then the lack of Doppler capabilities, as the mainstay of this technique.

4 Treatment of acute aortic dissection

The task of treating acute aortic dissection is the prevention of death and irreversible ischemic damage to the abdominal organs and lower extremities. Unlike patients with type A acute dissection, where surgical treatment is almost always necessary, treatment with medicaments, in the absence of serious complications, is in most cases applied in acute aortic dissection of type B (Elefteriades et al., 1999; Augoustides and Andritsos, 2010). Reasons for decision-making on the treatment of medicines based on numerous studies (Hagan et al., 2000; Golledge and Eagle, 2008) are

- nearly two-thirds of approximately 85% of patients who were discharged from the hospital after treatment with medication felt well without complications after aggressive antihypertensive drug therapy (Svensson et al., 2008),
- avoiding the risk of emergency surgery,
- a similar long-term outcome for survivors after any treatment with medications or surgical.

In all patients suspected of having acute aortic dissection, one of the diagnostic methods should be done to exclude the possibility of an error if the wrong therapeutic approach was made in the event of any other illness.

4.1 Treatment with medicaments

The main goal of drug therapy in the case of aortic dissection is to reduce the wall shear stress on the affected parts of the aorta by reducing blood pressure and cardiac contractility. A large number of patients with aortic disorders have comorbidity such as coronary artery disease, hypertension, and chronic kidney disease. Therefore, the prevention strategy and self-treatment must be similar to those of these diseases. Cessation of smoking is a very important step, as studies have shown that smoking induces significantly faster spread of the aorta (about 0.4 mm/year) (Brady et al., 2004). Moderate physical activity also slows down the progression of atherosclerosis. To prevent heart attacks, patients with an enlarged aorta must avoid competitive sports.

Patients with a suspected acute aortic dissection must be transferred to an intensive care unit, and a diagnostic assessment of their condition is urgently needed. Control of pain and blood pressure targeted to 110 mmHg is achieved using morphine sulfate and intravenous beta blockers (metoprolol, esmolol, or labetalol) or in combination with vasodilating agents such as sodium nitroprusside or angiotensin-converting enzyme inhibitors. Intravenous verapamil may be used in case beta blockers cause contraindications. Monotherapy with beta blockers can be good for controlling mild hypertension, and in combination with sodium nitroprusside at an initial dose of 0.3 μg/kg/min, it often acts as an effective agent and in severe hypertension states. In patients with lower blood pressure, a careful estimate of blood loss, pericardial effusion, or cardiac insufficiency is required before administration of volume. Patients with high hemodynamic instability often require intubation, mechanical ventilation, and urgent transesophageal echocardiography or CT as a confirmation of the previous technique of recording. In rare cases, an external ultrasound diagnosis of tamponade of the heart may justify emergency sternotomy and surgical approach in ascending aorta to prevent circulatory arrest, stroke, or ischemic brain damage.

4.2 Surgical treatment

The goal of surgical intervention in aortic dissection type A (type I and II) is to prevent rupture or development of pericardial effusion, which can lead to tamponade of the heart and death. Also, the sudden onset of aortic regurgitation (aortic valve disease) and coronary obstruction of flow requires urgent surgical intervention, with the aim of replacing the beginning of the dissection region with a composite graft. In the case of type B (type III) of acute aortic dissection, the goal is to prevent aortic rupture. There are many approaches to the management of acute aortic dissection, which provides a broad methodological application (Culliford et al., 1982; Miller, 1991).

4.3 Surgical intervention in acute aortic dissection of type A (type I and II)

The surgical treatment of the type A disorder (type I and II) of the proximal part of the aorta aims to reconstruct the ascending aorta with the restoration of the flow in the right lumen, and it depends on the size of the aortic root and the aortic valve state. Aortic valve is a trickle valve located between the left ventricle and the aorta. When ejecting blood from the heart, it opens and passes blood into the aorta. Its role is also to close it afterward and not allow the return of blood back to the heart from the aorta. If the valve turns blood to return to the heart, such a disease is called aortic insufficiency—regurgitation. If the aortic and aortic root diameters are normal and without the relapse of the coronary arteries and without the separation of aortic valve commissures, then the tubular graft is placed on the sinotubular ridge (Fig. 5.3).

When separation of one or more commissioners occurs, it is necessary to strengthen the valve before placing the graft. There are cases (Fraser et al., 1994; Gerber et al., 2010) when reconstruction of the valve runs risky and anomalies can occur, and in these cases, it is necessary to perform a complete reconstruction of the valves. Aortic valve replacement can be done with mechanical or biological prostheses. Mechanical dentures have the advantage of being long lasting, but their serious lack is the need to take anticoagulant therapy that has side effects. In contrast, biological valves have the advantage of not requiring the use of the said therapy. Most often, older patients receive biological dentures because they have a higher likelihood of complications from the use of anticoagulant therapy. The surgical

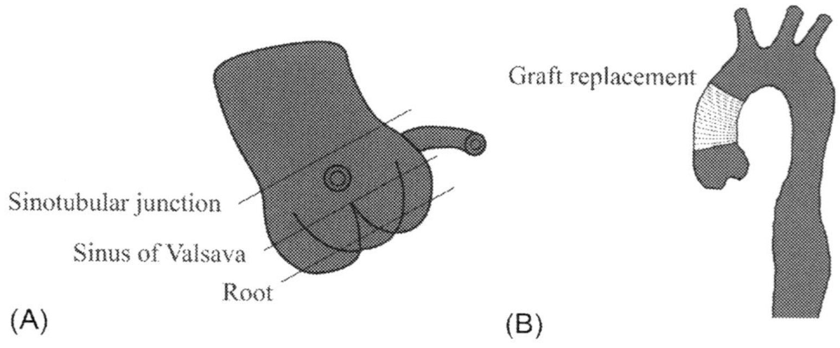

Fig. 5.3 (A) Components of the aortic root and (B) illustration of tubular graft implantation.

treatment of acute artery aortic dissection confronts the problem of when and to what extent the aortic arch must be replaced. If the aortic arch is not scattered, open distal anastomosis of graft and adjacent layers of the aortic wall is formed on the composition of the ascending aorta and arch (Ergin et al., 1982). Aortic arch is cleaved in about 30% of patients with acute aortic dissection (Bachet et al., 1991; Griepp et al., 1991). When the entry point of the split passes over the aortic arch, then the distal graft is placed to the aortic anastomosis by replacing the aortic arch below the entrance part. If there is an extensive cleft that continues below the joint of the transverse and descending segments of the aorta, a complete replacement of the aortic arch is necessary. This requires reconnection of individual or all supraaortic blood vessels for graft.

The technique applied to the aortic root has a lot of similarities with the technique used to reconstruct the aortic arch. In some cases, gelatin-resorcinol-formaldehyde or Teflon adhesive tape is used to close the dead spaces between the graft and the aorta.

4.4 Surgical intervention in acute aortic dissection type B (type III)

Patients with acute dissection of this type are subjected to surgical treatment in cases of severe pain and sudden enlargement of the diameter of the aorta and in the case of periacortic and metastatic hematomas as signs indicating aortic rupture. In the case of other complications, such as ischemia of the kidneys, extremities, or intestines, the solution of the problem is achieved by the installation of the catheter, thus achieving the decompression of the true lumen of the abdominal part of the aorta. Patients, whose acute type B disorders (type III) are not overly complicated, receive drug therapy. Surgical treatment of acute dissection of a descending aorta involves the replacement of the affected segments by a tubular graft of the appropriate length and diameter. The standard procedure of surgical intervention on a descending aorta is the opening of the chest at the level of the fifth rib, which allows the replacement of damaged parts of the aorta to the eighth intercostal space.

4.5 Interventional techniques

Solving aortic dissection can be a challenge. As already mentioned, most patients with type A acute dissection have been referred to a surgical replacement of the ascending aorta to restore the flow in the true lumen, while in the case of dissection of type B, in the absence of complicated dissection, most patients are treated with medicines. Patients of type A and B who are surgically treated have the potential to develop vascular complications such as mesenteric or peripheral ischemia, which cannot be resolved by drugs. Depending on the characteristics of the true and false lumen flow, the involvement of the aortic branch, the four major forms of ischemic complications are distinguished:

- Compression of the true lumen proximal from the main abdominal branches
- Compression of the true lumen in the region of the main abdominal branches
- The involvement of the main branches of the aorta dissection
- Increased false aneurysm due to proximal splitting of the aorta (Elefteriades et al., 1992)

There are two most commonly used interventional techniques: technique of balloon fenestration and stent placement. There are also combinations of these two techniques. Aortic fenestration is a method of decompression of hypertensive false lumen by creating an opening in the distal part of the aortic cleft. This increases the pressure in the true lumen. The next goal of fenestration is to ensure the communication of true and false lumen to prevent thrombosis of false lumen and avoid compression of blood vessels that emerge from the true and false lumen. The main indication for stent placement is the compression of the true lumen of the aorta distal to the main branches of the abdominal aorta. By placing the stent increases the diameter of the true lumen and therefore the flow of blood. The first use of aortic stents was designed to exclude true and false aneurysms in the abdominal and later breast sections (Dake et al., 1994; Slonim et al., 1996).

5 Solution of nonalinear problems with the final elements method

The finite element method (FEM) is a numerical method for solving engineering problems. It was created in the 1960s of the last century as a generalization of the matrix method of structural analysis after which it became the most common method in all engineering fields. The basic idea of this method is to divide the domain of a field of some physical size into the final number of subdomains that we call the finite elements. Thereafter, interpolation functions are introduced that serve to approximate the observed physical quantities (e.g., the field of pressure and velocity in viscous fluids).

The progress of computer technology has enabled and accelerated the application of this method in various fields of industry and science.

Many modern software packages, such as ANSYS and ABAQUS, have the ability to solve complex engineering problems and to visualize the obtained simulation results. In recent years, the method of finite elements has also been found to be used in medicine for, for example, cardiovascular analysis of blood flow, prediction of arterial wall rupture and the site of blood vessel narrowing, determination of wall shear stress, and pressure of walls during the flow of blood.

6 Basic equations of fluid flow

Fluid, by definition, is a matter that deforms under the influence of tangential forces. In contrast to the source gases, which are compressible, fluids can be considered incompressible because the same amount of fluid does not change its volume under the influence of pressure. The reason for this is the existence of less attractive forces among their molecules, so they move less. The state of fluids that flow is determined by density, pressure, and velocity. Within this chapter, a basic theory will be presented to solve the problem of flow of incompressible fluid.

6.1 Continuity equation

The equation of continuity is an analytic form of the law on the maintenance of mass. This law can be applied both to the elemental mass of the fluid particle dm and to the final mass m. According to this law, the mass of the fluid particle does not change during movement in an uninterrupted electric field. By applying Reynolds theorem to the law on the maintenance of mass, it is obtained as (Filipovic, 1999)

$$\frac{Dm}{Dt} = \frac{D}{Dt}\int_V \rho dV = \int_V \left(\frac{D\rho}{Dt} + \rho\frac{\partial v_i}{\partial x_i}\right) dV = 0 \tag{5.1}$$

For an arbitrary control volume V, the equation of continuity in the point is obtained as

$$\frac{D\rho}{Dt} + \rho\frac{\partial v_i}{\partial x_i} = 0 \tag{5.2}$$

After this, Reynolds theorem is reduced to the following form:

$$\frac{D}{Dt}\int_V (f\ \rho) dV = \int_V \left(\frac{\partial f}{\partial t} + v_i\frac{\partial f}{\partial x_i}\right)\rho dV = \int_V \rho\frac{Df}{Dt} dV \tag{5.3}$$

6.2 Navier-Stokes equations

The Navy-Stokes equations describe the motion of viscous incompressible fluids. They represent a system of partial differential equations derived from the second Newtonian law. Let the volume of the fluid be given at time t (Fig. 5.4). The external forces are represented as the surface forces $\mathbf{f}^S$ per unit area and the volume $\mathbf{f}^B$ per mass unit.

Based on the equation of the moment of change in the amount of motion, it is obtained as

$$\frac{D}{Dt}\int_V \rho v dV = \int_V \mathbf{b} dV + \int_S \mathbf{p} dS \tag{5.4}$$

By applying Eq. (5.3), which represents the equation of maintaining the mass, Eq. (5.4) is reduced to the following form:

$$\int_V \rho \frac{D\mathbf{v}}{Dt} dV = \int_V \mathbf{f}^B dV + \int_S \mathbf{f}^S dS \tag{5.5}$$

Using the Gaussian and Koshi theorem (Kojić et al., 1998) on the transformation of a surface into a volume integral, we obtain the following form:

$$\int_V \rho \frac{Dv_i}{Dt} dV = \int_V f_i^B dV + \int_V \frac{\partial \sigma_{ij}}{\partial x_j} dV \tag{5.6}$$

Subsequently, constituent relations for Newton's fluid are introduced:

$$\sigma_{ij} = -p\delta_{ij} + 2\mu \dot{e}_{ij} \tag{5.7}$$

where p represents the fluid pressure, μ the dynamic viscosity, and $\dot{e}$ the deformation rate tensor.

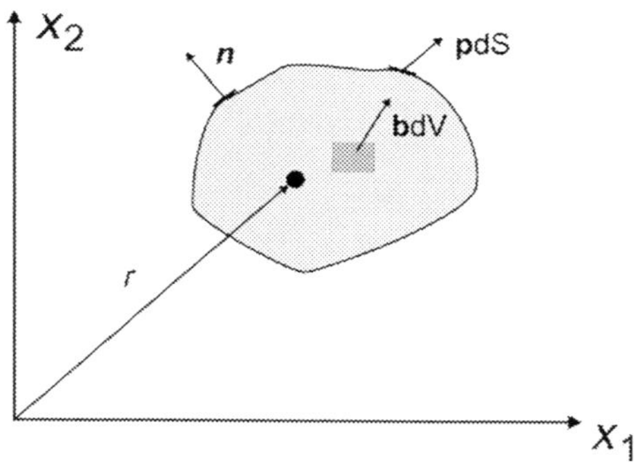

Fig. 5.4 Two-dimensional definitions of volume and surface forces.

The deformation rate tensor is determined as

$$\dot{e}_{ij} = \frac{1}{2}\left(\frac{\partial v_i}{\partial x_j} + \frac{\partial v_j}{\partial x_i}\right) \tag{5.8}$$

When Eqs. (5.7) and (5.8) are replaced by the Eq. (5.6), it is obtained as

$$\int_V \rho \frac{Dv_i}{Dt} dV = \int_V f_i^B dV + \int_V \left(-\frac{\partial p}{\partial x_i} + \mu\left(\frac{\partial^2 v_i}{\partial x_j \partial x_j} + \frac{\partial^2 v_j}{\partial x_j \partial x_i}\right)\right) dV \tag{5.9}$$

Since the control volume V is arbitrary, Eq. (5.9) can be written through the differential form, and finally the standard form of the Navi-Stokes equation for the incompressible viscous flow is obtained:

$$\rho\left(\frac{\partial v_i}{\partial t} + v_j \frac{\partial v_i}{\partial x_j}\right) = -\frac{\partial p}{\partial x_i} + \mu\left(\frac{\partial^2 v_i}{\partial x_j \partial x_j} + \frac{\partial^2 v_j}{\partial x_j \partial x_i}\right) + f_i^B \tag{5.10}$$

7 Basic equations of solid motion

The principle of virtual work is one of the basic principles in the mechanics of the continuum. It represents the basis for performing the necessary relationships in numerous numerical programs.

Let the body be balanced in terms of the external loads involved (Fig. 5.5), where δu is virtual displacement, and there are virtual deformations.

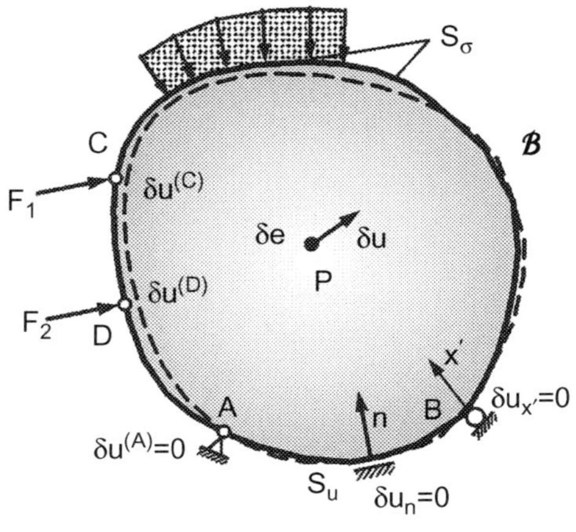

Fig. 5.5 Display of a deformable body under the influence of external forces (Kojic et al., 2008).

We will make the following assumptions:

- The virtual displacement field $\boldsymbol{\delta u}$ is assigned.
- External loads are immutable.
- Virtual movements at points in which the external forces operate are $\boldsymbol{\delta u}^{(C)}$ and $\boldsymbol{\delta u}^{(D)}$.
- Virtual movements in the relocation points are limited.
- Stress and displacements are set on the surfaces of S_σ and S_u.
- Virtual displacements are infinitesimal and satisfy specified boundary conditions.

If it starts from the equilibrium equations in the literature derived from (Kojic et al., 2008):

$$\frac{\partial \sigma_{ij}}{\partial x_i} + F_j^V = 0 \tag{5.11}$$

with the application of border conditions:

$$\sigma_{ij} n_i - F_j^S = 0 \tag{5.12}$$

the equality of virtual works of internal and external forces arises:

$$\sigma_{ij} n_i - F_j^S = 0 \tag{5.13}$$

wherein

$\delta W_{\text{int}} = \int_V \sigma_{ij} \delta e_{ij} dV$ virtual work of internal forces on virtual deformations

$\delta W_{ext} = \int_V F_k \delta u_k dV + \int_{S^\sigma} F_k^S \delta u_k^S dV + \sum_i F_k^{(i)} \delta u_k^{(i)},\ k = 1,3$ virtual work of external forces on virtual displacements

The virtual work of internal and external forces can be written in matrix form:

$$\delta W_{\text{int}} = \int_V \delta \mathbf{e}^T \boldsymbol{\sigma} dV \tag{5.14}$$

$$\delta W_{ext} = \int_V \delta \mathbf{u}^T \mathbf{F}^V dV + \int_{S^\sigma} \delta \mathbf{u}^T \mathbf{F}^S dV + \sum_i \delta \mathbf{u}^T \mathbf{F}^{(i)} \tag{5.15}$$

Applying the principles of virtual work and constituent relations for linear elastic material in matrix form:

$$\boldsymbol{\sigma} = \mathbf{C}\mathbf{e} \tag{5.16}$$

Using the isoparametric concept of interpolation (Bathe, 1996) within the final elements, on the basis of which the coordinates and displacements at any point within the element:

$$\mathbf{x} = \mathbf{N}\mathbf{X} \tag{5.17}$$

$$\mathbf{u} = \mathbf{N}\mathbf{U} \tag{5.18}$$

The equation of equilibrium of the finite element is obtained (Kojić et al., 1998):

$$\mathbf{K}\mathbf{U} = \mathbf{F}_{ext} \tag{5.19}$$

where

$\mathbf{B}$—moving in nodes, containing statements of interpolation functions
$\mathbf{C}$—elastic constituent matrix
$\mathbf{e} = \mathbf{B}\mathbf{U}$—matrix deformation
$\mathbf{U}$—movements in nodes
$\mathbf{X}$—coordinates of nodes
$\mathbf{N}$—matrix of interpolation functions
$\mathbf{F}_{ext}$—external forces in the element nodes

Eq. (5.19) represents the equality of internal and external forces. In general, the internal forces are nonlinear displacement functions. By linearizing the previous equation, the following form is obtained:

$$\mathbf{K}\mathbf{U} = \mathbf{F}_{ext} \tag{5.20}$$

where

$\Delta\mathbf{U}^{(i)}$ vector of incremental displacement vector
${}^{t+\Delta t}\mathbf{F}_{int}^{(i-1)} = \int_V {}^{t+\Delta t}\mathbf{B}^{T(i-1)t+\Delta t}\boldsymbol{\sigma}^{(i-1)}dV$ the internal force vector at the end of the load step

The equations of motion of a material system can be written using the principles of virtual work, taking into account the effects of inertial forces. The elementary volume inertial force is

$$d\mathbf{F}^{in} = -\ddot{\mathbf{u}}dm = -\ddot{\mathbf{u}}\rho dV \tag{5.21}$$

When taking into account the influence of inertial forces, the virtual work of external forces is

$$\delta W_{ext} = \int_V \delta\mathbf{u}^T\left(\mathbf{F}^V - \rho\ddot{\mathbf{u}}\right)dV + \int_{S^\sigma} \delta\mathbf{u}^T\mathbf{F}^S dV + \sum_i \delta\mathbf{u}^T\mathbf{F}^{(i)} \tag{5.22}$$

Time differentiation Eq. (5.18) gives us interpolations for velocity and acceleration of points:

$$\dot{\mathbf{u}} = \mathbf{N}\dot{\mathbf{U}} \tag{5.23}$$

$$\ddot{\mathbf{u}} = \mathbf{N}\ddot{\mathbf{U}} \tag{5.24}$$

where

$\dot{\mathbf{u}}$—the velocity of the material point in the element
$\ddot{\mathbf{u}}$—the acceleration of material point in the element
$\dot{\mathbf{U}}$—the velocity in nodes
$\ddot{\mathbf{U}}$—the acceleration in nodes

In matrix form, Eq. (5.23) can be represented in the form

$$\mathbf{M}\ddot{\mathbf{U}} + \mathbf{K}\mathbf{U} = \mathbf{F} \tag{5.25}$$

where $\mathbf{M} = \int_V \mathbf{N}^T \rho \mathbf{N} dV$ represents matrix of the mass of the finite element

Eq. (5.25) can be written in the following form:

$$\mathbf{M}\ddot{\mathbf{U}} + {}^{t+\Delta t}\mathbf{K}^{(i-1)}\Delta\mathbf{U}^{(i)} = {}^{t+\Delta t}\mathbf{F}_{ext} - {}^{t+\Delta t}\mathbf{F}_{int} \tag{5.26}$$

Assuming that the linear analysis of the displacement of the solids is infinitesimal small, that the material is linearly elastic, and that the nature of boundary conditions remains unchanged under the action of external loads, Eq. (5.19) is derived and refers only to linear analysis. Unlike linear, nonlinear analysis is applied when the displacements are not linearly dependent on the load. Nonlinearity arises from large displacements, deformations, and variable boundary conditions in solid due to mutual interaction of solid domains and fluid domains. To calculate the stress and deformation of such problems, the total Lagrangian formulation is used (Bathe, 1996).

8 Solid fluid interaction

The blood flow through blood vessels represents fluid flow through a tube of deformable walls. Modeling such a physical process requires the determination of the interaction of fluids and solids. During the last 30 years, a lot of effort has been invested in the development of computational mechanics and the field of analysis of solids and fluids. The development of computer technology expands the application of the method of solving the problem of fluids and solids not only in many engineering but also in other multidisciplinary areas. With this aim, special numerical disciplines for solid problems, computer computational structure dynamics (CSD), and fluid, computational fluid dynamics (CFD), have been developed.

The solid-fluid interaction requires simultaneous numerical resolution of both the fluid domain and the domain of solids. There are solvents that specifically solve the fluid, especially solid. There are two concepts of their unification—strong and poor clinging. Strong coupling aims to solve a complete system of equations in one step. At the same time, all sizes for both fluid and solid are solved along this route. The lack of this concept is reflected to the fact that, when 3D problems are present, the number of equations that are solved increases dramatically, which increases the time of obtaining the solution. In the case of poor coupling, the external program is managed by the solution of solid materials separately from fluids. At first glance, poor cohesion is a better alternative for solving multidisciplinary problems. Due to their specificities that do not normally occur in the method of strong coupling, poor coupling contains a series of problems that need to be addressed (the problem of time integration). Due to the diversity of physical properties of solids and fluids, the same time step cannot be used in general because the numerical stability domain is different in solving the problem of solids and in solving the fluid problem. Then, the next problem occurs when transferring data between the CFD and CSD programs, since different discretization further complicates the problem.

9 3D modeling of aortic dissection

9.1 Introduction

Computer tomography is a medical imaging technique that produces tomographic images or slices of specific body surfaces to X-rays. These cross-sectional views are used for diagnosis and for therapeutic purposes in various medical disciplines. Digital processing of these images provides the possibility of generating three-dimensional images of a scanned object recorded around an axis of rotation. The 3D modeling of the dissecting aorta model used in this study is done by creating volume models using 2D axial cross sections obtained by the CT scan method.

In this study, several software packages were used to obtain good, representative aortic dissection models. Several stages make up this whole process:

- Acquisition of CT images
- Segmentation of CT images and contour detection
- Generating a 3D model

- Processing the 3D model
- Creating a mesh of finite elements
- Setting boundary conditions

The first phase involves the application of a semiautomatic algorithm of the segmentation (thresholding) CT images to get an axial array of contours and then a creation of a 3D view using the Materialize Mimics 10.01 program. The next stage is the application of the Geomagic Studio 10 program to obtain the volume model presented as a mesh of triangular surfaces from noise and unnecessary parts and parts of interest as best prepared for generating a finite element mesh. The creation of a finite element mesh is done using the Femap v10 program, which represents the final phase in the preparation of the model for simulation. The last phase involves adjusting the final elements mesh, obtained by the previous step and the PAKSF solver, and setting boundary conditions. These programs are widely accepted, and for this reason, they are used in this study.

10 3D reconstruction using Materialize Mimics 10.01

Materialize Mimics, developed at the University of Leuven in Belgium, is a software that converts anatomical images into 3D models, and this process is known as segmentation. This software is specially developed for processing medical images, such as CT, microCT, MRI, and ultrasound. The accuracy of the patient's anatomy model will also depend on the quality of the mentioned images. The standard 3D reconstruction procedure involves several steps:

- Loading CT images in DICOM format (Digital Imaging and Communications in Medicine)
- Adjusting image contrast
- Segmentation of uploaded images
- Generating 2D surfaces of interest and 3D models
- Exporting geometry in standard formats

Fig. 5.6 shows the application of the segmentation algorithm and the final appearance of the 3D model.

Automatic contour detection has a disadvantage in that it cannot distinguish certain deformities that occur during segmentation. In this sense, manual repairs and deleting the pixels to get a fully accurate model are required.

By removing the noise, the contours that are intersecting, correcting the blurred images, and others, the 3D geometry reconstruction is reconstructed. The resulting 3D model is exported as a stl format (stereolithography) and continues to work in Geomagic Studio 10.0.

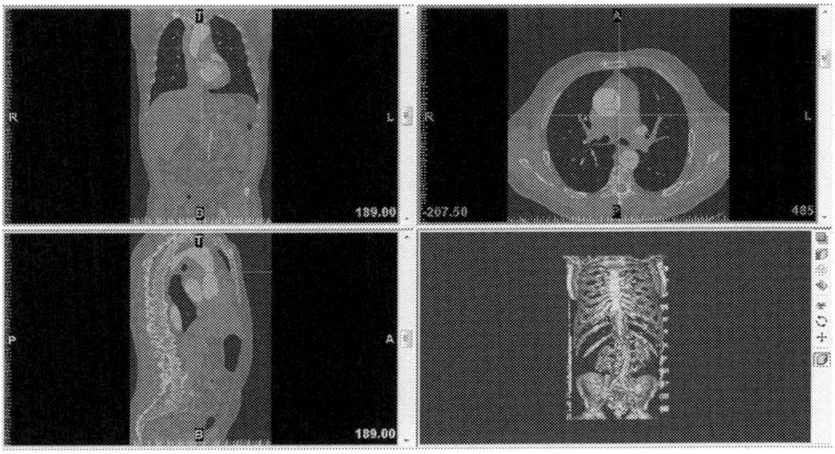

Fig. 5.6 Application of segmentation algorithm and 3D model representation.

11 Geometric 3D modeling using Geomagic Studio 10.0

Geomagic Studio 10.0 is a professional engineering software that is used to work with 3D objects. The company was founded in 1997 in North Carolina, in the city of Morrisville. It is primarily intended for computer-assisted design, with emphasis on 3D scanning and other nontraditional design methodologies. In this study, Geomagic Studio 10.0 was used for further reconstruction of polygonal models of dissecting aortic models obtained by Materialize Mimics 10.01. The reconstruction process begins by loading the stl model (Fig. 5.7). The mesh of the triangles obtained contains a lot of sharp triangles and triangles that are intersected, so it is necessary to correct them. For this purpose, a combination of the various useful tools that this software possesses is used.

For the purpose of creating a mesh of finite elements, a TetGen was used. TetGen is a software for tetrahedral mesh generation. Its goal is to generate good quality tetrahedral meshes suitable for numerical methods and scientific computing. It can be used as either a stand-alone program or a library component integrated in other software.

12 Virtual surgery

Structural wall weaknesses can lead to minor forms of aortic dissection. Authors (Svensson et al., 1999) describe a mild dissection as a partial splitting of the blood vessel covered by the thrombus. Partial splitting of the inner

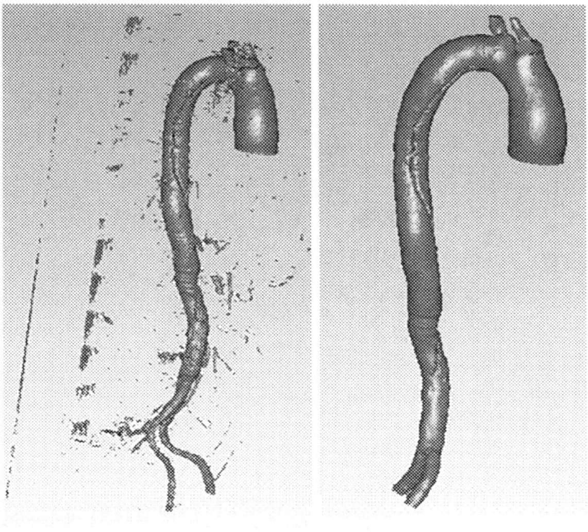

Fig. 5.7 Loaded mesh of triangles of true and false lumen with the noise of unnecessary objects (*left*) and the triangle mesh after cleaning (*right*).

layer of the aorta allows the blood to enter the already damaged medium and thus leads to dissection of the aortic wall and then the creation of false lumen in the wall and rupture.

In this chapter, the results of two patients with acute aortic dissection type I were analyzed. As already mentioned, surgical treatment of the dissection type A (type I or II) of the proximal part of the aorta is primarily aimed at reconstructing the ascending aorta with the restoration of the flow in the true lumen. Fig. 5.8 shows a process of applying computer simulation in performing a virtual replacement of a damaged ascending part of the aorta.

As Fig. 5.8 shows, the first task is to close the flow through the false lumen. After a real surgical intervention, surgeons close the flow in the false lumen without a complete replacement of the ascending aorta, as they are not sure to what extent the false lumen has taken over the overall flow.

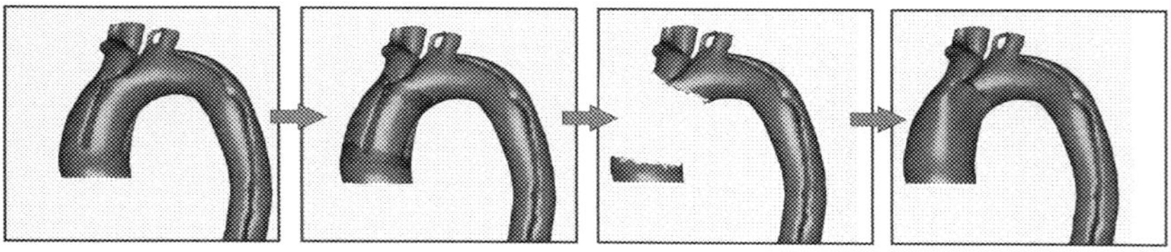

Fig. 5.8 Virtual surgical treatment of aortic dissection.

By computer simulation, it is possible to accurately determine the percentage of flow in all branches before actual surgical intervention. In this way, it is possible to reduce high mortality rates after surgery.

13 Results of numerical analysis

In this section of the chapter, the results of simulations performed on two preoperative models of patients with acute aortic dissection and the results of simulations of postoperative models of patients, made by computer manipulation, are presented. Also, the results of von Mises stresses on the false lumen walls are shown. It is very important to determine the potential sites of the rupture of the aorta to prevent its bursting, as well as those places where reverse flows can occur and the return of the flow from the false to the true lumen of the bloodstream.

Numerical solving of the relationship between true and false lumens involves determining and analyzing the obtained results, primarily referring to the wall shear stress, lumen pressure, and velocity. All numerical simulations were performed using the PAKSF solver (Kojic et al., 2008). A parabolic flow profile is assigned at the entrance of the ascending part of the aorta, while the constant value of the zero pressure is given at the output (Karmonik et al., 2011). The volume flow of the fluid from the heart to the aortic system was adopted from the literature (Joong et al., 2007). It has been assumed that patients have 70 heartbeats per minute, so that the mean flow of blood flow is 5 L/min. The simulations were performed with one whole heart cycle, and all the results presented in this paper were presented for the maximum systole (0.22 s). The blood is modeled as an incurable Newtonian fluid, suitable for larger arteries. The blood density was $\rho = 1.05\,\text{g/cm}^3$ and dynamic viscosity $\mu = 0.0037\,\text{Pa s}$. It is assumed that blood flow is laminar in large blood vessels, since it is assumed that the low mean velocity results in low Reynolds number values. In the literature, a solid-fluid interaction of the aorta can be found as a hyperelastic, homogeneous, isotropic material of a modulus of elasticity ranges from 0.4 MPa (Colciago et al., 2014) to 3 MPa (Nathan et al., 2011). The results of solid-fluid interactions shown in this chapter have adopted a modulus of elasticity of 2.5 MPa and a Poisson's coefficient of 0.45.

Fig. 5.9 shows the anatomy of the aorta and certain segments in whose crossings this analysis of aortic dissection will be carried out (Erbel et al., 2014).

The marked segments taken from the literature are the most important places for changing the diameter of the aorta. Cross section 1.1 represents the beginning of the ascending part of the aorta, the cross-section 2.2 of the middle of the ascending aorta, the cross-section 3.3 proximal aortic arch or the beginning of the brachiocephalic tree, the cross-section 4.4 of the aortic arch or cross-section between the common carotid and subclavian artery, the cross-section 5.5 the proximal descending aorta, the cross-section 6.6 of the center of the descending aorta, the cross-section 7.7 diaphragm, the cross-section 8-8 of the center of the abdominal aorta, while the cross-section 9.9 represents the cross-section immediately before the aortic bifurcation.

14 Results of the simulation of preoperative models

The results of the numerical simulations of preoperative models were carried out on models that present the state, without any computer modifications.

Patient #01 is a 60-year-old man, a nonsmoker, who is diagnosed with hypotension (low blood pressure) and acute aortic dissection of DeBakey type I. The size of the entry tear is 4.25 cm^2 at a distance of 81 mm from the aortic top. There is a reentry at a distance of 9.1 mm from aortic bifurcation of 8.77 cm^2. The volume of the true lumen is 389.64 cm^3, while the volume of the false lumen is 133.83 cm^3.

The results of the wall shear stress, pressure, and velocity are shown in Fig. 5.10. The figure shows that the false lumen from the ascending aorta has a higher pressure up to the diaphragm, that is, the end of the descending aorta. At the same time, the value of the wall shear stress of the false lumen in the area of the ascending aorta is up to 7.42 Pa. The highest recorded values of the wall shear stress are recorded on the branch of the brachiocephalic tree, 20.50 Pa, and along the left of the common subclavian artery, also the same value. As it is still not clear how the dissecting regions react to wall shear stress values, it is reasonable to expect to behave like aneurysms (Shojima, 2004). Also, the literature (Kondo et al., 1997; Filipovic et al., 2013) shows that the values of the wall shear stresses are correlated to the expansion of the tear, additional tears, and the formation of aneurysms. Through the brachiocephalic artery that remained in the true lumen, a flow of 15.77 cm^3/s (18.79% of the total flow) is achieved.

False lumen, through which flow 35.43% of the fluid flow, took the flow through the left common carotid artery through which flow 2.61 cm^3/s (3.12%) of the fluid flows, while the flow rate through the left subclavian

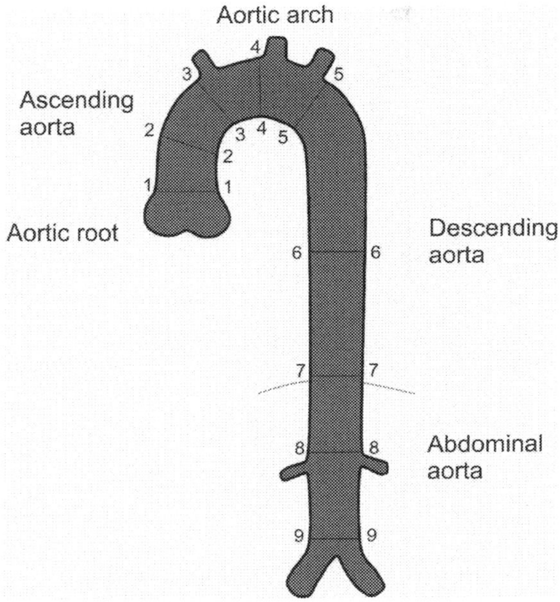

Fig. 5.9 Anatomy of the aorta.

artery is 9.22 cm^3/s (10.98%). The celiac branch, with a flow of 10.27 cm^3/s (12.31%), and upper mesenteric artery, with a flow of 7.81 cm^3/s (9.29%), were merged with the false lumen. The left renal artery is exposed to static obstruction because it is partially powered from the true and partly from the false lumen. The flow rate of the blood in the left renal branch is 0.16 m/s, corresponding to a flow of 7.58 cm^3/s (9.03%), and in the right renal, which is fed from the true lumen, the flow rate is 8.41 cm^3/s (10.02%). Both iliac branches belong to the true lumen, and they recorded flows of 10.61 cm^3/s (12.62%), in the right, and 11.62 cm^3/s (13.84%), in the left.

Patient #02 is a 51-year-old man, extremely obese, smoker who has been diagnosed with hypertension (high blood pressure), diabetes, blood fats, and acute aortic dissection of DeBakey type I. The size of the entry tear is 10.16 cm^2 at a distance of 95 mm from the top aortic port. The volume of the true lumen is 36.06 cm^3 and a false lumen of 102.32 cm^3. Fig. 5.11 shows the numerical results of the patient #02. It is clear that the true and false lumens have almost equal values of pressure down to the central part of the descending aorta. As in the case of the previous patient, here, in the case of a wall shear stress of the false lumen, very high values in the ascending part of the aorta reaching 29.33 Pa are observed. The high values of this stress are

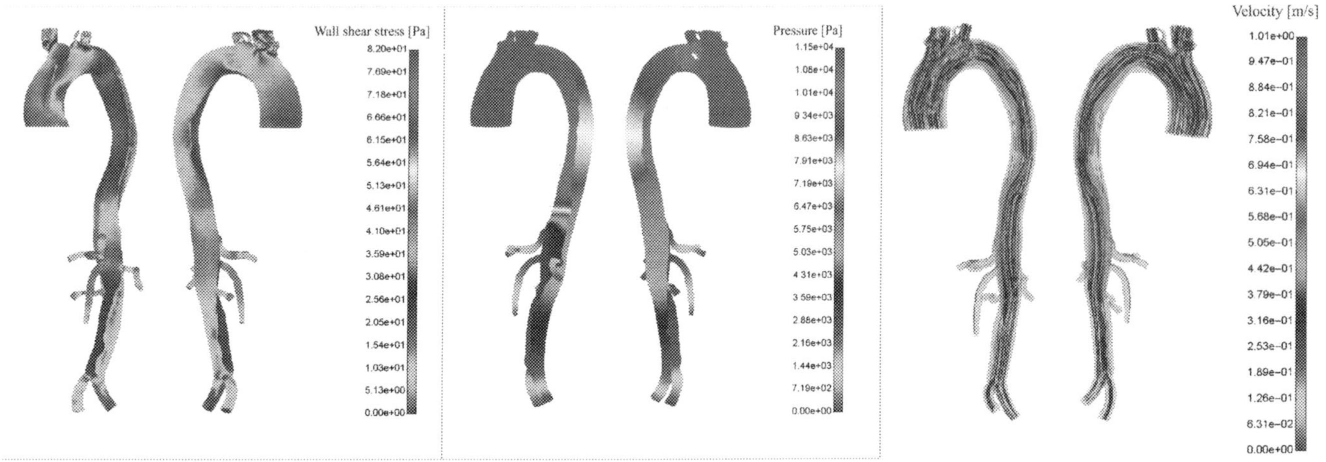

Fig. 5.10 Results of the numerical analysis of the preoperative patient model # 01: (A) the wall shear stress, (B) the pressure, and (C) the velocity.

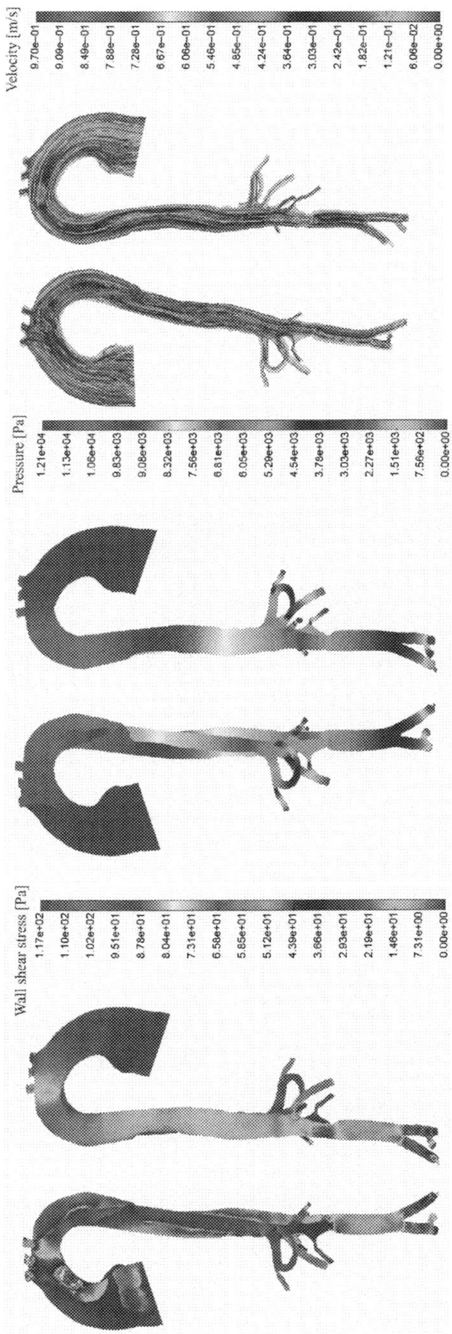

Fig. 5.11 Results of the numerical analysis of the preoperative patient model #02: (A) the wall shear stress, (B) the pressure, and (C) the velocity.

observed at the beginning of the branch of the brachiocephalic artery and left common carotid, which, as already been said, can lead to endothelial function disorders.

The false lumen took over 41.18% of the flow. The flow through the branches of the aortic arch was completely taken over by this lumen with the following flows: 12.52 cm^3/s (14.91%) through brachiocephalic, 9.64 cm^3/s (11.48%) through the left carotid, and 8.45 cm^3/s (10.06%) through the left subclavian artery. The celiac artery, also originating from the false lumen, had a flow rate of 8.31 cm^3/s (9.18%), while a flow of 6.81 cm^3/s (8.11%) was achieved through the upper mesentricular. The right renal artery attached to the right lumen had a flow rate of 7.01 cm^3/s (8.34%). An interesting phenomenon occurs in the left renal artery. One part of the flow was delivered through the true lumen, in the amount of 6.81 cm^3/s (8.12%), and the other part through the false lumen, in the amount of 1.97 cm^3/s (2.35%). The iliac arteries were branched from the true lumen with flows of 12.13 cm^3/s (14.44%), through the right, and 10.85 cm^3/s (12.93%), through the left branch.

15 Results of simulation of postoperative models

After surgery, a large percentage of patients die due to the lack of knowledge of how the operation performed affects the flow of blood in the aorta and its branches. Computer manipulation of preoperative models provides the ability to create virtual geometries of postoperative models of all patients. In this way, doctors provide a clear picture of the results of a particular surgical procedure (Filipovic et al., 2015).

Postoperative geometry model of patient #01 implied the replacement of the ascending aorta with stent graft and closure of the flow through the false lumen. After surgery, there was an increase in pressure in the true lumen and a fall in the false lumen. The results of numerical analysis are presented in Fig. 5.12. The maximum values of the pressure in the upright range of the true lumen range were up to 15812 Pa. After surgery, the flow through the brachiocephalic artery increased from 15.77 cm^3/s to 19.39 cm^3/s. The flow rate through the left common carotid artery was 2.61 cm^3/s, and now, it is 0.74 cm^3/s, while the flow rate through the left subclavian artery is now 1.97 cm^3/s (drop by 78.63%). The celiac artery now has a new flow rate of 7.17 cm^3/s (drop by 30.18%) and upper mesentricular 4.21 cm^3/s (46.02% decreases). The left renal artery of a new flow of 10.18 cm^3/s records an increase of 34.3%, while the right renal, with a flow of 8.41 cm^3/s, has a leap at 11.30 cm^3/s. The iliac branches have new values of

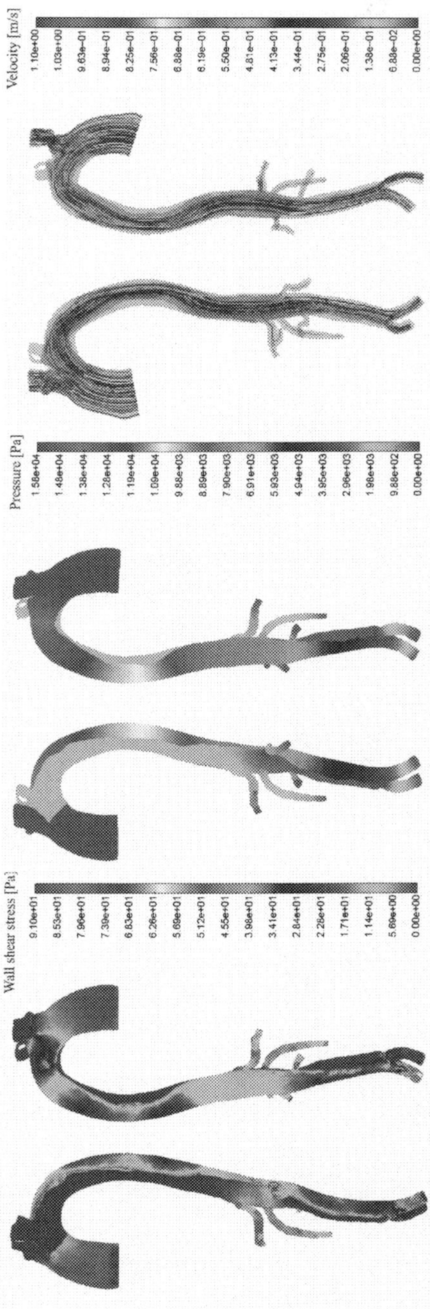

Fig. 5.12 Results of the numerical analysis of the postoperative patient model #01: (A) the wall shear stress, (B) the pressure, and (C) the velocity.

14.13 cm^3/s, in the right (33.3% increase), and 14.72 cm^3/s (increase by 26.67%), in the left iliac artery.

The proposed postoperative procedure for the replacement of the ascending aorta, and at the same time the closure of the false lumen, gives relatively good results in terms of the final outcome of the operation. In addition to the fall in the flow in the left subclavian branch, as in the left common carotid artery, an increase in flow in almost all other branches is observed. By virtually manipulating the preoperative patient model #01, a much clearer picture of what happens to the flows of the aortic arch and distal branches is provided, as doctors can thus predict whether the flow will be able to satisfy the needs of other organs of the body.

Postoperative geometry model of patient #02 also implied the replacement of the ascending aorta with stent graft and closure of the flow through the false lumen. In this case, as the branches of the aortic arch belonged to the false lumen, it was necessary to completely switch all three branches to the flow of the true lumen and to observe what happens afterward with the flow. The results of numerical analysis are presented in Fig. 5.13.

From Fig. 5.13, it can be noticed that there has been a redistribution of pressure, and the true lumen again gets more values (from 12140 to 16106 Pa). The flow through the branches of the aortic arch is characterized by the following flows: 14.14 cm^3/s through brachiocephalic (up by 12.93%), 10.71 cm^3/s through left carotid (elevation by 11.09%), and 10.14 cm^3/s through the left subclavian artery (20% increases). The celiac, upper mesenteric artery, and the right renal artery, after postoperative procedure, remain free of flow due to the absence of a reentry in the reconstructed part of the aorta. In the real state of the patient, there would probably have been a small flow through these branches due to the closed circulatory system, but certainly, such a procedure would not yield the expected results. The left renal artery, which is fed by the flow of true lumen, with values of 6.81 cm^3/s, has a jump to 10.65 cm^3/s (an increase of 56.38%). The iliac arteries of new flows of 14.88 cm^3/s, through the right, and 12.71 cm^3/s, through the left, record an increase of 22.67% and 17.14%, respectively. It can be concluded that the performed postoperative procedure partially fulfills its goal. The flow through the branches of the aortic arch increases, leading to the unobstructed operation of the upper parts of the body. However, due to the nonexistence of the reversal reentry of the reconstructed part of the aorta, branches of the abdominal aorta remain without a significant blood flow, which can profoundly affect kidney function, accounting for a high mortality rate after surgery (Suzuki et al., 2003).

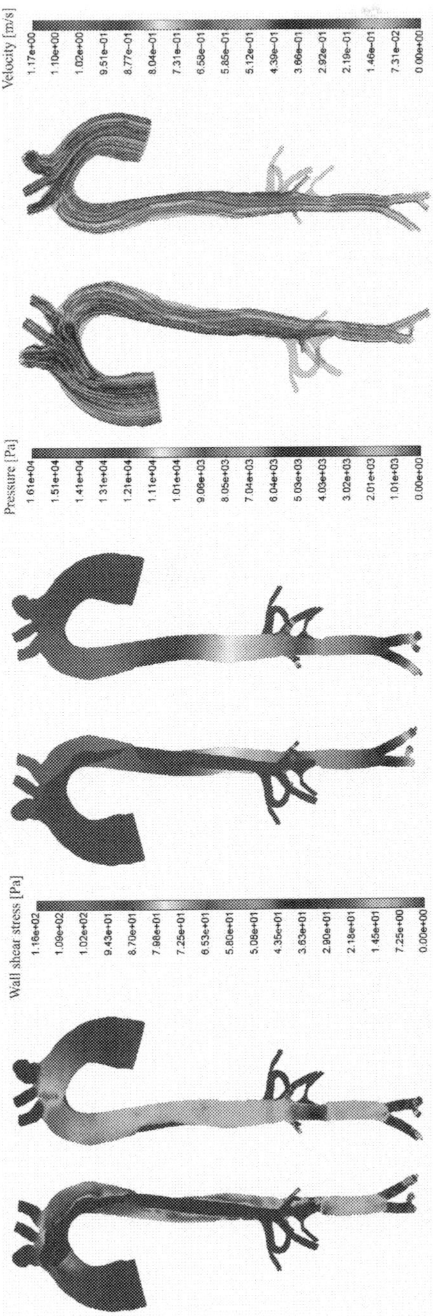

Fig. 5.13 Results of the numerical analysis of the postoperative patient model #02: (A) the wall shear stress, (B) the pressure, and (C) the velocity.

16 Results of simulation of the wall shear stress on the false lumen of preoperative models

Good knowledge of more physical quantities that can affect the occurrence and further development of acute aortic dissection can greatly assist in treating these conditions better. In addition to the numerical results of the pressure distribution along the aorta, the distribution of wall shear stress, and the results of the achieved velocities through the observed model, much attention should be paid to determining the stress on the wall of the arteries. It is especially important to determine what distribution of the von Mises stress has a false lumen during a single heartbeat cycle. Knowing these values can help in determining the potential points of reentry flow of the false lumen or places on the aorta where rupture may occur.

Fig. 5.14 shows the result of the distribution of the von Mises stress along the geometry of the false lumen of patient #01. The first potential point of reentry is located in the middle of the ascending part of the aorta.

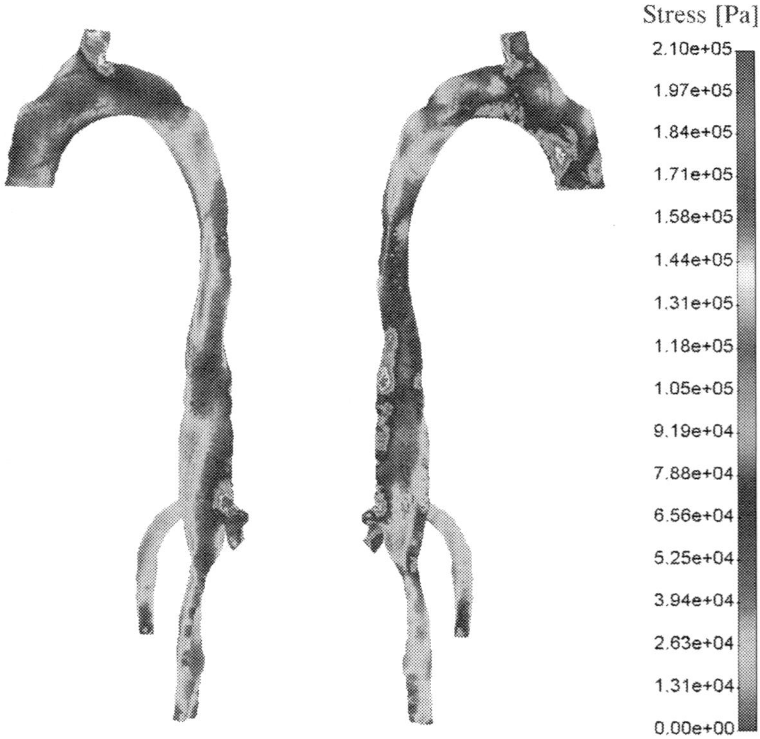

Fig. 5.14 The result of numerical analysis of the stress on the false lumen wall of the preoperative patient model #01.

The maximum value of the wall stress recorded at that site is 1.64e05 Pa. The shape of the geometry and fluid flow velocity in the initial part of the brachiocephalic tree result in a higher value of the stress (2.10E05 Pa), as in the case of left renal artery (2.10E05 Pa), where there is already a reentry tear of false lumen. In this reconstructed model, the left renal branch is completely attached to the false lumen, although it was originally part of the true lumen, to verify that the actual zones of higher stress values can be a prediction of the formation of reversible circuits.

Fig. 5.15 is really confirmed that in this way, these potentially dangerous places can be determined with great certainty.

Another high stress zone, a value of 2.07E5 Pa, can be seen at a height of 64 mm, measured from the left renal artery. There is no clear confirmation of the DICOM image on that cross section, but this site can be treated as a high-percentage aortic rupture.

Numerical analysis of the distribution of the stress on the false lumen of the patient #02 (Fig. 5.16) shows a pair of regularly connected high stress zones along the ascending aorta, as seen in the figure on the left side.

However, there is a greater likelihood that a rupture occurs in the stress zone in the middle of the ascending part of the aorta, with a maximum value of 2.32E05 Pa (right side in Fig. 5.15). Further, distally, at a height of 78 mm from the celiac branch, it is possible to notice the zone of the mean values of the stress (5.1E04 Pa), but they are not at this moment too significant, except that it can only be stated that in the future, it can be of significance for consideration.

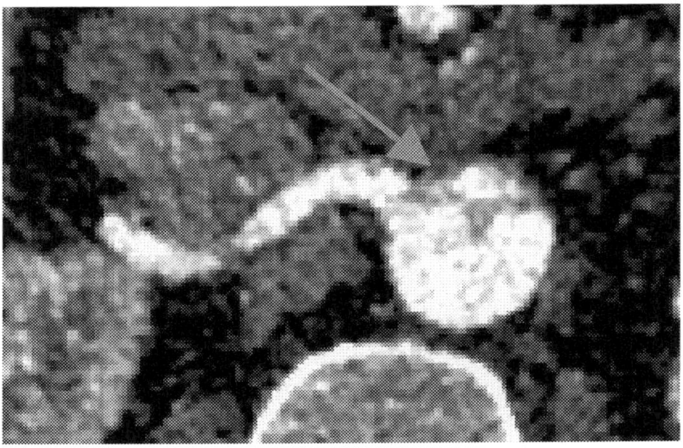

Fig. 5.15 DICOM image branching of the left renal artery of the patient #01.

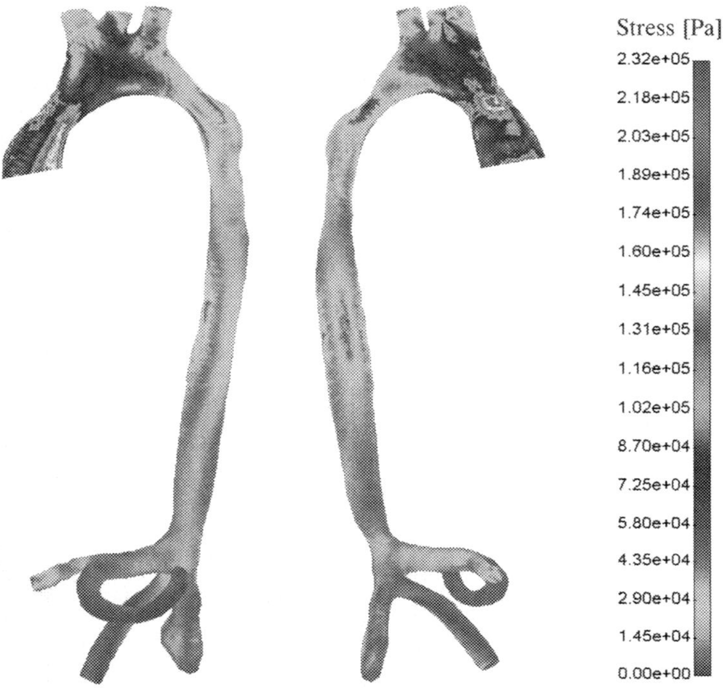

Fig. 5.16 The result of numerical analysis of the stress on the false lumen wall of the preoperative patient model #02.

17 Conclusion

Cardiovascular diseases are the leading cause of death in all developed countries. Modern lifestyle, reduced physical activity, daily stress exposure, irregular nutrition, and obesity are some of the causes of hypertension, one of the most significant risk factors for the development of cardiovascular diseases. Dissection of the aorta is one of the most serious diseases that begins with a crack in the intima or the first layer of the aorta wall. The blood under pressure is penetrating through the newly emerging cleavage, which causes further layering of the media layer, as a result of which a new, false lumen creates. This new lumen compresses the true lumen due to which the pressure drops and later, as the false lumen acquires a higher proportion of the flow, leads to obstruction of the aortic branch flow. As the mortality rate is high in this case, it was necessary to find a way to better understand this phenomenon and predict its outcome.

In this chapter, three-dimensional simulations were conducted to determine the hemodynamic behavior of the blood in aorta with the presence of acute dissection. The aim of the simulated exercises was to determine the numerical relationship between the true and the false lumens of acute aortic dissection. It is very important to predict the behavior of false lumen and its propagation in the further course of disease development and to identify potential sites for the occurrence of reversal or rupture. The reverse patterns allow for the communication of true and false lumens again.

One part of this chapter was dedicated to virtual surgery that includes a numerical analysis of the models obtained on the basis of the doctor's recommendations for the implementation of a particular surgical procedure. In the case where the ascending part of the aorta is affected, it is necessary to cut off that part of the aorta and replace the tubus graft while the flow through the false lumen closes at that point. When it comes to the more complex development of the disease, when the aortic arches are affected, a complete replacement of the ascending aorta and aortic arch with its branches is required.

At the end of the chapter, answers to questions about the possible rupture of the aorta are given. By solid-fluid interaction, for the given flow rate at the inlet of the aorta, in addition to determining the physical sizes of fluid domains such as velocity, wall shear stress, and pressure, it is possible to determine the von Mises stress on the wall of the aorta—domain of the solid. In this way, the potential positions of the aortic rupture are determined. It is of great importance to know which sites have high stress values because they represent the points with the highest potential for developing a reentry or aortic rupture, which can have a fatal outcome.

References

Alfonso, F., Goicolea, J., Aragoncillo, P., Hernandez, R., Macaya, C., 1995. Diagnosis of aortic intramural hematoma by intravascular ultrasound imaging. Am. J. Cardiol. 76, 735–738.

Augoustides, J.G.T., Andritsos, M., 2010. Innovations in aortic disease: the ascending aorta and aortic arch. J. Cardiothorac. Vasc. Anesth. 24, 198–207.

Bachet, J., Guilmet, D., Goudot, B., 1991. Cold cerebroplegia. J. Thorac. Cardiovasc. Surg. 102, 85–93.

Bathe, K.J., 1996. Finite Element Procedures. Prentice-Hall, Inc., Englewood Cliffs, N.J.

Brady, A.R., Thompson, S.G., Fowkes, F.G., Greenhalgh, R.M., Powell, J.T., 2004. Abdominal aortic aneurysm expansion: risk factors and time intervals for surveillance. Circulation 110, 16–21.

Colciago, C.M., Deparis, S., Quarteroni, A., 2014. Comparisons between reduced order models and full 3D models for fluid–structure interaction problems in haemodynamics. J. Comput. Appl. Math. 265, 120–138.

Creswell, L.L., Kouchoukos, N.T., Cox, J.L., Rosenbloom, M., 1995. Coronary artery disease in patients with type A aortic dissection. Ann. Thorac. Surg. 59, 585–590.

Culliford, A.T., Ayvaliotis, B., Shemin, R., Colvin, S.B., Isom, O.W., Spencer, F.C., 1982. Aneurysms of the ascending aorta and transverse arch: surgical experience in 80 patients. J. Thorac. Cardiovasc. Surg. 82, 701–710.

Dake, M.D., Miller, D.C., Semba, C.P., Mitchell, R.S., Walker, P.J., Liddell, R.P., 1994. Transluminal placement of endovascular stentgrafts for the treatment of descending thoracic aortic aneurysms. N. Engl. J. Med. 331, 1729–1734.

Deeb, G.M., Williams, D.M., Bolling, S.F., 1997. Surgical delay for acute type A dissection with malperfusion. Ann. Thorac. Surg. 64, 1669–1675.

Deutsch, H.J., Sechtem, U., Meyer, H., Theissen, P., Schicha, H., Erdmann, E., 1994. Chronic aortic dissection: comparison of MR imaging and transesophageal echocardiography. Radiology 192, 645–650.

Elefteriades, J.A., Hartleroad, J., Gusberg, R.J., 1992. Long-term experience with descending aortic dissection: the complication-specific approach. Ann. Thorac. Surg. 53, 11–20.

Elefteriades, J.A., Lovoulos, C.J., Coady, M.A., Tellides, G., Kopf, G.S., Rizzo, J.A., 1999. Management of descending aortic dissection. Ann. Thorac. Surg. 67, 2002–2005.

Erbel, R., Aboyans, V., Boileau, C., Bossone, E., Bartolomeo, R.D., Eggebrecht, H., Evangelista, A., Falk, V., Frank, H., Gaemperli, O., Grabenwöger, M., Haverich, A., Iung, B., Manolis, A.J., Meijboom, F., Nienaber, C.A., Roffi, M., Rousseau, H., Sechtem, U., Sirnes, P.A., Allmen, R.S., Vrints, C.J., 2014. ESC Guidelines on the diagnosis and treatment of aortic diseases: document covering acute and chronic aortic diseases of the thoracic and abdominal aorta of the adult. The Task Force for the Diagnosis and Treatment of Aortic Diseases of the European Society of Cardiology (ESC). Eur. Heart J. 35 (41), 2873–2926.

Erbel, R., Engberding, R., Daniel, W., Roelandt, J., Visser, C.M., Rennollet, H., 1989. Echocardiography in diagnosis of aortic dissection. Lancet 1, 457–461.

Erbel, R., Oelert, H., Meyer, J., 1993. Influence of medical and surgical therapy on aortic dissection evaluated by transesophageal echocardiography. Circulation 87, 1604–1615.

Ergin, M.A., O'Connor, J., Guinto, R., Griepp, R.B., 1982. Experience with profound hypothermia and circulatory arrest in the treatment of aneurysms of the aortic arch. Aortic arch replacement for acute aortic arch dissections. J. Thorac. Cardiovasc. Surg. 84, 649–655.

Filipovic, N., 1999. Numerical Analysis of Coupled Problems: Deformable Body and Fluid Flow. Ph.D. Thesis, Faculty of Mech. Engrg., University of Kragujevac, Serbia.

Filipovic, N., Nikolic, D., Saveljic, I., Djukic, T., Adjic, O., Kovacevic, P., Cemerlic-Adjic, N., Velicki, L., 2013. Computer simulation of thromboexclusion of the complete aorta in the treatment of chronic type B aneurysm. Comput. Aided Surg. 18 (1-2), 1–9.

Filipovic, N., Saveljic, I., Nikolic, D., Milosevic, Z., Kovacevic, P., Velicki, L., 2015. Numerical simulation of blood flow and plaque progression in carotid–carotid bypass patient specific case. Comput. Aided Surg. 20 (1), 1–6.

Fraser, C.D., Wang, N., Mee, R.B., 1994. Repair of insufficient bicuspid aortic valves. Ann. Thorac. Surg. 58, 386–390.

Gerber, R.T., Osborn, M., Mikhail, G.W., 2010. Delayed mortality from aortic dissection post transcatheter aortic valve implantation (TAVI): the tip of the iceberg. Catheter. Cardiovasc. Interv. 76 (2), 202–204.

Golledge, J., Eagle, K.A., 2008. Acute aortic dissection. Lancet 372 (9632), 55–66.

Griepp, R.B., Ergin, M.A., Lansman, S.L., Galla, J.D., Pogo, G., 1991. The physiology of hypothermic circulatory arrest. Semin. Thorac. Cardiovasc. Surg. 3, 188–193.

Hagan, P.G., Nienaber, C.A., Isselbacher, E.M., Bruckman, D., Karavite, D., Russman, P., 2000. The International Registry of Acute Aortic Dissection (IRAD): new insights into an old disease. JAMA 283, 897–903.

Iliceto, S., Ettore, G., Francisco, G., Antonelli, G., Biasco, G., Rizzon, P., 1984. Diagnosis of aneurysm of the thoracic aorta. Comparison between two non invasive techniques: twodimensional echocardiography and computed tomography. Eur. Heart J. 5, 545–555.

Joong, Y.P., Chan, Y.P., Chang, M.H., Kyung, S., Byoung, G.M., 2007. Pseudo-organ boundary conditions applied to a computational fluid dynamics model of the human aorta. Comput. Biol. Med. 37 (8), 1063–1072.

Karmonik, C., Bismuth, J., Shah, D.J., Davies, M.G., Purdy, D., Lumsden, A.B., 2011. Computational study of haemodynamic effects of entry and exit-tear coverage in a DeBakey type III aortic dissection: technical report. Eur. J. Vasc. Endovasc. Surg. 42, 172–177.

Khan, I.A., Nair, C.K., 2002. Clinical, diagnostic, and management perspectivesof aortic dissection. Chest 122, 311–328.

Khandheria, B.K., 1993. Aortic dissection: the last frontier. Circulation 87, 1765–1768.

Kojic, M., Filipovic, N., Stojanovic, B., Kojic, N., 2008. Computer Modeling in Bioengineering—Theoretical Background, Examples and Software. John Wiley and Sons, Chichester, England.

Kojić, M., Slavković, R., Živković, M., Grujović, N., 1998. Metod Konačnih Elemenata I, Mašinski fakultet u Kragujevcu, Kragujevac, Serbia.

Kondo, S., Hashimoto, N., Kikuchi, H., Hazama, F., Nagata, I., Kataoka, H., 1997. Cerebral aneurysms arising at nonbranching sites. An experimental study. Stroke 28, 398–403.

Krinsky, G.A., Rofsky, N.M., DeCorato, D.R., 1997. Thoracic aorta: comparison of gadolinium-enhanced threedimensional MR angiography with conventional MR imaging. Radiology 202, 183–193.

Mehta, R.H., O'Gara, P.T., Bossone, E., Nienaber, C.A., 2002. Acute type A aortic dissection in the elderly: clinical characteristics, management, and outcomes in the currentera. J. Am. Coll. Cardiol. 40, 685–692.

Miller, D.C., 1991. Surgical management of acute aortic dissection: new data. Semin. Thorac. Cardiovasc. Surg. 3, 225–237.

Mintz, G.S., Kotler, M.N., Segal, B.L., Parry, W.R., 1979. Twodimensional echocardiographic recognition of the descending thoracic aorta. Am. J. Cardiol. 44, 232–238.

Nathan, D.P., Xu, C., Gorman, J.H., Fairman, R.M., Bavaria, J.E., Gorman, R.C., Chandran, K.B., Jackson, B.M., 2011. Pathogenesis of acute aortic dissection: a finite element stress analysis. Ann. Thorac. Surg. 91 (2), 458–463.

Prince, M.R., Narasimham, D.L., Jacoby, W.T., 1996. Threedimensional gadolinium-enhanced MR angiography of the thoracic aorta. Am. J. Roentgenol. 166, 1387–1397.

Reul, G.J., Cooley, D.A., Hallman, G.L., Reddy, S.B., Kyger 3rd, E.R., Wukasch, D.C., 1975. Dissecting aneurysm of the descending aorta. Arch. Surg. 110, 632–640.

Sanders, C., 1990. Current role of conventional and digital aortography in the diagnosis of aortic disease. J. Thorac. Imaging 5, 48–59.

Shojima, M., 2004. Magnitude and role of wall shear stress on cerebral aneurysm: computational fluid dynamic study of 20 middle cerebral artery aneurysms. Stroke 35 (11), 2500–2505.

Slonim, S.M., Nyman, U.R., Semba, C.P., Miller, D.C., Mitchell, R.S., Dake, M.D., 1996. True lumen obliteration in complicated aortic dissection: endovascular treatment. Radiology 201, 161–166.

Sommer, T., Fehske, W., Holzknecht, N., 1996. Aortic dissection: a comparative study of diagnosis with spiral CT, multiplanar transesophageal echocardiography, and MR imaging. Radiology 199, 347–352.

Suzuki, T., Mehta, R.H., Ince, H., Nagai, R., Sakomura, Y., Weber, F., Suniyoshi, T., Bossone, E., Trimarchi, S., Cooper, J.V., Smith, D.E., Isselbacher, E.M., Eagle, K.A., Nienaber, C.A., 2003. Clinical profiles and outcomes of acute type B aortic dissection in the current era: lessons learned from the International Registry of Aortic Dissection (IRAD). Circulation 108, 312–317.

Svensson, L.G., Kouchoukos, N.T., Miller, D.C., 2008. Expert consensus document on the treatment of descending thoracic aortic disease using endovascular stent-grafts. Ann. Thorac. Surg. 85, S1–S41.

Svensson, L.G., Labib, S.B., Eisenhauer, A.C., Butterly, J.R., 1999. Intimal tear without hematoma. Circulation 99, 1331–1336.

Yamada, E., Matsumura, M., Kyo, S., Omoto, R., 1995. Usefulness of a prototype intravascular ultrasound imaging in evaluation of aortic dissection and comparison with angiographic study, transesophageal echocardiography, computed tomography, and magnetic resonance imaging. Am. J. Cardiol. 75, 161–165.

CHAPTER 6

The biomechanics of lower human extremities

Contents

1 Introduction

The skeletal system (skeleton) is the foundation of the human body. It consists of bones (dense, calcified connective tissue) connected to each other by

Computational Modeling in Bioengineering and Bioinformatics
https://doi.org/10.1016/B978-0-12-819583-3.00006-0

cartilage, tendons, or ligaments. Each of these elements serves a particular function and enables the skeletal system to (OpenStax College, 2013):
- provide the shape for our bodies that changes as we age,
- provide support for the body,
- preserve the adequate positions and shield internal organs,
- produce blood cells,
- store and release minerals and fat,
- provide the movement of the body (together with the muscular system).

Based on the shape, bones can be divided into five categories (OpenStax College, 2013):
- *Long bones*—they are cylindrical with the length of the bone being greater than the radius. Their role is to move when there is a muscle contraction. Long bones can be found in the legs (femur, tibia, and fibula), arms (radius, humerus, and ulna), fingers (metacarpals and phalanges), and toes (metatarsals and phalanges).
- *Short bones*—they are shaped like a cube with almost equal length, width, and height. Their roles are to provide support, stability, and limited motion. Short bones are located in the ankles (tarsals) and wrists (carpals).
- *Flat bones*—they are usually thin and often curved. Their roles are to protect the internal organs and to provide points for muscle attachment. They are located in the pelvis (iliac bone, ischium bone, and pubic bone), thoracic cage (ribs and sternum), and skull (occipital bone, parietal bone, frontal bone, nasal bone, lacrimal bone, and vomer bone).
- *Sesamoid bones*—they are small and round. Their role is to protect tendons from compressive forces. Sesamoid bones are embedded in tendons, and their number varies from person to person. The only sesamoid bone that every person has is patella (located in the knee joint).
- *Irregular bones*—they have complex shapes, and they do not fit in any of the previous categories. Their role is to protect the internal organs. The examples of the irregular bones are facial bones and the vertebrae.

The adult skeleton includes 206 bones that form axial and appendicular skeleton. The axial skeleton consists of 80 bones that include the thoracic cage (main role is to protect the heart and lungs), the vertebral column (main roles are to protect the spinal cord; support the head, neck, and body; and allow their movements), and the skull (main roles are to support the face and protect the brain) (OpenStax College, 2013). The appendicular skeleton includes 126 bones that are divided into:

- bones of the extremities (upper and lower),
- the girdle bones that connect extremities to the axial skeleton.

In the appendicular skeleton, we can differentiate between two girdles (OpenStax College, 2013):

- Pectoral girdle that connects the upper extremities to the thoracic cage of the axial skeleton
- Pelvic girdle that connects lower extremity to the vertebral column

The lower human extremity starts from the pelvis (Fig. 6.1) and contains 30 bones.

These bones can be divided into three regions:

- Thigh—located between the hip and knee joint
- Leg—located between the knee and ankle joint
- Foot—distal to the ankle joint

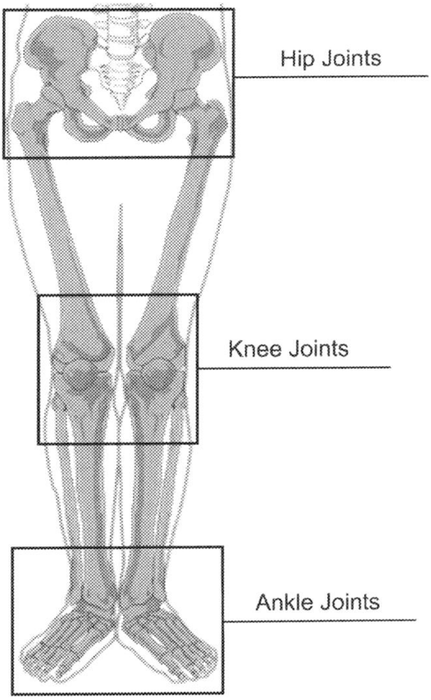

Fig. 6.1 Joints of the lower human extremities.

The main functions of the lower extremity are to bear the body weight and provide mobility to the body. Besides these main functions, lower extremities allow us some specific activities such as jumping or kicking a ball.

2 Anatomy of lower extremities

2.1 Hip joint anatomy

The hip joint (acetabulofemoral joint) is a ball-and-socket joint. It provides stability during daily activities (e.g., standing, walking, and running) and supports the body weight through them. The joint consists of the femoral head that is connected with the acetabulum (Griffiths and Khanduja, 2012).

Acetabulum is the socket part of the joint that has a substantial articulation area for the femoral head. Both of these articulating surfaces are covered by joint cartilage. Acetabulum is formed by merging of the three bones: iliac bone, ischium bone, and pubic bone (Fig. 6.2).

The ball of the hip joint is the head located at the proximal end of the femur (Fig. 6.3). It is connected to acetabulum with ligaments, which are surrounding it on all sides. Femoral bone is the strongest and the longest bone in the human body. It is the weight-bearing bone whose weakest part is the femoral neck. Fractures of the femoral neck are common in the elder population (Ly and Swiontkowski, 2008).

The ligaments are responsible for the joint stability (to hold a femoral head inside the socket) and limitation of the motion range of the leg in the hip joint. There are three ligaments that spiral around the head and the neck of the femur and connect pelvic bone to the femoral neck

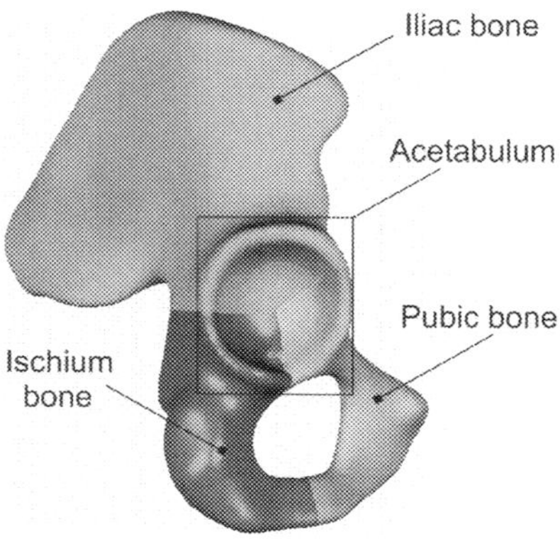

Fig. 6.2 Acetabulum.

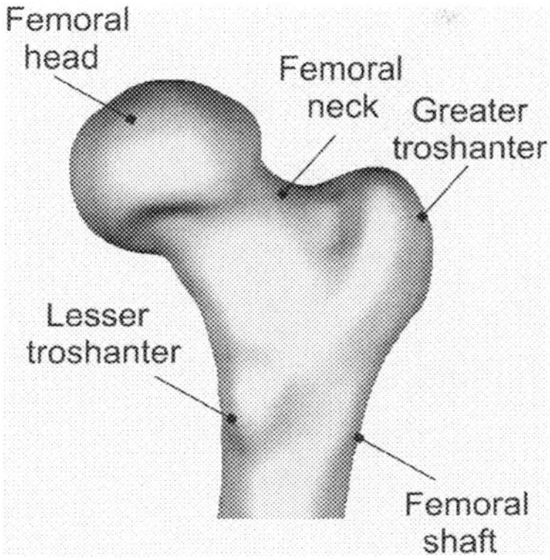

Fig. 6.3 Femoral bone.

(OpenStax College, 2013). Those are ischiofemoral ligament, iliofemoral ligament, and the pubofemoral ligament (Hall, 2011).

The ischiofemoral ligament starts from the ischium bone (lower posterior part of the pelvis) and extends laterally from the back of the hip joint over the neck of the femur. Its role is to contain the femoral head in the socket and limit the degree of medial and internal rotation in the socket. On the anterior side of the joint, there are two ligaments: pubofemoral ligament and iliofemoral ligament. The pubofemoral ligament connects the pubic bone (lower anterior part of the pelvis) with the femur. It supports the bottom of the joint while limiting lateral and external rotation in the socket and holding the femoral head in the socket. The iliofemoral ligament connects the iliac bone to the femoral bone. It is one of the strongest ligaments in the human body responsible for the force absorptions and transfer of the forces and for structural integrity of the joint.

2.2 Knee joint anatomy

The knee joint (Fig. 6.4) is the largest and one of the most complex joints in the human body. It provides stability while at the same time allowing the body to move. It is a hinge joint that supports flexion and rotation and consists of the femorotibial and the patellofemoral joint (Goldblatt and Richmond, 2003). The bones that form the knee joint are femur, tibia, fibula, and patella. Beside bones, the knee consists of ligaments (connect bone to bone) and tendons (connect muscle to bone).

The tibiofemoral joint is the largest joint in the body. This is a condyloid joint formed between the femur and tibia. The distal end of the femur has

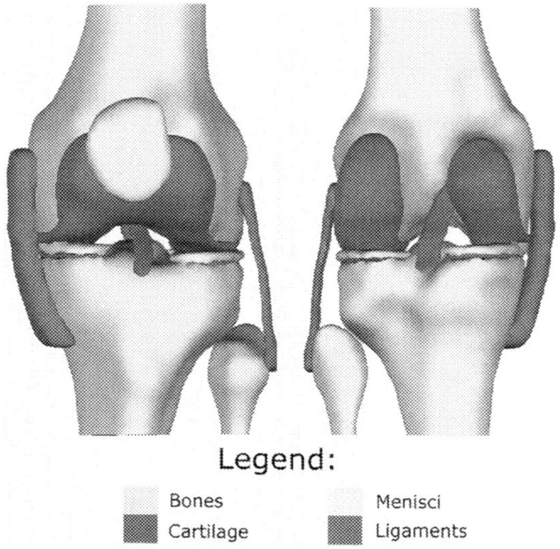

Fig. 6.4 Knee joint.

two condyles, a medial and a lateral. These condyles articulate with the corresponding tibial plateaus (Goldblatt and Richmond, 2003). The main role of the tibiofemoral joint is to provide the body weight transmission from the femur to the tibia while at the same time allowing small rotation of the joint (Oliveira et al., 2017).

The patellofemoral joint consists of two bones: patella and femur. The patella is a small, sesamoid bone embedded in the patellar tendon (Oliveira et al., 2017), whose posterior surface is in contact with the anterior side of the femur condyles. The friction between the femur and patella is reduced by articular cartilage that covers the posterior surface of the patella (Hall, 2011).

Other surrounding structures have an important effect on this joint. The menisci (medial and lateral) are fibrocartilage discs that are located between the femoral condyles and tibial plateaus. The menisci allow better shock absorption and load transmission in the knee joint (Hall, 2011). They are the thickest at their outer side where they are connected to tibia, while they are paper thin on the inner side where they are unattached. The menisci have different dimensions (Makris et al., 2011), and it is noticeable that the lateral meniscus has more shape, size, and thickness variety (Greis et al., 2002). The lateral meniscus is almost circular, while the medial meniscus is shaped as a crescent (Brindle et al., 2001).

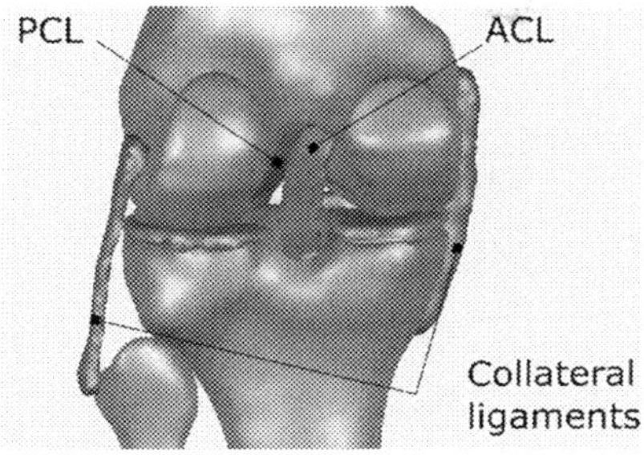

Fig. 6.5 Cruciate and collateral ligaments.

The ligaments help align the bones of the knee joint. The location of the ligament defines the direction in which it restricts the knee movements. Main ligaments of the joint are cruciate (anterior and posterior) and collateral (medial and lateral) ligaments (Fig. 6.5). The cruciate ligaments (PCL and ACL) are responsible for limiting the knee hyperextension and the femur sliding (forward and backward) on the tibial plateaus during the knee flexion and extension (Hall, 2011). On the other side, the collateral ligaments limit the knee lateral motion.

Besides cruciate and collateral ligaments, several other ligaments help to keep the knee from dislocating. The coronary ligaments are responsible for attachments between the menisci and tibial plateau. The anterior (ligament of Humphrey) and posterior (ligament of Wrisberg) meniscofemoral ligaments are responsible for the attachment between femur and lateral meniscus, while the medial meniscus mobility is limited by its attachment to the medial collateral ligament. The ligaments are attached to each other with the transverse ligament (Brindle et al., 2001).

2.3 Ankle joint anatomy

The ankle joint (Fig. 6.6) consists of the lower part of the leg and the foot, allowing for the interaction between the lower extremities and the ground. It transmits the force between the lower extremity and the ground and acts as a flexible shock absorber during the gait cycle (Dawe and Davis, 2011). This joint combines several smaller ones: talocalcaneal, tibiotalar, and transverse tarsal joint (Brockett and Chapman, 2016).

The talocalcaneal joint is formed between the talus and calcaneus bone. The talus is located on the anterior part of the calcaneus. This joint allows for ankle rotary motions (eversion and inversion) (Michael et al., 2008). While there are several ligaments that support this joint, the main connection

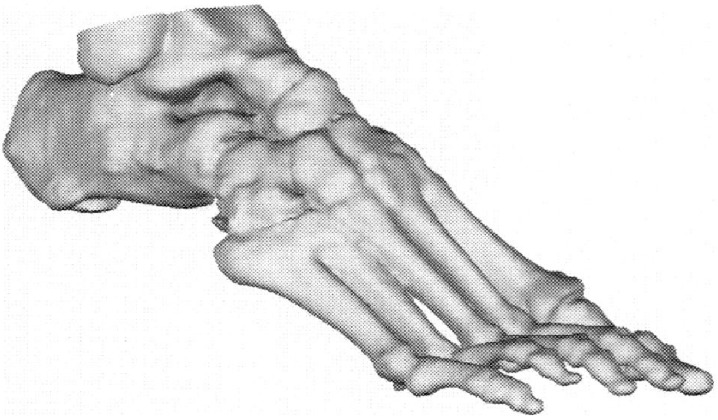

Fig. 6.6 Ankle joint.

between the talus and calcaneus bone is the interosseous talocalcaneal ligament that connects the inferior talus to the superior surface of the calcaneus (Brockett and Chapman, 2016).

The tibiotalar joint consists of the distal tibia and fibula of the lower leg and the talus. The tibial-talar interface is responsible for the load bearing. During the stance phase of the gait cycle, the geometry of the tibiotalar joint provides stability, while during the other phases, soft tissues are responsible for providing stability (Brockett and Chapman, 2016). There are three groups of ligaments responsible for the stability: tibiofibular syndesmosis, medial collateral ligaments, and lateral collateral ligaments. The tibiofibular syndesmosis provides stability between the tibia and fibula by limiting the motion between the bones. Medial collateral ligaments are responsible for reducing eversion motion and valgus stresses. The lateral collateral ligaments are responsible for limiting the joint inversion and rotation and reducing varus stress.

The transverse tarsal joint (Chopart's joint) is formed by two articulations, one of the talus bone with the navicular bone and other of the calcaneus bone with the cuboid bone. The joint help with eversion and inversion motion of the foot (Brockett and Chapman, 2016).

3 Finite element method application in biomechanics

Finite element method is a numerical method that has been used for the past couple of decades for the problems in bone biomechanics. In the beginning, the finite element method was used for structural stress analysis in the engineering mechanics (Huiskes and Chao, 1983). Due to the success of this methodology, in 1972, it was introduced in the orthopedic biomechanics for the stress analysis of the human bones. This method divides the complex problem into a greater number of smaller problems easier to solve. The solution to the problem obtained this way is an approximation that depends on the number of the elements used for the simulation. The greater number of elements used leads to more accurate results; however, this means that it will take more time to analyze the problem. Due to the advantages of this methodology such as ability to calculate the stress distribution for the complex geometries, it was applied for the analysis of the fractures and artificial joints. Since the 1972, when the first application of the finite element method in the field of biomechanics started, the complexity of the models has significantly increased.

Considering the number of papers available that have used finite element method for the analysis of hip, knee, or ankle joint, it is noticeable that more research has been dedicated toward the analysis of hip and knee joint.

3.1 Hip joint

The finite element approach has been used to investigate the femur during the gait cycle. Seo et al. (Seo et al., 2014) have used FEA to obtain information about the femur during the stance phase of the gait cycle. In the study presented in (Sverdlova and Witzel, 2010), the obtained results indicate that during the normal gait cycle and sitting down, the bone is loaded predominantly in compression, and the compressive stress was concentrated in the proximal and diaphyseal areas of the femur. Nishiyama et al. (2013) have used the finite element models to estimate the failure load of the femur. Their results indicate that the structural stiffness and failure load of femur obtained by FEA are in agreement with results obtained by the mechanical testing. The finite element method has also been used to understand the effects of the impingement and dysplasia on stress distributions in the hip joint during sitting and walking (Chegini et al., 2009). A significant number of finite element analysis of the hip joint are related to the analysis of the hip implants (Vulović and Filipović, 2018; Senalp et al., 2007; Bennett and Goswami, 2008; Abdul-Kadir et al., 2008).

3.2 Knee Joint

Peña et al. (2006) have created a 3D finite element model of the healthy knee joint to analyze the role that ligaments and menisci have in the load transmission and stability of the human knee. The healthy knee joint has also been analyzed with goal not only to understand tibiofemoral contact

(Haut Donahue et al., 2002) but also to assess knee cartilage stress distribution and deformation (Mijailović et al., 2015). Some authors analyzed only the specific parts of the knee joint, such as the ligaments (Filipović et al., 2013; Song et al., 2004) or articular cartilage. Limbert et al. (2004) have used finite element analysis to simulate clinical procedures that were performed to assess the existence and severity of anterior cruciate ligament injury, while Weiss et al. (2005) have analyzed the technical aspects of the 3D finite element modeling of ligaments. Their goal was to describe the methods for obtaining ligament geometry for computational models, FE modeling of ligament mechanics, and the verification and validation of ligament FE models. The effects of meniscectomy on the knee biomechanics have also been analyzed using the finite element method (Peña et al., 2005; Bae et al., 2012; Zielinska and Haut Donahue, 2006). Shriram et al. (2017) have used finite element analysis to evaluate the effects of material properties of artificial meniscal implant in the human knee joint. For this study, they have analyzed five different situations: intact knee, knee with medial meniscectomy, and the knee joint with the meniscal implant for which three distinct material stiffness were used. Although CFD has been used mostly for the problems related to the cardiovascular system (Zhong et al., 2018), drug delivery (Passos et al., 2018), respiratory system, and inhalers (Vulović et al., 2018), it has also found application for the analysis of the cartilage (Wu and Ferguson, 2016). Besides the analysis of the healthy knee joints, many authors have analyzed knee replacement (Halloran et al., 2005; Baldwin et al., 2012; Harrysson et al., 2007).

3.3 Ankle joint

Gefen et al. (2000) were the first in the literature to present the 3D foot model with realistic geometry and material properties of skeletal and soft tissue, with a goal to analyze the biomechanics of the foot during gait. Before them, a lot of papers have used 2D finite element model (Nakamura et al., 1981; Patil et al., 1996) or 3D models with simplified foot structure or material properties (Chu et al., 1995; Jacob et al., 1996; Kitagawa et al., 2000; Chen et al., 2001). Besides analyzing the ankle model, researchers have also been analyzing the footwear using the finite element method (Cheung and Zhang, 2005; Chen et al., 2003; Lemmon et al., 1997). This method has also been used for the analysis of the total ankle replacement during the gait cycle (Reggiani et al., 2006). Although the material properties of ligaments have been simplified, this study is an improvement compared with the previous published results. Recently, Wang et al. (2017) have developed a 3D finite element model of the lateral ankle ligaments to compare three surgical techniques for lateral ankle ligament reconstruction.

4 3D model development

This section is an overview of the methodology commonly used for the development of the patient-specific model. The model development requires several steps:

- Obtaining the patient's scans
- Segmentation of the domains of interest
- Creating the surface mesh for the previously segmented domains
- Manual refinement of the surface mesh
- Creating the volumetric mesh based on the refined segmented mesh

The first step is obtaining the appropriate scans. Most often, either computed tomography (CT) or magnetic resonance imaging (MRI) scans are used. This depends on the organs that are being analyzed. Both types of scans can be used for the development of the bone models; however, if the soft tissue is also modeled, then the MRI scans are used. Depending on the problem being analyzed, researchers can obtain scans from either a healthy person or a person that has a disease (e.g., osteoarthritis) or an injury (e.g., femur fracture). The obtained scans often correspond to the neutral and nonweight-bearing position.

Obtained 2D scan images are then loaded into a software where the segmentation process is performed to obtain the boundaries of the domains of interest (such as femur and cartilage) that will be analyzed. For this process, commercial software Mimics (Materialise, Leuven, Belgium) is commonly used. The quality of the scans has tremendous effect on the quality of the model that will be created. Higher resolution and larger number of slices improve the anatomical correctness of the model.

For segmentation process, thresholding algorithm is used. Segmentation can be performed

- manually,
- semimanually,
- automatic.

Automatic segmentation is the quickest, but it cannot always provide the completely anatomically correct model. An example of the automatic segmentation of left and right tibia and fibula is shown in Fig. 6.7. In the right down corner, it can be noticed that besides the tibia and fibula, the

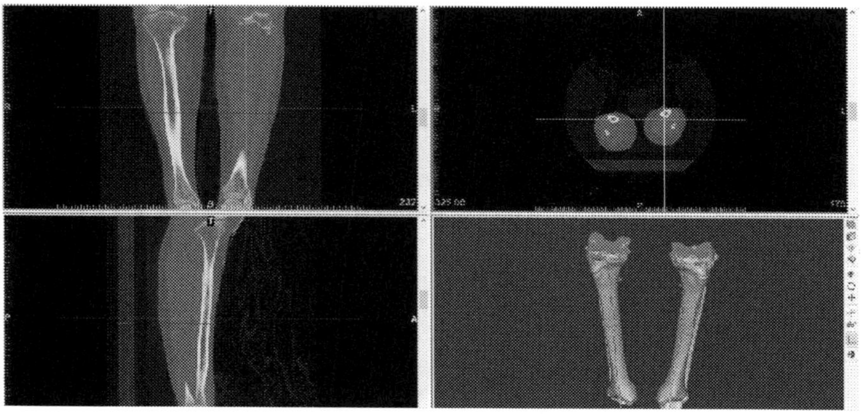

Fig. 6.7 Example of the automatic segmentation.

segmentation algorithm has singled out additional tissue that does not belong to either of these bones.

Often, as it was the case in the example earlier, it is necessary to make some manual adjustments to the segmentation as either there is more or not enough tissue segmented. The least used is the manual segmentation as it takes too much time. However, in cases where the scans have lower quality, this could be the only option. An example of the manually defined boundaries of the tibia and fibula is shown in Fig. 6.8. In this case, the generated model (right down corner) is not as smooth as the model created using automatic segmentation, which means that the model will require a lot of time for the refinement process.

The result of the segmentation process is the initial 3D model, which needs to be additionally refined. This initial 3D model is exported as the stereolithography (STL) format and contains information about the model's surface mesh.

The surface mesh of the initial 3D model is usually additionally refined. The process of the model refinement is often performed using Geomagic Studio (Raindrop Geomagic Inc., Research Triangle Park, NC, USA) software. In this part of the 3D model creation, several modifications are usually performed:

- Remove the additional segmented tissue
- Smooth the surface mesh
- Remove the spikes in the model mesh
- Increase/decrease the number of triangles (elements of the surface mesh)
- Remesh to obtain a more uniform surface mesh

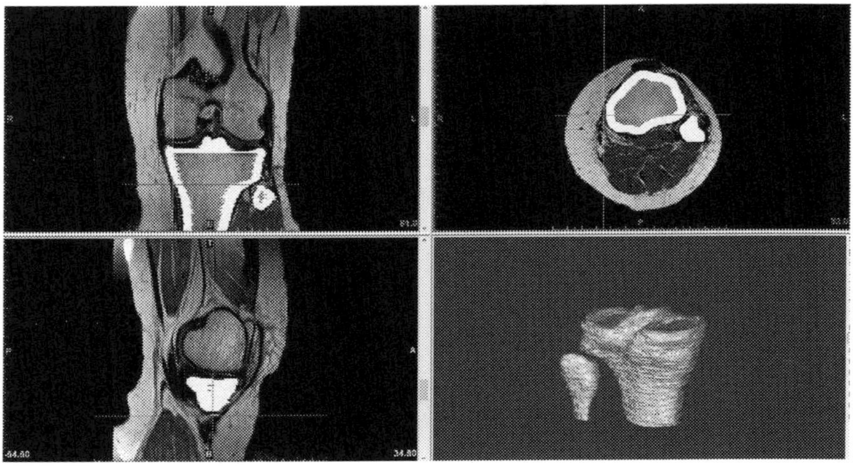

Fig. 6.8 Example of the manual segmentation.

The end result of this process is an anatomically correct 3D model. The last step of 3D model development is the creation of the volumetric mesh based on the surface mesh. This step can be performed using the software Femap (Siemens PLM Software, USA). The 3D model with surface mesh is imported, and the quality of the mesh is checked. It the quality is not satisfying, additional refinement is needed. After the mesh quality is satisfied, the volumetric mesh will be created by filling the model with tetrahedral elements.

The volumetric mesh can consist of hexahedral or tetrahedral elements. Although hexahedral elements provide better results, tetrahedral elements are used more often due to the less effort needed to create a volumetric mesh.

These steps are repeated for each model that needs to be created.

5 Material properties

One of the main difficulties of any numerical simulation that deals with an anatomical model is the application of appropriate biological material properties.

5.1 Bone

Bone is a composite material that consists of a mineral components, hydroxyapatite; an organic components; and water. The properties of the bone depend on the quantity, arrangement, and characteristics of each of these components (Boskey, 2013). The quantity depends on the several factors such as

- age (Boskey and Coleman, 2010),
- gender (Gregson et al., 2013),
- health status (Boskey and Mendelsohn, 2005),
- ethnicity (Leslie, 2012).

There are two types of bones (Pal, 2013):

- Cortical (also known as compact)
- Cancellous (also known as trabecular or spongy)

The difference between these two types is in the degree of porosity and organization, as the material forming both bones is identical. Cortical bone has lower porosity (5%–30%) compared with the cancellous bone (30%–90%). These values are not fixed, and depending on the disease, the aging process or loading conditions can be changed (Pal, 2013).

Both structural and material properties of the bone can be determined using different tests. Bone material properties are obtained using mechanical testing and ultrasonic techniques used for other structural materials. However, it is important to note that the bone properties depend on the freshness of the tissue used for tests. The organic components provide bone with flexibility, while the inorganic components (mainly hydroxyapatite) give the bone its resilience. Higher mineralization lowers the toughness but increases the strength and stiffness. Cortical bone is 100 times stiffer than cancellous bone, which means that it can sustain greater stress, but lower strain before the fracture (Pal, 2013).

Bone is usually assumed to be one of the following types (Krone and Schuster, 2006):

- Isotropic
- Transversely isotropic
- Orthotropic

Isotropic is the simplest type, where only two independent parameters are needed—Young's modulus and Poisson's ratio. The more complex is transversely isotropic, which requires five parameters—two Young's moduli, two shear moduli, and one Poisson's ratio. The orthotropic material has material properties that are different in three mutually perpendicular directions at any point. Orthotropic material requires nine parameters—three Young's moduli (for each direction), three shear moduli (for each direction), and three Poisson's ratios (for each direction) (Krone and Schuster, 2006).

5.1.1 Cortical bone

In the majority of the papers, cortical bone is usually considered to be homogeneous, linear elastic, and isotropic to simplify the problem. This is especially the case that the complex geometry is used. However, some authors consider cortical bone to be orthotropic or anisotropic material, which improves the obtained results. Table 6.1 presents a short overview of the cortical bone material properties used for the finite element analysis.

Table 6.1 Overview of the cortical bone material properties

Reference	Type of properties	Material property	Value
Prochor (2017)	Orthotropic material properties	Young's modulus (GPa)	Radial: 12.0 Transverse: 13.4 Longitudinal: 20.0
		Poisson's ratio	Radial: 0.376 Transverse: 0.222 Longitudinal: 0.235
		Shear modulus (GPa)	Radial: 4.53 Transverse: 5.61 Longitudinal: 6.23
Reilly and Burnstein (1974)	Orthotropic material properties	Young's modulus (GPa)	Radial: 3.76 Transverse: 4.19 Longitudinal: 8.69
Krone and Schuster (2006)	Orthotropic material properties	Young's modulus (GPa)	Radial: 6.30 Transverse: 6.88 Longitudinal: 16.0
		Poisson's ratio	Radial: 0.3 Transverse: 0.45 Longitudinal: 0.3
		Shear modulus (GPa)	Radial: 3.20 Transverse: 3.60 Longitudinal: 3.20
Krone and Schuster (2006)	Isotropic material properties	Young's modulus (GPa) Poisson's ratio	16.0 0.36

Continued

Table 6.1 Overview of the cortical bone material properties—cont'd

Reference	Type of properties	Material property	Value
Perez et al. (2008)	Orthotropic material properties	Young's modulus (GPa)	Radial: 11.6 Transverse: 14.6 Longitudinal: 21.9
		Poisson's Ratio	Radial: 0.11 Transverse: 0.30 Longitudinal: 0.21
		Shear modulus (GPa)	Radial: 6.29 Transverse: 5.29 Longitudinal: 6.99
Soni et al. (2017)	Isotropic material properties	Young's modulus (GPa)	14371
		Poisson's ratio	0.315
Soni et al. (2017)	Isotropic material properties	Young's modulus (GPa)	20033
		Poisson's ratio	0.315
Tseng et al. (2014)	Isotropic material properties	Young's modulus (GPa)	17000
		Poisson's ratio	0.3

5.1.2 Cancellous bone

Cancellous bone is an extremely anisotropic and heterogeneous; however, it is also considered to be homogeneous, linear elastic, and isotropic based on the same reasons as for the cortical bone. Table 6.2 presents a short overview of the material properties of cancellous bone used for the finite element analysis.

5.2 Cartilage

Articular cartilage is a multiphasic material consisting of two major phases—a fluid and a solid phase. Fluid phase is composed of water (68%–85%) and electrolytes, while a solid phase is composed of collagen fibrils (primarily type II collagen) (10%–20%), proteoglycans, other glycoproteins (5%–10%), and the chondrocytes. Articular cartilage covers the bone surfaces within the joint capsule, and its thickness depends on the joint and location within the joint. It is the thickest over the ends of the femur and tibia, ranging from 2 to 4 mm (Pal, 2013).

Table 6.2 Overview of the cancellous bone material properties

Reference	Type of properties	Material property	Value
Krone and Schuster (2006)	Orthotropic material properties	Young's modulus (MPa)	Radial: 676 Transverse: 968 Longitudinal: 1352
		Poisson's ratio	Radial: 0.3 Transverse: 0.3 Longitudinal: 0.3
		Shear modulus (MPa)	Radial: 505 Transverse: 370 Longitudinal: 292
Krone and Schuster (2006)	Transversely isotropic material properties	Young's modulus (MPa)	Radial: 822 Transverse: 822 Longitudinal: 1352
		Poisson's ratio	Radial: 0.3 Transverse: 0.3 Longitudinal: 0.3
		Shear modulus (MPa)	Radial: 399 Transverse: 370 Longitudinal: 399
Krone and Schuster (2006)	Isotropic material properties	Young's modulus (MPa)	1.0
		Poisson's ratio	0.30
Soni et al. (2017)	Isotropic material properties	Young's modulus (MPa)	295
		Poisson's ratio	0.315
Perez et al. (2008)	Isotropic material properties	Young's modulus (MPa)	600
		Poisson's ratio	0.2

Depending on the paper, articular cartilage is assumed to have either three (superficial, middle, and deep) or four distinct layers (superficial, middle, deep, and calcified) along its depth. This indicates that mechanical properties of articular cartilage depend on its depth (Pal, 2013).

Table 6.3 Overview of the articular cartilage material properties

Reference	Type of properties	Material property	Value
Hopkins et al. (2010)	Isotropic material properties	Young's modulus (MPa) Poisson's ratio	12 0.45
Anderson et al. (2007)	Isotropic material properties	Young's modulus (MPa) Poisson's ratio	12 0.42
Cheung et al. (2005)	Isotropic material properties	Young's modulus (MPa) Poisson's ratio	1 0.4
Gefen et al. (2000)	Isotropic material properties	Young's modulus (MPa) Poisson's ratio	1 0.1
Peña et al. (2006)	Isotropic material properties	Young's modulus (MPa) Poisson's ratio	5 0.46

Cartilage is considered to be an anisotropic viscoelastic material with porous structure. This structure allows the fluid to move in and out of the cartilage. The mechanical properties of cartilage depend not only on the fluid content but also on the pathology (Pal, 2013).

Table 6.3 presents a short overview of the material properties of articular cartilage used for finite element analysis.

Using linearly elastic material properties for cartilage was found to be appropriate for the short loading times (Haut Donahue et al., 2002).

5.3 Ligaments

Ligaments consist of collagen, elastin, glycoproteins, protein polysaccharides, glycolipids, water, and cells (mostly fibrocytes). Collagen constitutes 70%–80% of the dry weight of a ligament, the majority being type I collagen. Water makes up about 60%–80% of the wet weight of ligaments. Collagen fiber bundles are parallel and closely packed, which provides motion and stability of the joints (Pal, 2013).

Ligaments are considered to be composite, anisotropic structures with nonlinear time- and history-dependent viscoelastic properties that can change based on (Pal, 2013):

- strain rate,
- temperature,
- aging,
- hydration,
- maturation,

- immobilization,
- exercise,
- healing.

Table 6.4 presents a short overview of the isotropic material properties of ligaments used for finite element analysis.

In Wang et al. (2017), ligaments were assumed to be nonlinear hyperelastic. The constitutive relation of the ligaments followed the neo-Hookean model with formula:

$$\psi = \frac{1}{2D} \ln(J)^2 + C_1(\bar{I}_1 - 3) + F_2(\lambda) \tag{6.1}$$

where C_1 is the neo-Hookean constant (value of the constant = 1.44) and $\frac{1}{D}$ is the ligament of the bulk modulus (value of the bulk modulus = 0.00126).

Peña et al. (2005) considered ligaments to be isotropic and hyperelastic. This was represented using incompressible neo-Hookean behavior with the following energy density function:

$$\psi = C_1(\bar{I}_1 - 3) \tag{6.2}$$

where C_1 is the initial shear modulus and $\bar{I}_1$ is the first modified invariant of the right Cauchy-Green strain tensor. The C_1 values for lateral collateral ligament, medial collateral ligament, anterior cruciate ligament, and posterior cruciate ligament are presented in Table 6.5.

Table 6.4 Overview of the ligament material properties

Reference	Type of properties	Material property	Value
Kubiček and Florian (2009)	Isotropic material properties	Young's modulus (MPa)	400
		Poisson's ratio	0.45
Soni et al. (2017)	Isotropic material properties	Young's modulus (MPa)	345
		Poisson's ratio	0.22

Table 6.5 Initial shear modulus for the ligaments of the knee (Peña et al., 2005)

Ligament	Initial shear modulus (MPa)
Lateral collateral ligament	6.06
Medial collateral ligament	6.43
Anterior cruciate ligament	5.83
Posterior cruciate ligament	6.06

5.4 Menisci

The menisci are usually described as the biphasic tissue with the liquid and solid phase. As well as the articular cartilage, menisci are also considered to be anisotropic viscoelastic material (Aufderheide and Athanasiou, 2004). Table 6.6 presents a short overview of the isotropic material properties of menisci used for finite element analysis.

Besides isotropic material properties, papers such as, Shriram et al. (2017), have used transversely isotropic hyperelastic neo-Hookean material properties with the strain-energy function:

$$\Phi = C_{10}\left(\overline{G}_1 - 3\right) + \frac{1}{D_1}(J_F - 1)^2 + S(\lambda) \tag{6.3}$$

where C_{10} is the bulk material constant associated with the shear modulus, J_F is the Jacobian of the deformation gradient, and $\overline{G}_1$ is the first invariant of the left Cauchy-Green tensor.

Table 6.6 Overview of the menisci material properties

Reference	Type of properties	Material property	Value
Bae et al. (2012)	Isotropic material properties	Young's modulus (MPa)	59
		Poisson's ratio	0.49
Hopkins et al. (2010)	Isotropic material properties	Young's modulus (MPa)	80
		Poisson's ratio	0.3
Kubiček and Florian (2009)	Isotropic material properties	Young's modulus (MPa)	112
		Poisson's ratio	0.45
Zielinska and Haut Donahue (2006)	Transversely isotropic material properties	Young's modulus (MPa)	Axial/radial: 20 Circum: 150
		Poisson's ratio	In-plane: 0.2 Out-of-plane: 0.3
		Shear modulus (MPa)	57.7

6 Boundary conditions

The boundary conditions depend on the problem being analyzed. More precise boundary conditions will lead to obtaining more accurate results. As the geometry of the lower extremities is very complex, boundary conditions used for numerical simulations are quite simplified. These conditions are usually obtained using experiments. Researchers are mostly interested in analyzing the gait (walking) cycle. The walking cycle or the gait cycle is an interval that starts when one heel comes in contact with the ground and ends when the same heel comes in contact with the ground again (Perry and Jon, 1992).

The ankle joint has not been analyzed as much as the knee or the hip joint; however, interest in this joint has been increasing. In this section, we will present boundary conditions commonly used for the finite element analysis of the hip, knee, and ankle joint.

6.1 Hip joint

The majority of the simulations that are concerning the hip joint are related to the hip arthroplasty. Senalp et al. (2007) have performed both static and dynamic analysis of the hip implant. For the static analysis, they have used a load of 3 kN (correspond to a person of 70 kg) that was applied on the surface of the implant and abductor muscle load of 1.25 kN that was applied to the proximal area of the greater trochanter. On the bottom of the femur in the longitudinal direction, iliotibial tract load was applied. For the dynamic analysis, the force applied on the surface of the implant was time-dependent on the load during the gait cycle. More complex boundary conditions are presented in the paper (Chalernpon et al., 2015) where both walking and climbing stair were analyzed.

6.2 Knee joint

For the analysis of the healthy knee joint model, frictionless nonlinear contact was assumed for all contact surfaces. To simulate the full extension position of the femur, the flexion-extension and the varus-valgus rotations were fixed. The lower surface of the tibia was fixed as well. To simulate full extension position, vertical compression force of 1150 N (force at full extension position during the gait cycle) was applied at the upper femur surface (Peña et al., 2005). Similar boundary conditions were used in the papers (Shriram et al., 2017; John et al., 2013). For the analysis of the medial effects meniscectomy, the following boundary conditions were used: A vertical load of 570 N was applied on the sacrum bone, under the assumption of the single-leg stance under full extension; frictionless contact condition was assumed for all contact surfaces; distal surfaces of the fibula and tibia were fixed; and the displacement of the nodes on the sacrum and coxal bones symmetric surfaces were fixed due to the symmetry (Bae et al., 2012). For the analysis of the knee implant, Harrysson et al. (2007) have considered contact between the distal femur and implant to be standard augmented Lagrange contact without separation after contact. They have analyzed standing position, end of a normal walking gait cycle, and climbing stairs by using different forces (2200 N, 3200 N, and 2800 N) on the upper surface of the femur, which corresponds to the change of angle during the previous activities.

6.3 Ankle joint

Wang et al. (2018) have used the Vicon system to obtain boundary conditions of the ankle joint during the gait cycle. They have put seven reflective markers on the lower limb that were used to obtain the motion of the segments of interest while the participant walked and stepped on the force platform. This resulted in obtaining the curve for ground reaction forces during the gait cycle. Although they have obtained the curve, they decided to use three characteristic values that correspond to the maximum force impacting the hind (first peak), forefoot (midstance), and full body weight (second peak) during single-foot support.

For the finite element analysis used for the evaluation of the calcaneus and talus fracture, contact between the ground and the soft tissue surrounding the ankle joint was considered with coefficient of friction that was set to a value of 0.6 (Wong et al., 2016). The goal was to analyze the axial loading, which is an important injury mechanism for the lower extremities (Funk et al., 2002). This was done by fixing the proximal tibia and fibula and the surrounding soft tissue in all six degrees of freedom, while the impactor was moving at 5.0 m/s until it came into the contact with the foot after which the compression force of 270 N was applied.

For the simulation of the neutral standing on the ground, Taha et al. (2016) used automatic surface-to-surface contact algorithm available in ANSYS. This was used for the simulation of the interaction between the bony surfaces, which was assumed to be frictionless. The same algorithm was used for the contact between the foot and the ground; however, in this case, it was assumed to be frictional with coefficient of friction set to be 0.6.

The force acting on the tibia and fibula was 300 N, as the authors considered a normal male person with weight of 60 kg evenly distributed.

For the insole design, a person with an average weight of 70 kg was considered (Cheung and Zhang, 2005). In this case, during balanced stance, a vertical force of 350 N on each foot was applied. Also, the reaction of the Achilles' tendon was considered. The magnitude was 175 N, based on the study by Simkin (1982), where this force was calculated to be approximately 50% of the force applied on the foot during balanced standing.

7 Finite element analysis of the knee joint with ruptured anterior cruciate ligament

This example shows the finite element analysis of the knee joint with ruptured anterior cruciate ligament.

7.1 Step 1: Three-dimensional model development

The first step included segmentation of the area of interest, which in this case were femur (cortical and cancellous), tibia (cortical and cancellous), fibula, menisci (medial and lateral meniscus), articular cartilage (femoral cartilage, medial tibial cartilage, and lateral tibial cartilage), and ligaments (medial collateral ligament, lateral collateral ligament, and posterior cruciate ligament). Since the influence of the ruptured anterior cruciate ligament was analyzed, this ligament was not segmented.

Available DICOM images were used for the development of the knee model. Resolution of the used scans was 320×320 pixels with slice increment of 0.589mm. Segmentation and the generation of the knee model were performed using the software Mimics 10.01 (Mimics 10.01, Materialise, Leuven, Belgium). Fig. 6.9 shows loaded knee scans.

For each of the previously mentioned elements of the knee joint, a mask was created. Each mask contained information about the boundaries of regions of interest for the specific knee joint element. This information was needed for the 3D model development. The thresholding algorithm was used to segment elements of the knee joint. However, some manual modifications were necessary as the algorithm was not able to perform complete segmentation. These segmentations were exported as surface models with triangular mesh and had to be manually refined to remove additional segmented tissue and to smooth the surface mesh. The refinement process was done using Geomagic 2013 (Geomagic Studio 2013, Raindrop

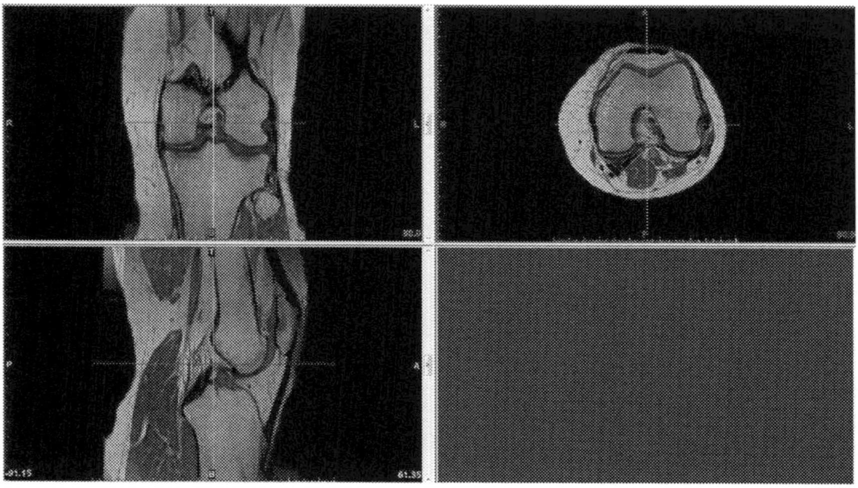

Fig. 6.9 Imported knee DICOM images.

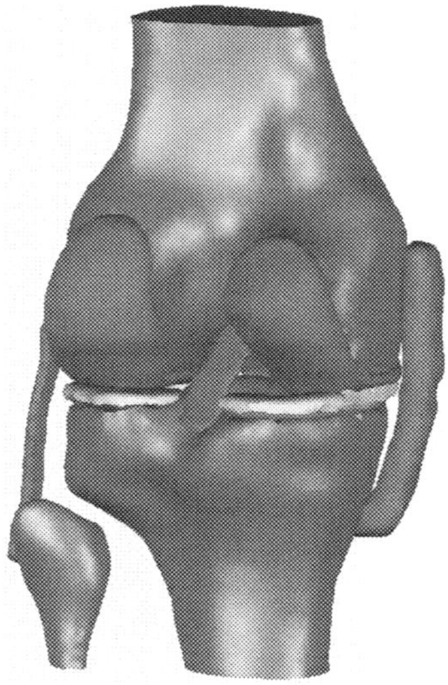

Fig. 6.10 Three-dimensional knee joint model.

Geomagic Inc., Research Triangle Park, NC, USA). Fig. 6.10 shows the knee joint model after the refinement process was finished.

This model with surface mesh was used for creating the volumetric mesh. The result of the volumetric mesh creation was that each triangle on the surface mesh became the face of at least one tetrahedral element. The number of nodes and elements in the volumetric mesh of the knee joint model was 997455 and 665153, respectively.

7.2 Step 2: Material properties

As already explained, material properties of the knee joint elements are very complex, especially articular cartilage, menisci, and ligaments. However, they are often simplified and considered to be linear elastic, homogeneous, and isotropic. Usually, material properties are obtained from the literature as a result of different experiments. Table 6.7 lists all the materials used for calculation and their properties (Young's modulus and Poisson's ratio).

7.3 Step 3: Boundary conditions

For this simulation, considered boundary conditions corresponded to the physiological conditions when a person is standing on the fully extended leg. Boundary conditions included constraints and loading. Applied constraints included fixing the nodes on the lower tibia and fibula surface. These nodes were fixed in x-, y-, and z-directions, while other nodes were allowed to move along all three (x, y, and z) axes. The force was applied perpendicularly to the femur's upper surface. The force value corresponds to the value of the force during the gait cycle when the leg is in the full extension (Kutzner et al., 2010).

Fig. 6.11 shows a graphical representation of the applied boundary conditions.

Table 6.7 Material properties used for simulation

Material	Young's modulus (MPa)	Poisson's ratio	Reference
Femur-cortical	14317	0.315	Soni et al. (2017)
Femur-cancellous	295	0.315	Soni et al. (2017)
Tibia-cortical	20033	0.315	Soni et al. (2017)
Tibia-cancellous	295	0.315	Soni et al. (2017)
Fibula	17000	0.3	Tseng et al. (2014)
Articular cartilage	12	0.45	Hopkins et al. (2010)
Menisci	80	0.3	Hopkins et al. (2010)
Ligaments	345	0.22	Soni et al. (2017)

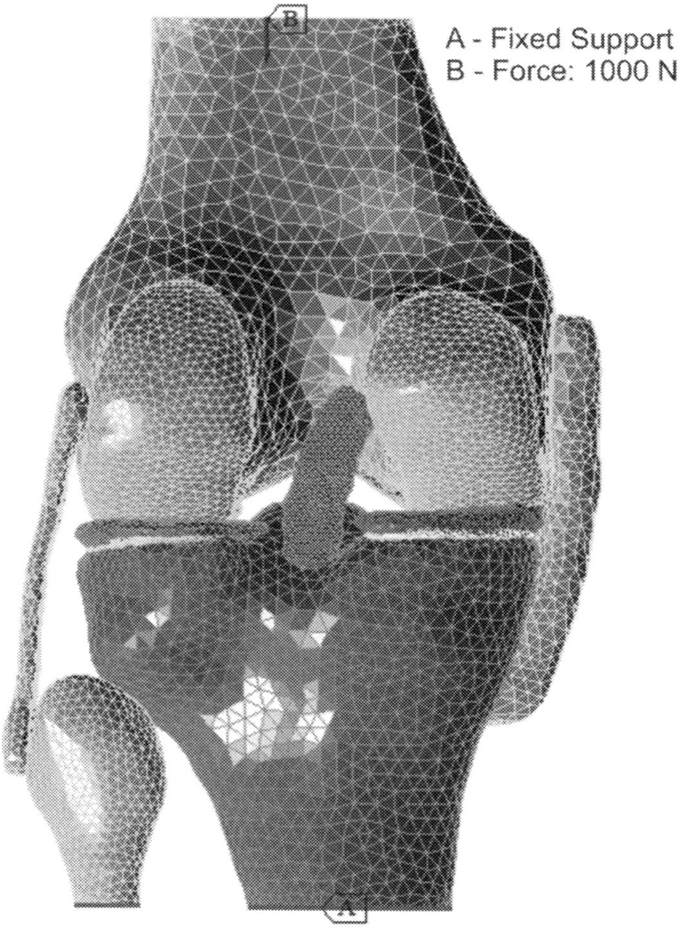

Fig. 6.11 Applied boundary conditions.

Bonded contact was assumed between the knee joint elements. Contact was set up automatically by the software, with the assumption that it should be defined where the overlap between materials exists.

7.4 Step 4: Results and discussion

FE analysis was performed using ANSYS 14.5.7 software. The goal of the study was to analyze the influence of the ruptured ACL on the stress distribution in the knee joint. The results presented here are von Mises stress (Vulović et al., 2016) and maximum principal stress. Fig. 6.12 shows von

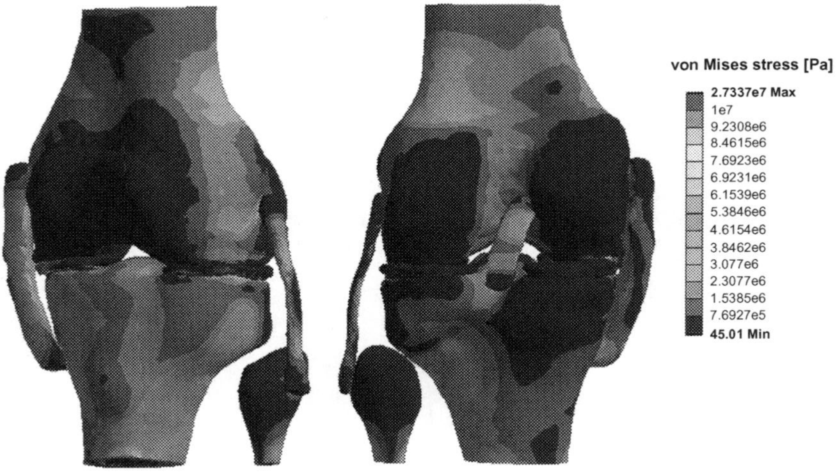

Fig. 6.12 Von Mises stress distribution.

Mises stress distribution for the knee joint model based on the previously explained material properties and boundary conditions.

The medial and lateral menisci have the highest stress value, calculated to be around 10 MPa. The obtained results are significantly higher than the values when only the healthy knee joint is analyzed. This can be partially explained not only by the used boundary conditions but also with the fact that the analyzed situation included ruptured ACL. Bone (femur, tibia, and fibula) stress values are in the range from 45 Pa to 7 MPa. Near the lower tibia surface is one area with stress values above 7 MPa, which can be seen in the figure. This is the result of the used boundary conditions (node fixing). Ligament stress values are lower than the bone stress values, and they are in the range from 45 Pa to 6.5 MPa.

Maximum principal stress distribution is shown in Fig. 6.13.

The lowest maximum principal stress value was near the lower tibia surface, and it was calculated to be −5 MPa, which was the result of the boundary conditions. The highest value was calculated to be on the place of the connection between the PCL and the femur, and it was around 10 MPa. This is the area of the high interest, and it will require further analysis.

This static analysis indicated that in the case of the ACL rupture, a person could be able to stand on his/her legs, without worrying that it will have any effect on the other parts of the joint. However, the obtained results require additional studies, where both dynamic analyses will be not only performed (e.g., gait cycle) but also improved material properties taken into consideration.

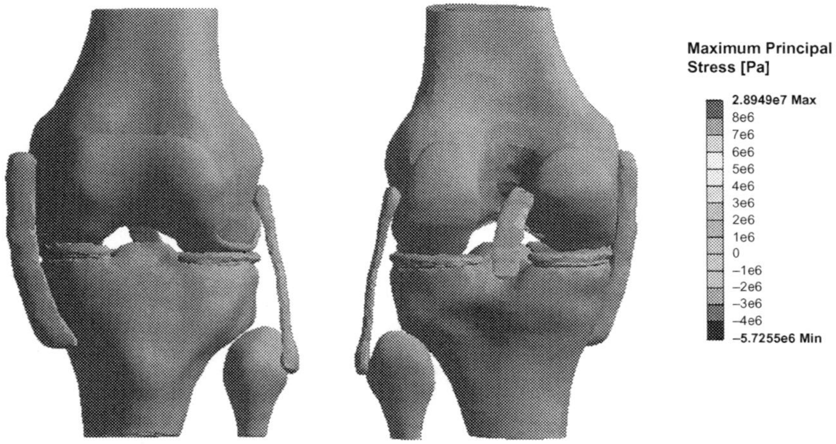

Fig. 6.13 Maximum principal stress distribution.

8 Conclusion

The use of numerical methods, especially the finite element method, for the patient-specific analysis of the biomechanics of the lower human extremities, provides us with possibility to assess the biomechanics of the human joint. This approach is noninvasive and can provide us with information that cannot be obtained with experimental studies. The geometries of the lower human extremities can be easily modified to represent not only healthy joints but also the joints with diseases or bones with fracture. Also, it can be custom made to represent any individual. Another big advantage is that one model can be used multiple times with different boundary conditions, which requires less time compared with starting the whole process from the beginning.

On the other side, there are several disadvantages, and they are related to the simplification of the material properties and boundary conditions that are necessary to scale down the complexity of the model and simulation. The simplification is especially present in the finite element analysis of the knee joint, where used material properties (linear/elastic isotropic and homogeneous materials instead of nonlinear and anisotropic behavior) cannot accurately describe the behavior of the knee joint. However, the constant development of the technology and the additional material testing will lead to the numerical simulation that will be able to eventually substitute the experimental studies.

Acknowledgment

The research has been carried out with the support of the Ministry of Education, Science and Technological Development, Republic of Serbia with projects III41007 and OI174028.

References

Abdul-Kadir, M.R., Hansen, U., Klabunde, R., Lucas, D., Amis, A., 2008. Finite element modelling of primary hip stem stability: the effect of interference fit. J. Biomech. 41 (3), 587–594.

Anderson, D.D., Goldsworthy, J.K., Li, W., James Rudert, M., Tochigi, Y., Brown, T.D., 2007. Physical validation of a patient-specific contact finite element model of the ankle. J. Biomech. 40 (8), 1662–1669.

Aufderheide, A.C., Athanasiou, K.A., 2004. Mechanical stimulation toward tissue engineering of the knee meniscus. Ann. Biomed. Eng. 32 (8), 1163–1176.

Bae, J.Y., Park, K.S., Seon, J.K., Kwak, D.S., Jeon, I., Song, E.K., 2012. Biomechanical analysis of the effects of medial meniscectomy on degenerative osteoarthritis. Med. Biol. Eng. Comput. 50 (1), 53–60.

Baldwin, M.A., Clary, C.W., Fitzpatrick, C.K., Deacy, J.S., Maletsky, L.P., Rullkoetter, P.J., 2012. Dynamic finite element knee simulation for evaluation of knee replacement mechanics. J. Biomech. 45 (3), 474–483.

Bennett, D., Goswami, T., 2008. Finite element analysis of hip stem designs. Mater. Des. 29 (1), 45–60.

Boskey, A.L., 2013. Bone composition: relationship to bone fragility and antiosteoporotic drug effects. Bonekey Rep. 2, 447.

Boskey, A.L., Coleman, R., 2010. Aging and bone. J. Dent. Res. 89 (12), 1333–1348.

Boskey, A., Mendelsohn, R., 2005. Infrared analysis of bone in health and disease. J. Biomed. Opt. 10, 031102–031108.

Brindle, T., Nyland, J., Johnson, D.L., 2001. The meniscus: review of basic principles with application to surgery and rehabilitation. J. Athl. Train. 36 (2), 160–169.

Brockett, C.L., Chapman, G.J., 2016. Biomechanics of the ankle. J. Orthop. Trauma 30 (3), 232–238.

Chalernpon, K., Aroonjarattham, P., Aroonjarattham, K., 2015. Static and dynamic load on hip contact of hip prosthesis and thai femoral bones. Int. J. Mech. Mechatron. Eng. 9 (3), 251–255.

Chegini, S., Beck, M., Ferguson, S.J., 2009. The effects of impingement and dysplasia on stress distributions in the hip joint during sitting and walking: a finite element analysis. J. Orthop. Res. 27 (2), 195–201.

Chen, W., Ju, C., Tang, F., 2003. Effects of total contact insoles on the plantar stress redistribution: a finite element analysis. Clin. Biomech. 18, S17–S24.

Chen, W.P., Tang, F.T., Ju, C.W., 2001. Stress distribution of the foot during mid-stance to push-off in barefoot gait: a 3-D finite element analysis. Clin. Biomech. 16, 614–620.

Cheung, J.T.-M., Zhang, M., 2005. A 3-dimensional finite element model of the human foot and ankle for insole design. Arch. Phys. Med. Rehabil. 86 (2), 353–358.

Cheung, J.T.-M., Zhang, M., Leung, A.K.-L., Fan, Y.-B., 2005. Three-dimensional finite element analysis of the foot during standing—a material sensitivity study. J. Biomech. 38 (5), 1045–1054.

Chu, T.M., Reddy, N.P., Padovan, J., 1995. Three-dimensional finite element stress analysis of the polypropylene ankle-foot orthosis, static analysis. Med. Eng. Phys. 17, 372–379.

Dawe, E.J., Davis, J., 2011. (vi) Anatomy and biomechanics of the foot and ankle. Orthop. Trauma 25 (4), 279–286.
Filipović, N., Isailović, V., Nikolić, D., Peulić, A., Mijailović, N., Petrović, S., Ćuković, S., Vulović, R., Matić, A., Zdravković, N., Devedžić, G., Ristić, B., 2013. Biomechanical modeling of knee for specific patients with chronic anterior cruciate ligament injury. Comput. Sci. Inf. Syst. 10, 525–545.
Funk, J.R., Crandall, J.R., Tourret, L.J., MacMahon, C.B., Bass, C.R., Patrie, J.T., Khaewpong, N., Eppinger, R.H., 2002. The axial injury tolerance of the human foot/ankle complex and the effect of achilles tension. J. Biomech. Eng. 124 (6), 750–757.
Gefen, A., Megido-Ravid, M., Itzchak, Y., Arcan, M., 2000. Biomechanical analysis of the three-dimensional foot structure during gait: a basic tool for clinical applications. J. Biomech. Eng. 122 (6), 630–639.
Goldblatt, J.P., Richmond, J.C., 2003. Anatomy and biomechanics of the knee. Oper. Tech. Sports Med. 11 (3), 172–186.
Gregson, C.L., Paggiosi, M.A., Crabtree, N., Steel, S.A., McCloskey, E., Duncan, E.L., Fan, B., Shepherd, J.A., Fraser, W.D., Smith, G.D., Tobias, J.H., 2013. Analysis of body composition in individuals with high bone mass reveals a marked increase in fat mass in women but not men. J. Clin. Endocrinol. Metab. 98 (2), 818–828.
Greis, P.E., Bardana, D.D., Holmstrom, M.C., Burks, R.T., 2002. Meniscal injury: I. Basic science and evaluation. J. Am. Acad. Orthop. Surg. 10 (3), 168–176.
Griffiths, E.J., Khanduja, V., 2012. Hip arthroscopy: evolution, current practice and future development. Int. Orthop. 36 (6), 1115–1121.
Hall, S.J., 2011. Basic Biomechanics, sixth ed. McGraw-Hill, New York.
Halloran, J.P., Petrella, A.J., Rullkoetter, P.J., 2005. Explicit finite element modeling of total knee replacement mechanics. J. Biomech. 38 (2), 323–331.
Harrysson, O.L., Hosni, Y.A., Nayfeh, J.F., 2007. Custom-designed orthopedic implants evaluated using finite element analysis of patient-specific computed tomography data: femoral-component case study. BMC Musculoskelet. Disord. 8, 91.
Haut Donahue, T.L., Hull, M.L., Rashid, M.M., Jacobs, C.R., 2002. A finite element model of the human knee joint for the study of tibio-femoral contact. J. Biomech. Eng. 124 (3), 273–280.
Hopkins, A.R., New, A.M., Rodriguez-y-Baena, F. &., 2010. Finite element analysis of unicompartmental knee arthroplasty. Med. Eng. Phys. 32 (1), 14–21.
Huiskes, R., Chao, E.Y., 1983. A survey of finite element analysis in orthopedic biomechanics: the first decade. J. Biomech. 16 (6), 385–409.
Jacob, S., Patil, K.M., Braak, L.H., Huson, A., 1996. Stresses in a 3-D two arch model of a normal human foot. Mech. Res. Commun. 387–393.
John, D., Pinisetty, D., Gupta, N., 2013. Image based model development and analysis of the human knee joint. In: Plissiti, M.E., Nikou, C., Andreaus, U., Iacoviello, D. (Eds.), Biomedical Imaging and Computational Modeling in Biomechanics. Springer, pp. 55–79.
Kitagawa, Y., Ichikawa, H., King, A.I., Begeman, P.C., 2000. Development of a human ankle/foot model. In: Kajzer, J., Tanaka, E., Yamada, H. (Eds.), Human Biomechanics and Injury Prevention. Springer, Tokyo, pp. 117–122.
Krone, R., Schuster, P., 2006. An Investigation on the Importance of Material Anisotropy in Finite-Element Modeling of the Human Femur. SAE Technical Paper Series, No. 2006-01-0064.
Kubiček, M., Florian, Z., 2009. Stress strain analysis of knee joint. Eng. Mech. 16 (5), 315–322.
Kutzner, I., Heinlein, B., Graichen, F., Bender, A., Rohlmann, A., Halder, A., Beier, A., Bergmann, G., 2010. Loading of the knee joint during activities of daily living measured in vivo in five subjects. J. Biomech. 43 (11), 2164–2173.

Lemmon, D., Shiang, T., Hashmi, A., Ulbrecht, J., Cavanagh, P., 1997. The effect of insoles in therapeutic footwear: a finite-element approach. J. Biomech. 30, 615–620.
Leslie, W., 2012. Clinical review: ethnic differences in bone mass—clinical implications. J. Clin. Endocrinol. Metab. 97, 4329–4340.
Limbert, G., Middleton, J., Taylor, M., 2004. Finite element analysis of the human ACL subjected to passive anterior tibial loads. Comp. Methods Biomech. Biomed. Eng. 7 (1), 1–8.
Ly, T.V., Swiontkowski, M.F., 2008. Management of femoral neck fractures in young adults. Ind. J. Orthop. 42 (1), 3–12.
Makris, E.A., Hadidi, P., Athanasiou, K.A., 2011. The knee meniscus: structure function, pathophysiology, current repair. Biomaterials 32 (30), 7411–7431.
Michael, J.M., Golshani, A., Gargac, S., Goswami, T., 2008. Biomechanics of the ankle joint and clinical outcomes of total ankle replacement. J. Mech. Behav. Biomed. Mater. 1 (4), 276–294.
Mijailović, N., Vulović, R., Milanković, I., Radaković, R., Filipović, N., Peulić, A., 2015. Assessment of knee cartilage stress distribution and deformation using motion capture system and wearable sensors for force ratio detection. Comput. Math. Methods Med. 2015, 963746.
Nakamura, S., Crowninshield, R.D., Cooper, R.R., 1981. An analysis of soft tissue loading in the foot: a preliminary report. Bull. Prosth. Res. 18, 27–34.
Nishiyama, K.K., Gilchrist, S., Guy, P., Cripton, P., 2013. Proximal femur bone strength estimated by a computationally fast finite element analysis in a sideways fall configuration. J. Biomech. 46 (7), 1231–1236.
Oliveira, I., Goncalves, C., Reis, R.L., Oliveira, J.M., 2017. Synovial knee joint. In: Oliveira, M., Reis, R.L. (Eds.), Regenerative Strategies for the Treatment of Knee Joint Disabilities. Springer International Publishing, pp. 21–28.
OpenStax College, 2013. Anatomy and Physiology. OpenStax College.
Pal, S., 2013. Mechanical properties of biological materials. In: Pal, S. (Ed.), Design of Artificial Human Joints & Organs. Springer, USA, pp. 23–40.
Passos, A.D., Tziafas, D., Mouza, A.A., Paras, S.V., 2018. Computational modelling for efficient transdentinal. Fluids 3, 4.
Patil, K.M., Braak, L.H., Huson, A., 1996. Analysis of stresses in two-dimensional models of normal and neuropathic feet. Med. Biol. Eng. Comput. 34, 280–284.
Peña, E., Calvo, B., Martínez, M.A., Doblaré, M., 2006. A three-dimensional finite element analysis of the combined behavior of ligaments and menisci in the healthy human knee joint. J. Biomech. 39 (9), 1686–1701.
Peña, E., Calvo, B., Martínez, M.A., Palanca, D., Doblaré, M., 2005. Finite element analysis of the effect of meniscal tears and meniscectomies on human knee biomechanics. Clin. Biomech. 20 (5), 498–507.
Perez, A., Mahar, A., Negus, C., Newton, P., Impelluso, T., 2008. A computational evaluation of the effect of intramedullary nail material properties on the stabilization of simulated femoral shaft fractures. Med. Eng. Phys. 30 (6), 755–760.
Perry, J., Jon, D.J., 1992. Gait analysis: normal and pathological function. J. Pediatr. Orthop. 12 (6), 815.
Prochor, P., 2017. Finite element analysis of stresses generated in cortical bone during implantation of a novel Limb cortical bone during implantation of a novel Limb. Biocybern. Biomed. Eng. 37 (2), 255–262.
Reggiani, B., Leardini, A., Corazza, F., Taylor, M., 2006. Finite element analysis of a total ankle replacement during the stance phase of gait. J. Biomech. 39 (8), 1435–1443.
Reilly, D.T., Burnstein, A.H., 1974. The mechanical properties of cortical bone. J. Bone Joint Surg. 56 (5), 1001–1022.

Senalp, A.Z., Kayabasi, O., Kurtaran, H., 2007. Static, dynamic and fatigue behavior of newly designed stem shapes for hip prosthesis using finite element analysis. Mater. Des. 28 (5), 1577–1583.

Seo, J.-W., Kang, D.-W., Kim, J.-Y., Yang, S.-T., Kim, D.-H., Choi, J.-S., Tack, G.-R., 2014. Finite element analysis of the femur during stance phase of gait based on musculoskeletal model simulation. Biomed. Mater. Eng. 24 (6), 2485–2493.

Shriram, D., Praveen Kumar, G.P., Cui, F., Lee, Y.H., Subburaj, K., 2017. Evaluating the effects of material properties of artificial meniscal implant in the human knee joint using finite element analysis. Sci. Rep. 7 (6011), 1–11.

Simkin, A., 1982. Structural Analysis of the Human Foot in Standing. Tel Aviv University, Tel Aviv.

Song, Y., Debski, R.E., Musahl, V., Thomas, M., Woo, S.L.-Y., 2004. A three-dimensional finite element model of the human anterior cruciate ligament: a computational analysis with experimental validation. J. Biomech. 37 (3), 383–390.

Soni, A., Chawla, A., Mukherjee, S., 2017. Effect of muscle contraction on knee loading for a standing pedestrian in lateral impacts. In: Proceedings of 20th ESV conference.

Sverdlova, N.S., Witzel, U., 2010. Principles of determination and verification of muscle forces in the human musculoskeletal system: muscle forces to minimise bending stress. J. Biomech. 43 (3), 387–396.

Taha, Z., Norman, M.S., Omar, S.F., Suwarganda, E., 2016. A finite element analysis of a human foot model to simulate neutral standing on ground. Procedia Eng. 147, 240–245.

Tseng, J.-G., Huang, B.W., Liang, S.-H., Yen, K.T., Tsai, Y.C., 2014. Normal mode analysis of a human fibula. Life Sci. J. 11 (12), 711–718.

Vulović, A., Filipović, N., 2018. Computational analysis of hip implant surfaces. J. Serbian Soc. Comput. Mech. 13(1).

Vulović, A., Šušteršič, T., Cvijić, S., Ibrić, S., Filipović, N., 2018. Coupled in silico platform: computational fluid dynamics (CFD) and physiologically-based pharmacokinetic (PBPK) modelling. Eur. J. Pharm. Sci. 113, 171–184.

Vulović, A., Vukićević, A., Jovičić, G., Ristić, B., Filipović, N., 2016. The influence of ruptured anterior cruciate ligament on the biomechanical weakening of knee joint and posterior cruciate ligament. J. Serbian Soc. Comput. Mech. 10 (2), 1–8.

Wang, Y., Li, Z., Wong, D.W.-C., Cheng, C.-K., Zhang, M., 2018. Finite element analysis of biomechanical effects of total ankle arthroplasty on the foot. J. Orthop. Transl. 12, 55–65.

Wang, C.-W., Muheremu, A., Bai, J.-P., 2017. Use of three-dimensional finite element models of the lateral ankle ligaments to evaluate three surgical techniques. Int. Med. Res. 46 (2), 699–709.

Weiss, J.A., Gardiner, J.C., Ellis, B.J., T, L., & Phatak, N. S., 2005. Three-dimensional finite element modeling of ligaments: technical aspects. Med. Eng. Phys. 27 (10), 845–861.

Wong, D.W.-C., Niu, W., Wang, Y., Zhang, M., 2016. Finite element analysis of foot and ankle impact injury: risk evaluation of calcaneus and talus fracture. PLoS One 11(4) e0154435.

Wu, Y., Ferguson, S.J., 2016. The influence of cartilage surface topography on fluid flow in the intra-articular gap. Comp. Methods Biomech. Biomed. Eng. 20 (3), 250–259.

Zhong, L., Zhang, J.-M., Su, B., Tan, R.S., Allen, J.C., Kassab, G.S., 2018. Application of patient-specific computational fluid dynamics in coronary and intra-cardiac flow simulations: challenges and opportunities. Front. Physiol. 9, 742.

Zielinska, B., Haut Donahue, T.L., 2006. 3D finite element model of meniscectomy: changes in joint contact behavior. J. Biomech. Eng. 128 (1), 115–123.

CHAPTER 7

Different theoretical approaches in the study of antioxidative mechanisms

Contents

1 Prevention of oxidative stress

Although oxygen is crucial for aerobic organisms, it also represents one of the main intergradients for the production of the reactive oxygen species

Computational Modeling in Bioengineering and Bioinformatics
https://doi.org/10.1016/B978-0-12-819583-3.00007-2

(ROS) (Augustyniak et al., 2010). Free radicals and other reactive species are constantly generated in the human body. Oxidative stress (OS) is caused by an imbalance between the production of ROS and ability of biological systems to readily detoxify the reactive intermediates or easily repair the resulting damage. OS has been proposed to play an important role in the pathogenesis of many (if not all) diseases, such as inflammation, cancer, hypertension, cardiovascular disease, diabetes mellitus, atherosclerosis, ischemia/reperfusion injury, neurodegenerative disorders, rheumatoid arthritis, and ageing (Halliwell and Gutteridge, 2015; Fridovich, 1978; Sies, 1997; Thomas, 1998; Halliwell, 2001). The organisms have developed a variety of the internal defense mechanisms that include endogenous enzymes (such as superoxide dismutase and catalase), copper and iron transport proteins, and water-soluble and lipid-soluble antioxidants to counteract the damaging effect of free radicals. Also, there are some external factors including dietary substances (flavonoids, phenolic acids, vitamins C and E, hydroquinones, and various sulfhydryl compounds), which help in the prevention of damage caused by free radical species. All these substances constitute complex antioxidant defense systems.

A possible association between the consumption of foods containing phenolic compounds and a reduced risk of developing disorders, such as cancer and cardiovascular diseases, has been evaluated in several epidemiological investigations (Ness and Powles, 1997; Jacob and Burri, 1996; Block et al., 1992; Huang and Ferraro, 1992). Both natural and synthetic phenolic compounds have been characterized by their antioxidant activity (Rice-Evans et al., 1996; Halliwell et al., 1995; Anouar et al., 2009). Phenolic compounds are plant secondary metabolites commonly found in herbs and fruits. The term phenolics are used for more than several thousands of naturally occurring compounds. One of them is phenolic compounds, which are plant secondary metabolites commonly found in herbs and fruits. The common structural feature of all phenolics is an aromatic ring bearing one or more hydroxyl substituents.

2 Characteristics of good antioxidants in general

There are several desirable characteristics that every good antioxidant should have, regardless of their sources. Indeed, even though there are many molecules that show antioxidant properties, not all of them are equally efficient for that purpose. There is a series of requirements proposed, which are helpful in the identification of good antioxidants (Rose and Bode, 1993). They are listed later:

1) Toxicity: Good antioxidant should not be toxic before and after the antioxidant action. In addition, it is also important to be aware of the influence of possible interactions with any drug that may be concurrently consumed.
2) Availability: Antioxidant should be available when needed.
3) Location and concentration: Not only should an efficient antioxidant be present, but also it should be in adequate concentration in cells, due to the fact that most free radicals have short half-lives within biological systems.
4) Versatility: A good antioxidant should be able to easily react with different free radicals, since actually there are a wide variety of them in biological systems.
5) Fast reactions: Based on the many definitions of antioxidants, it is obvious that good antioxidants must react faster with free radicals than the molecules that they protect, and in that way, they can efficiently protect biological targets.
6) Crossing physiological barriers: It is expected that a good antioxidant be able to cross physiological barriers and be rapidly transported into the cells, where it is needed the most.
7) Regeneration: In this context, the term *regeneration* refers to antioxidants that are able to scavenge different free radicals. Antioxidants that have physiological mechanisms that regenerate their original form are expected to be particularly efficient in reducing OS because it is expected that they could scavenge more than one free radical.
8) Minimal loss: The concentration of any chemical compound is reduced in physiological environments by metabolic routes. Therefore, those antioxidants that still possess antioxidant activity after they undergo metabolic paths are expected to be particularly efficient (e.g., melatonin).

The antioxidative mechanisms can be examined using in silico approaches, such as quantum mechanical calculations. Thermodynamically favorable mechanisms of the action of the investigated molecules can be quantitatively expressed through physicochemical descriptors such as bond dissociation enthalpy (BDE), ionization potential (IP), proton dissociation enthalpy (PDE), proton affinity (PA), and electron-transfer enthalpy (ETE). The transition-state theory (TST) can be applied for the study of the reaction mechanisms in describing the geometry of the corresponding transitional states and calculating the reaction rate constants (k).

However, elucidation of the main reaction mechanisms involved in the antioxidant activities of chemical compounds may be a challenge. For these reasons, some of the most important reaction mechanisms involved in antioxidant protection, in both experimental and theoretical approaches, are investigated in many cases (Ivanović et al., 2016; Petrović et al., 2015). In this chapter, some theoretical results will be presented. For this purpose, quercetin, gallic acid, and dihydroxybenzoic acids are chosen.

Quercetin (Q) is naturally occurring and the most abundant dietary flavonol commonly found in onions, grape, and other vegetables and fruits. According to the DPPH, ABTS, and VCEAC values, as well as structure-activity relationships, quercetin is ranked as one of the most powerful antioxidants in the flavonoid class of compounds (Cai et al., 2006; Kim and Lee, 2004). Constant scientific interest in quercetin arises from its multiple effects on human health including antiviral protection (against parainfluenza virus type 3, herpes simplex virus type 1, and poliovirus type 1), cardiovascular and anticancer protection, inhibition of oxidation of LDL cholesterol in vitro, and inhibitory effects on inflammation-producing enzymes like cyclooxygenase and lipoxygenase, which in their metabolic cycles cause edema, dermatitis, arthritis, gout, and other pathological states. It is found as glycoside in high concentrations in many types of fruits and vegetables (*Ginkgo biloba*, *Hypericum perforatum*, *Sambucus canadensis*, apples, onion, berries, red wine, barks, nuts, flowers, leaves, and seeds) (Cao et al., 1997; Rice-Evans and Miller, 1996; Havsteen, 1983; Bors et al., 1990; Benavente-García et al., 1997; Hertog and Hollman, 1996; Kaul et al., 1985). It has been shown that quercetin glucosides are hydrolyzed by lactase phlorizin hydrolase (LPH), a β-glucosidase found on the brush border membrane of the mammalian small intestine (pH 7–9). Subsequently, the liberated aglycone can be absorbed across the small intestine (Day et al., 2000; Hollman, 2004; Van Dorsten et al., 2012). The majority of theoretical investigations of Q is focused on all rings where OH groups are located.

Phenolic acids, as hydroxylated derivatives of cinnamic and benzoic acids, bear many functions in plants. One of the compounds from this group is gallic acid (GA). In the plant world, this acid is widely distributed as a free compound, as well as part of hydrolyzable tannins (Bianco et al., 1998). GA can be found in gallnuts, witch hazel, and tea leaves (Pettersen et al., 1993) and also as a part of some hardwood species such as eucalyptus wood (e.g., *Eucalyptus microcrys* f. muell., *E. triantha* link., and *E. regnans* f. muell.) and oak trees (e.g., *Quercus robur*, *Q. alba*, and *Q. rubra*) (Bianco et al., 1998). It is known that GA shows fungicidal and fungistatic properties, as well as strong affinity to form complexes (Koga et al., 1999). Derivatives of gallic acid are biologically active components of other plants and plant products: grape, different berries, fruits, different juices, and wine. GA shows extensive application in medicine, pharmaceutical and chemical industry, and foodstuff, as well as light industry. Because of its good antioxidant properties, GA has great application in medicine in protection of human cells against oxidative damage. Apart from that, GA and its derivatives are particularly effective in treating albuminuria and diabetes, psoriasis and external hemorrhoids, coronary heart disease, cerebral thrombosis, gastric ulcer, snail fever, viral hepatitis, senile dementia, and other diseases where oxidative stress is involved and inhibit insulin degradation (Sakagami et al., 2001). On the other hand, it shows cytotoxicity against cancer cells and antiallergic, antifungal, antiinflammatory, antiseptic, antivirus, and antiasthmatic effects in living organisms.

Dihydroxybenzoic acids (DHBA) as a subclass of hydroxybenzoic acids possess two hydroxyl groups whose mutual position determines their chemical properties. There is some experimental evidence that generally confirms good antioxidant activity of DHBAs (Cai et al., 2006; Kim and Lee, 2004), especially indicating good scavenging potency of 3,4-DHBA and 2,3-DHBA. Protocatechuic acid (3,4-DHBA) is a strong antiradical and antioxidant agent that inhibits the chemical carcinogenesis and show protection against the hydroperoxide-induced toxicity (De Graff et al., 2003). Also, a few positive health attributes of 3,4-DHBA such as antibacterial (Chao and Yin, 2009), antimutagenic (Stagos et al., 2006), antiinflammatory, anticoagulatory (Lin et al., 2009), and antihyperglycemic (Kwon et al., 2006) actions have been reported. Although all these compounds can be found in natural products, there are certain findings that implicate the nonenzymatic production of 2,3-DHBA (pyrocatechuic acid), which proceeds upon trapping the hydroxyl radical by salicylic acid (Prasad and Laxdal, 1994). Moreover, pyrocatechuic acid may act as a dioxygenases metabolite (Bugg, 2003).

3 The proposed reaction mechanisms

The scavenging of free radicals seems to play an important role in the antioxidant activity of phenolic compounds. The antiradical properties of phenolic compounds are related to their ability to transfer their phenolic H-atom to a free radical. The reactive radical species ($R^\bullet$) in the radical-scavenging mechanisms are inactivated by accepting a hydrogen atom from OH and NH groups of different antioxidants (Ivanović et al., 2016; Petrović et al., 2015). It is known that this reaction proceeds via several mechanisms: hydrogen atom transfer (hydrogen atom transfer [HAT]/proton-coupled electron transfer [PCET], Eq. 7.1), single-electron transfer (SET Eqs. 7.2–7.4), single-electron transfer followed by proton transfer (SET-PT, Eq. 7.5), sequential proton loss electron transfer (SPLET, Eqs. 7.6 and 7.7) sequential proton loss hydrogen atom transfer (SPLHAT, Eqs. 7.8 and 7.9), and radical adduct formation (RAF, Eq. 7.10) (Klein et al., 2007; Litwinienko and Ingold, 2007; Galano, 2015; Galano et al., 2016; Mazzone et al., 2016). These mechanisms may coexist, and they depend on polarity and other properties of solvents as well as on radical characteristics.

3.1 Hydrogen-atom transfer versus proton-coupled electron transfer

Understanding the leading reaction mechanism involved in the antioxidant action is a challenging task. It is well known that reactions ($RXH + X \rightarrow RX + XH$) involving HAT mechanism between two oxygen atoms have much lower activation energies and higher rate constants than HAT mechanism between two carbon atoms (Min and Boff, 2002).

Mayer and coworkers used the DFT method to examine the self-exchange reactions of the phenoxy radical/phenol, methoxy radical/methanol, and the benzyl radical/toluene systems (Mayer et al., 2002). They identified the geometrical differences in transition states of the HAT and PCET mechanisms. The identification of the transition state was based on the analysis of the singly occupied molecular orbital (SOMO). The HAT mechanism is characterized by a significant SOMO density along the donor$\cdots$H$\cdots$acceptor transition vector. On the other hand, the SOMO of PCET transition state involves *p*-orbitals, which are orthogonal to the transition vector.

In the HAT mechanism, radical species ($R^{\bullet}$) remove a hydrogen atom from the antioxidant molecule (A–OH) that itself converts to a radical ($A–O^{\bullet}$):

$$A - OH + R^{\bullet} \rightarrow A - O^{\bullet} + RH \tag{7.1}$$

This mechanism plays a significant role in the process of analyzing mechanisms of antioxidant activity of different classes of organic compounds, for example, polyphenols, Schiff base, and triazoles. The HAT was proposed as a key reaction mechanism of antioxidative action for different classes of

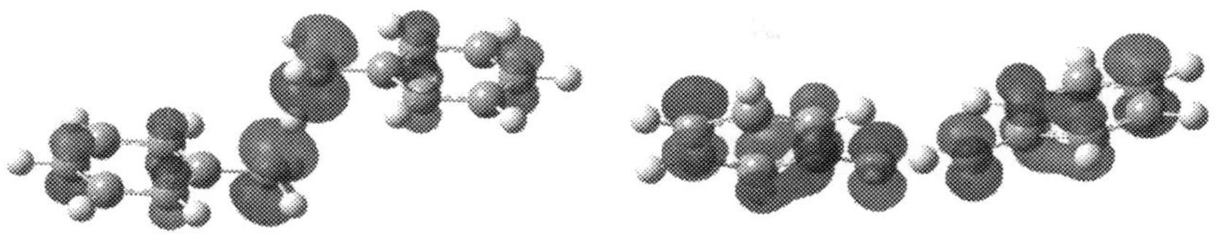

Fig. 7.1 SOMO orbitals in HAT mechanism (*left*) and SOMO orbitals in PCET mechanism (*right*).

organic compounds, from different literature data (Marković et al., 2014a, b, c; Dimić et al., 2017; Tošović et al., 2017). This mechanism, as mentioned earlier, involves transfer of a proton with one of its bonding electrons.

In the PCET mechanism, the same product is obtained as in the HAT mechanism, but the electron and proton are transferred in a single kinetic step via different routes. The proton is transferred from the phenolic compound to the radical's lone pair, while the electron moves from the 2*p* lone pair of the phenolic compound to the SOMO of phenoxy radical (Rose and Bode, 1993; Mayer, 2004) (Fig. 7.1). It should be pointed out that the PCET mechanism is present in many biological and biochemical processes (DiLabio and Johnson, 2007; Mayer et al., 2002).

3.2 Single electron transfer

The SET is the second one-step mechanism, and it can take place following three different pathways:

$$A-OH+{}^{\bullet}R \rightarrow A-OH^{\bullet +}+R^{-} \quad (7.2)$$

$$A-OH+{}^{\bullet}R \rightarrow A-OH^{\bullet -}+R^{+} \quad (7.3)$$

$$A-O^{-}+R^{\bullet} \rightarrow A-O^{\bullet}+R^{-} \quad (7.4)$$

Eq. (7.2) has been described as important for the free radical-scavenging activity of the enol isomer of curcumin (Barzegar, 2012) and for the reactions of carotenoids with ${}^{\bullet}NO_2$ (Mortensen et al., 1997) and $CCl_3OO^{\bullet}$ (Hill et al., 1995), catechin analogs with peroxyl radicals ($ROO^{\bullet}$) (Nakanishi et al., 2004), edaravone derivatives with ${}^{\bullet}OH$, ${}^{\bullet}OCCl_3$ and $CH_3COO^{\bullet}$ (Pérez-González and Galano, 2012), and resveratrol with oxygen radicals (Nakanishi et al., 2007). However, the second pathway (Eq. 7.3) is involved in the reactions of the superoxide anion radical ($O_2^{\bullet -}$) with xanthones (Martínez et al., 2012) and carotenoids (Galano et al., 2010) and in the reactions of the NO radical with trolox, caffeic acid, uric acid, and genistein (Sueishi et al., 2011). The relative importance of the second pathway increases with the electron donor capability of the reacting free radical.

3.3 Single electron transfer-proton transfer

The SET-PT mechanism implies two steps. The first step is transfer of an electron to the radical, while the primary antioxidant is transformed into the radical cation (Eq. 7.2). The next step is the heterolytic O–H bond dissociation of A–OH$^{\bullet+}$ (Eq. 7.5):

$$\mathrm{A-OH^{\bullet+}+R^- \rightarrow A-O^{\bullet}+RH} \quad (7.5)$$

This mechanism is less represented in comparison with the HAT and PCET mechanisms because the first step of this mechanism is very slow. On the other hand, once formed radical cation easily loses one proton in the second step of this mechanism, for example, in the case of baicalein (Marković et al., 2012a) and quercetin (Marković et al., 2013a, b, c). Furthermore, the SET-PT mechanism plays an important role in the oxidative damage of biomolecules by highly reactive radicals such as hydroxyl radical. For instance, the SET-PT mechanism is the main reaction path for the reaction of the guanosine and hydroxyl radical (Galano and Alvarez-Idaboy, 2009).

3.4 Sequential proton-loss electron transfer

In the SPLET mechanism, antioxidant losses one proton and converts to anion, A–O$^-$ (Eq. 7.6). Further, electron transfer from the antioxidant anion to a radical leads to the formation of the antioxidant radical, A–O$^{\bullet}$, and the corresponding anion R$^-$ (Eq. 7.4), which is further protonated (Eq 7.7):

$$\mathrm{A-OH \rightarrow A-O^- + H^+} \quad (7.6)$$

$$\mathrm{R^- + H^+ \rightarrow RH} \quad (7.7)$$

This mechanism is particularly important for the explanation of the antioxidant activity of phenolic compounds under physiological conditions (Foti, 2007). Analyzing the antioxidant activity, two important chemical characteristics of the antioxidant should be mentioned. The first one is its pK_a, which is responsible for the determination of the proportion of the deprotonated species in water solution and at each pH value, for instance, at pH 7.4 (physiological conditions). Moreover, the role of the solvents in these reactions also should not be forgotten. The solvent should be polar and protic and be able to provide good solvation of the formed anion.

Therefore, it is expected that the SPLET mechanism is dominant in water, but not in the lipid phase, characteristic of the biological systems. It should be emphasized that the SPLET mechanism has been identified as a crucial mechanism in the scavenging activity exerted by numerous compounds in polar environments.

3.5 Sequential proton loss hydrogen atom transfer mechanism

This mechanism also consists of two steps. The first one is identical to the first step of the SPLET process and yields the deprotonated antioxidant (Eq. 7.8). The second step is completely different compared with SPLET, where instead of the electron transfer, hydrogen atom transfer takes place (Eq. 7.9):

$$A-(OH)_2 \rightarrow A-(OH)O^- + H^+ \tag{7.8}$$

$$A-(OH)O^- + {}^{\bullet}R \rightarrow A-(O^{\bullet})O^- + HR \tag{7.9}$$

As a special mechanistic pathway, SPLHAT mechanism is mentioned in the study of the free radical-scavenging action of anthocyanidins (Estévez et al., 2010). However, its importance in the free radical-scavenging activity for other compounds also has been described. This mechanism is playing a very important role in the reactions of esculetin with ${}^{\bullet}OOCH_3$ and ${}^{\bullet}OOCHCH_2$ radicals (Medina et al., 2014) and in the reaction of gallic acid with ${}^{\bullet}OH$ (Marino et al., 2014). The SPLHAT mechanism is significant for the free radical-scavenging activities of α-mangostin (Martínez et al., 2011), ellagic acid (Galano et al., 2014), propyl gallate, and different phenolic acids (Medina et al., 2013).

Both processes, SPLHAT and SPLET, are mutually competitive, since the first step is common. Therefore, the second step, transferring a hydrogen atom (H) or an electron from the deprotonated antioxidant, plays a very important role in the evaluation of the mechanistic pathway of the action of investigated compound. Accordingly, any factor contributing to the increase in deprotonation (pH, polarity of solvent, reactivity of free radical, etc.) would be in favor of both processes. On the other hand the higher electron donor ability of the deprotonated antioxidant influences the higher probability of SPLET, while species with more labile H atoms would favor the SPLHAT mechanism.

4 Radical adduct formation

RAF is a simple mechanism of radical adduct formation between free radical and antiradical species (Eq. 7.10). This mechanism is important for systems containing the π conjugated systems (Galano and Alvarez-Idaboy, 2013; Galano and Francisco-Marquez, 2009). In contrast to the HAT and PCET mechanisms, the antioxidant does not provide its hydrogen atom, but forms the radical adduct with the free radical. This mechanism depends on the structure of the investigated antioxidant and the free radical. If the investigated antioxidant has multiple bonds, then RAF is a possible reaction path. In addition, the properties of the free radical play an important role, and the electrophilic free radicals have the greatest potential for participation in this type of reactions. Generally speaking the reaction center of the investigated antioxidant should be easily accessible, and a free radical should be small or medium sized to avoid potential steric hindrance:

$$\mathrm{A-OH+R^{\bullet}\rightarrow[A-OH-R]^{\bullet}} \quad (7.10)$$

4.1 Thermodynamical parameters for evaluation of antioxidative mechanisms

The most investigated antioxidative mechanisms are HAT, SET-PT, and SPLET. Important thermodynamic parameters that determine dominant mechanisms of free radical scavenging are as follows:

1) Bond dissociation enthalpy (BDE) of A–OH molecule for the HAT mechanism. The importance of BDE is due to the antioxidant activity estimation. The lower BDE value indicates the easier O–H bond rupture.
2) Ionization potential (IP) of A–OH molecule and proton dissociation enthalpy (PDE) of radical cation A–OH$^{\bullet+}$ for the SET-PT mechanism. The lower value of ionization potential of A–OH indicates easier electron transfers from antioxidant to radical species.
3) Proton affinity (PA) of molecule A–OH and electron energy transfer (ETE) correspondence to anion A–O$^-$ for SPLET mechanism.

The reaction enthalpies related to the studied free radical-scavenging mechanisms can be calculated by the following equations (Marković et al., 2012a, b, c; Galano and Alvarez-Idaboy, 2009; Milenković et al., 2017):

$$\mathrm{BDE}=H(\mathrm{A-O^{\bullet}})+H(\mathrm{H^{\bullet}})-H(\mathrm{A-OH}) \quad (7.11)$$

$$\mathrm{IP}=H(\mathrm{A-OH^{\bullet+}})+H(\mathrm{e^{-}})-H(\mathrm{A-OH}) \quad (7.12)$$

$$\mathrm{PDE}=H(\mathrm{A-O^{\bullet}})+H(\mathrm{H^{+}})-H(\mathrm{A-OH^{\bullet+}}) \quad (7.13)$$

$$\mathrm{PA}=H(\mathrm{A-O-})+H(\mathrm{H^{+}})-H(\mathrm{A-OH}) \quad (7.14)$$

$$\mathrm{ETE}=H(\mathrm{A-O^{\bullet}})+H(\mathrm{e-})-H(\mathrm{A-O^{-}}) \quad (7.15)$$

where H(A–OH), H(A–O$^{\bullet}$), H(A–OH$^{\bullet+}$), H(A–O$^-$), H(H$^{\bullet}$), H(e$^-$), and H(H$^+$) are the enthalpies of parent molecule, radical, radical cation, and anion of the examined compound, hydrogen atom, electron, and proton, respectively. For calculation of the mentioned thermodynamical parameters, values of $H(e^-)$ and $H(H^+)$ are used, which are taken from literature (Marković et al., 2013a; Marković et al., 2016).

4.2 Influence of different free radicals on scavenging potency of various antioxidants

From a chemical point of view, free radicals are species containing one or more unpaired electrons. This particular characteristic is responsible for their high reactivity and triggering chain reaction mechanisms. Most of the radicals found in vivo are oxygen-centered free radicals (ROS) or reactive nitrogen species (RNS). The ROS are the superoxide anion radical ($O_2^{\bullet -}$), hydroxyl ($^\bullet OH$), alkoxyl ($RO^\bullet$), peroxyl ($ROO^\bullet$), and hydroperoxyl ($HOO^\bullet$) radicals. The RNS are peroxynitrite ($ONOO^-$), nitric oxide ($NO^\bullet$), and nitrogen dioxide ($NO_2^\bullet$). Among the oxygen-centered radicals, $^\bullet OH$ is the most electrophilic (Pryor, 1988) and reactive, with a half-life of $\sim 10^{-9}$ s. It can react through a wide variety of mechanisms, and its reactions with a large variety of chemical compounds take place at, or close to, diffusion-controlled rates (rate constants $\geq 10^9$ M/s). It has been estimated that $^\bullet OH$ is responsible for 60%–70% of tissue damage caused by ionizing radiation (Vijayalaxmi et al., 2004). A hydroxyl radical is so reactive that it is capable of immediate reacting, after formation, with almost any molecule in its vicinity and with little selectivity toward the various possible sites of attack. It has been held responsible for the most important oxidative damage to DNA (Chatgilialoglu et al., 2009). With respect to $^\bullet OH$, $ROO^\bullet$ radicals are less reactive species, capable of diffusing to remote cellular locations. Their half-lives are of the order of seconds (Pryor, 1986), and their electrophilicity is significantly lower than that of $^\bullet OH$ (Pryor, 1988). However, $ROO^\bullet$ can also react with other chemical species through different mechanisms. Peroxy radicals are, in general, less reactive than $HOO^\bullet$ when R is an alkyl or an alkenyl group (Galano, 2011). However, if R is a more efficient electron-accepting group, such as CCl_3, the reactivity of $ROO^\bullet$ toward organic molecules significantly increases (Milenković et al., 2017). Indeed, the rate constants for the electron-transfer reactions involving $ROO^\bullet$ strongly depend on the chemical nature of R. The rate constant significantly increases with the electron-withdrawing character of the substituents (Neta et al., 1989). $RO^\bullet$ radicals are formed from the reduction of peroxides and are significantly more reactive than $ROO^\bullet$ radicals, provided that R is the same in both species, but they are less reactive than $^\bullet OH$ (León-Carmona et al., 2012).

On the other hand, RNS possess lower reactivity. The chemical reactivity of $NO^{\bullet}$ is rather limited, and as a consequence of that, its direct toxicity is less than that of ROS (Squadrito and Pryor, 1998). However, it reacts with $O_2^{\bullet-}$, yielding peroxynitrite (Radi et al., 2005), which is a very damaging species, as it is able to react with proteins, lipids, and DNA (Douki and Cadet, 1996). Nitrogen dioxide is a moderate oxidant, and its reactivity is somewhere between those of $NO^{\bullet}$ and $ONOO^{-}$ (Yan et al., 2014). It reacts with organic molecules at rates ranging from $\sim 10^4$ to 10^6 M/s, depending on the pH (Prütz et al., 1985).

The scavenging mechanisms are highly influenced by the electronic properties of the scavenged free radical species (Xie and Schaich, 2014). Bearing that in mind, the enthalpies of the reactants and products as well as of the reactions with described free radicals are calculated. It is well known that the values of the reaction enthalpies can significantly contribute to the understanding of the investigated reaction mechanisms. The reaction with the free radical ($RO^{\bullet}$) can occur via three mentioned mechanisms. In HAT/PCET mechanism, the reaction can be presented by Eq. (7.16):

$$A - OH + RO^{\bullet} \rightarrow A - O^{\bullet} + ROH \tag{7.16}$$

The SET-PT mechanism takes place in two steps, as it is described earlier in Eqs. (7.2) and (7.5). In interaction with free radicals ($RO^{\bullet}$), it can be presented by Eqs. (7.17) and (7.18):

$$A - OH + RO^{\bullet} \rightarrow A - OH^{\bullet +} + RO^{-} \tag{7.17}$$

$$A - OH^{\bullet +} + RO^{-} \rightarrow A - O^{\bullet} + ROH \tag{7.18}$$

The SPLET mechanism can be presented as follows:

$$A - OH + RO^{-} \rightarrow A - O^{-} + ROH \tag{7.19}$$

$$A - O^{-} + RO^{\bullet} \rightarrow A - O^{\bullet} + RO^{-} \tag{7.20}$$

The reaction of the examined compound with the particular free radical is considered thermodynamically favorable if it is exothermic:

$$\Delta_r H = [H(\text{products}) - H(\text{reactants})] < 0 \tag{7.21}$$

In radical inactivation, the HAT mechanism (Eq. 7.16) is characterized by the H-atom transfer from the examined compounds to the free radical ($RO^{\bullet}$). The value of $\Delta_r H_{BDE}$ can be calculated using the following equation:

$$\Delta_r H_{BDE} = [H(A - O^{\bullet}) + H(ROH)] - [H(A - OH) + H(RO^{\bullet})] \tag{7.22}$$

The SET-PT mechanism is described with Eqs. (7.17) and (7.18). The first step of this mechanism is determined by $\Delta_r H_{IP}$, while the second step is determined by $\Delta_r H_{PDE}$ (Eqs. 7.23 and 7.24, respectively):

$$\Delta_r H_{IP} = [H(A-OH^{\bullet+}) + H(RO^-)] - [H(A-OH) + H(RO^{\bullet})] \quad (7.23)$$

$$\Delta_r H_{PDE} = [H(A-O^{\bullet}) + H(ROH)] - [H(A-OH^{\bullet+}) + H(RO^-)] \quad (7.24)$$

$\Delta_r H_{PA}$ and $\Delta_r H_{ETE}$ are the reaction enthalpies related to the SPLET mechanism (Eqs. 7.19 and 7.20), and they are calculated using Eqs. (7.25) and (7.26), respectively:

$$\Delta_r H_{PA} = [H(A-O^-) + H(ROH)] - [H(A-OH) + H(RO^-)] \quad (7.25)$$

$$\Delta_r H_{ETE} = [H(A-O^{\bullet}) + H(RO^-)] - [H(A-O^-) + H(RO^{\bullet})] \quad (7.26)$$

5 Mechanistic approach

In the case of the transition states, it was verified, by Intrinsic Coordinate Calculations (IRC), that the imaginary frequency corresponds to the expected motion along the reaction coordinate. These calculations proved that each transition state (TS) connects two corresponding energy minima: reactant complex (RC) and product complex (PC). Natural bond orbital (NBO) analysis was performed for all participants in simulated reaction (Carpenter and Weinhold, 1988). Transition state theory (TST) affords one of the simplest theoretical approaches for estimating the rate constants (*k*), which requires only structural, energetic, and vibrational frequency information for reactants and transition states (Galano et al., 2014). The main advantage of using conventional TST is that it requires very limited potential energy information (only on reactants and the transition state), which makes it practical for a wide range of chemical reactions. Despite its relative simplicity, this theory has been proven to be sufficient to reproduce experimental rate constants of free radical-scavenging reactions (Galano and Alvarez-Idaboy, 2013). The rate constants were calculated using TST, implemented in TheRate program (Duncan et al., 1998), and 1 M standard state is calculated as follows:

$$k_{TST} = \frac{k_B T}{h} \exp\left(\frac{-\Delta G^{\neq}}{RT}\right) \quad (7.27)$$

where k_B and h stand for the Boltzman and Planck constants and $\Delta G^{\neq}$ is the free energy of activation, which is calculated as the difference in energies between transition states and reactants. Reaction path degeneracy (σ) and transmission coefficient $\gamma(T)$were taken into account, implying that the Eyring equation was transformed into the following:

$$k_{ZCT} = \sigma\gamma(T)\frac{k_B T}{h}\exp\left(\frac{-\Delta G^{\neq}}{RT}\right) \tag{7.28}$$

The transmission coefficient γ, corrections for tunneling effects (defined as the Boltzman average of the ratio between the quantum and classical probabilities), was calculated using the zero-curvature tunneling (ZCT) approach (Marcus, 1964).

5.1 Electron-transfer reaction rate constant calculation

One of the viable mechanisms to scavenge free radicals is electron transfer (ET), as the second step of the SPLET mechanism (Eq. 7.4) (Burton and Ingold, 1984).

Transition states are necessary for calculating the rate constant k_{TST}, Eq. (7.27), for HAT reactions. However, for electron-transfer reaction, transition state cannot be located using electronic structure methods, as it is not possible to describe mechanistic pathway of electron motion. To estimate the reaction barrier (the term $\Delta G^{\neq}$) in such cases, the Marcus theory was used (Marcus, 1997). Within this transition-state formalism, the SPLET activation barrier ($\Delta G^{\neq}_{ET}$) is defined in terms of the free energy of reaction (ΔG^{0}_{ET}) and the nuclear reorganization energy (λ):

$$\Delta G^{\neq}_{ET} = \frac{\lambda}{4}\left(1 + \frac{\Delta G^{0}_{ET}}{\lambda}\right)^2 \tag{7.29}$$

λ is the energy associated with the nuclear rearrangement involved in the formation of products in an ET reaction, which implies not only the nuclei of the reacting species but also those of the surrounding solvent. For λ calculation, a very simple approximation was used:

$$\lambda \approx \Delta E_{ET} - \Delta G^{0}_{ET} \tag{7.30}$$

where ΔE is the nonadiabatic energy difference between reactants and vertical products, that is, A–O$^{\bullet}$ and RO^{-} in geometries of A–O^{-} and RO$^{\bullet}$:

$$\Delta E_{ET} = E(\text{A}-\text{O}^{\bullet}) + E(\text{RO}^{-}) - E(\text{A}-\text{O}^{-}) - E(\text{RO}^{\bullet}) \tag{7.31}$$

The adiabatic Gibbs free energies of reaction were calculated as follows:

$$\Delta G^{0}_{ET} = G(\text{A}-\text{O}^{\bullet}) + G(\text{RO}^{-}) - G(\text{A}-\text{O}^{-}) - G(\text{RO}^{\bullet}) \tag{7.32}$$

This approach is similar to that that was used by Nelsen and coworkers (Nelsen et al., 1987) for a large set of self-exchange reactions.

The theory of diffusion-controlled reaction was originally utilized by Alberty, Hammes, and Eigen to estimate the upper limit of enzyme-substrate reactions (Alberty and Hammes, 1958). According to their estimation, the upper limit of enzyme-substrate reactions was $10^9\,M^{-1}\,s^{-1}$ (Eigen and Hammes, 1963). This fact has influence on the final value of steady-state Smoluchowski rate constant, k_D (Smoluchowski, 1918). If calculated rate constant is close to the diffusion limit, appropriate corrections are considered (the Collins-Kimball theory) as proposed by Galano and Alvarez-Idaboy (2013).

The apparent rate constant is then calculated as the correction to this value, according to the Collins-Kimball theory (Collins and Kimball, 1949):

$$k_{app} = \frac{k_D k_{TST}}{k_D + k_{TST}} \tag{7.33}$$

The steady-state Smoluchowski rate constant (k_D) (Smoluchowski, 1918), under the assumption of the irreversible bimolecular diffusion-controlled reaction, is calculated to encounter diffusion as a parameter. The diffusion-limited rate constant depends on the reactant distance, the diffusion coefficient of reactants, and Avogadro's constant:

$$k_D = 4\pi R D_{AB} N_A \tag{7.34}$$

where R denotes the reaction distance, N_A is the Avogadro number, and D_{AB} is the mutual diffusion coefficient of the reactants, A (anion of antioxidant A–O$^-$) and B free radical (RO$^\bullet$).

The Stokes-Einstein theory allows the calculation of the diffusion coefficient from Boltzmann's constant, temperature, viscosity of solvent, and radius of the solvent (Eq. 7.35) (Stokes, 1903). All of the reactions were investigated at 298 K:

$$D = \frac{k_B T}{6\pi\eta a} \tag{7.35}$$

The thermal rate constant (k) for hydroxyl radical is of the order of magnitude 10^9, and this value is at the upper limit of enzyme-substrate reaction, and because of that, one can say that reaction with hydroxyl radical is faster and diffusion controlled. When obtained rate constants are lower than 10^9, reaction is slower, and it is not controlled by diffusion.

6 Thermodynamical parameters for quercetin, gallic acid, and DHBA

Antioxidative activity of phenolic compounds is usually examined by analyzing the thermodynamic properties of the parent molecules, the corresponding radicals, radical cations, and anions. Here, antioxidative properties were analyzed for different natural compounds, which are presented in Fig. 7.2. The generally accepted approach based on the thermodynamic parameters (BDE, IP, and PA), related to the HAT, SPLET, and SET-PT mechanisms, was applied. The calculations were performed in the gaseous phase (for Q), benzene and pentyl ethanoate (for GA, DHBAs) as nonpolar solvents, and in water (for these three considered compounds)

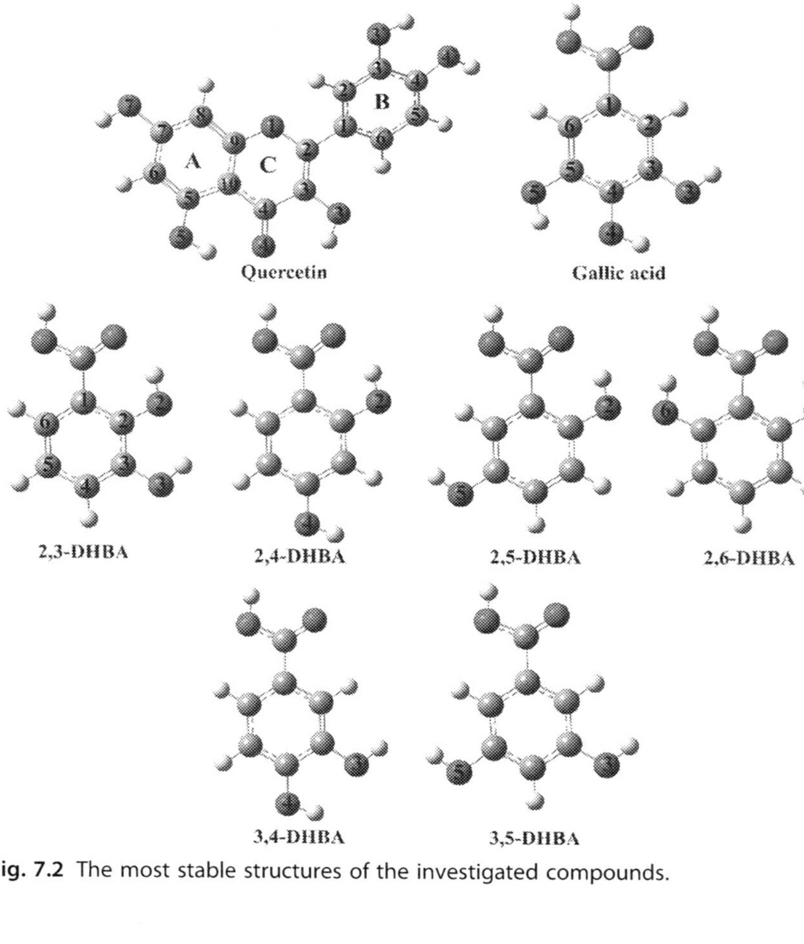

Fig. 7.2 The most stable structures of the investigated compounds.

and DMSO (for GA) as polar solvents. To estimate the effects of solvent, the SMD solvation model was used (Marenich et al., 2009). BDE, PA, and IP values provided insight into the thermodynamically most favorable reaction pathway. It was found that IP values are the highest for all investigated compounds, on the bases of the mutual comparison of the thermodynamical parameters, indicating that SET-PT mechanism is not the operative mechanism for all examined compounds, both in polar and nonpolar solvents. On the other hand, it can be concluded that HAT and PCET are preferred antioxidative mechanisms in the gas phase for Q, due to the fact that the BDE values of Q are significantly lower than the corresponding IP and PA values (Marković et al., 2013b). Taking into account BDE and PA values for GA and DHBAs in benzene as a common nonpolar solvent, it is evident that BDE values are significantly lower than the corresponding PA values, which indicates that reaction in benzene probably proceeds via HAT mechanism. Also, the HAT mechanism is a predominant reaction pathway for some other natural compounds in benzene, such as baicalein (Marković et al., 2014a), morin (Marković et al., 2012b), and morin $2'$-O^- phenoxide anion (Marković et al., 2012c). Regarding the obtained results for phenolic acids, it is evident that in polar solvent (water) and in nonpolar solvent (pentyl ethanoate), PA values are significantly lower than the corresponding BDE values for DHBAs and GA. These results unequivocally favor SPLET mechanism as the most probable reaction pathway. Further analysis of the thermodynamic parameters for DHBAs indicates that 3,4-DHBA and 2,3-DHBA have the best antioxidant activity in all solvents under investigation. The SPLET has been characterized as a crucial mechanism in the scavenging activity of numerous compounds in polar environments. Some examples are alizarin and alizarin red S (Jeremić et al., 2014), hydroxybenzoic and dihydroxybenzoic acids (Marković et al., 2014b; Milenković et al., 2017), kaempferol (Marković et al., 2014c), gallic acid (Đorović et al., 2014), erodiol (Marković et al., 2013c), baicalein (Marković et al., 2014a, b, c), and purpurin (Jeremić et al., 2012).

It should be pointed out that the influence of the polarity of the environment plays an important role in the feasibility of HAT and SPLET reaction mechanisms of investigated compounds. In addition to the polarity of the reaction medium, its acidity also plays an important role in antioxidative processes. In a polar protic solvent, such as aqueous solution, the environmental pH of the solution plays an important role in the changes of the reaction pathway of the antioxidant action. In acidic environment (when the pH is lower than the pK_a of the investigated compound), neutral form of the investigated molecule is predominant one. It should be expected that HAT or/and PCET are the main reaction pathways of an antioxidant action, for the investigated antioxidants, in this case. In the other case, when the pH is higher than the pK_a (basic environment), the anion species will prevail, indicating SPLET as the main reaction pathway in this case. The first step of the SPLET mechanism is the formation of the corresponding anion that is strongly solvated in the aqueous medium. On the other hand, free radicals are not solvated in the same way as ions. It is crucial for a solvent to have an unpaired electron to solvate free radicals. However, the interactions between free radicals are weak, because the solvents have all their electrons paired. In this case, water (as a polar solvent) increases molecular mobility and rate of reaction, reducing radical lifetime. Obtained results for all analyzed compounds in water revealed that SPLET mechanism is a predominant reaction pathway. The stronger synergistic antioxidant activity should be expected at pH values close to 7. Then, it should be expected the reaction to take place via both reaction paths.

From presented results, it is clear that C4′–OH and C4–OH positions should be more reactive OH sites of Q and GA, respectively. These groups have the lowest BDE values in all phases, so they represent the most probable sites of H-atom abstraction (Marković et al., 2013b; Đorović et al., 2014). Also, PA values of all presented OH groups of Q and GA indicate that proton transfer from C4′–OH and C4–OH groups is easier compared with other OH groups, in all phases. On the other hand, the C3–OH position of 3,4- and 2,3-DHBAs is the most favorable site for homolytic O–H cleavage in all solvents. The obtained results for PA imply the proton transfer from C4–OH of 3,4-DHBA as favored, while the C3–OH group of 2,3-DHBA is preferably the site for heterolytic cleavage of the O–H bond, in all solvents.

7 Antiradical mechanisms in the presence of different free radicals

The preferred mechanism of the antiradical activity of the considered compounds (Fig. 7.2) can be estimated from ΔH_{BDE}, ΔH_{IP}, and ΔH_{PA} values, namely, the lowest of these values indicates which mechanism is favorable.

The values of enthalpies of the reactions related to the HAT, SPLET, and SET-PT mechanisms of GA with three free radicals, $O_2^{\bullet -}$, $^{\bullet}OH$, and $CH_3OO^{\bullet}$ (methylperoxyl), imply 4-OH group as the most favorable site for homolytic and heterolytic O–H cleavage in all solvents (Đorović et al., 2014). The SET-PT mechanism is not a suitable pathway for the reactions of GA with three examined radicals in all solvents. On the basis of the procedure based on the assessment of the reaction enthalpies of GA with selected radical species, it can be said that HAT mechanism is the preferable reaction pathway only for the reaction of GA with $^{\bullet}OH$ in water. Also, it should be pointed out that there is no mechanism suitable for the reaction of GA with $O_2^{\bullet -}$, in water. On the other hand, the SPLET mechanism is more favorable in nonpolar solvents, benzene and pentyl ethanoate. It should be emphasized that ΔH_{PA} value is lower by 46 kJ mol^{-1} than ΔH_{BDE}, when $CH_3OO^{\bullet}$ reacts with GA in the aqueous medium. The careful analysis of the results showed that change of the solvent polarity considerably influences the enthalpy of the investigated reactions. Namely, ΔH_{PA} values decrease when solvent polarity decreases, while ΔH_{BDE} values remain almost constant. All these facts indicate that SPLET is the prevailing mechanism in nonpolar solvents, while HAT is favorable one in polar solvents such as water. In addition, both of these mechanisms are competitive in DMSO.

The enthalpies for the reactions of $HO^{\bullet}$ with GA and Q show that these reactions in water solution are considerably more exothermic when it obeys HAT mechanism than when it takes place via SPLET mechanism (Table 7.1). Similarly, as in the case of GA, superoxide anion radical cannot

Table 7.1 Calculated reaction enthalpies (kJ mol^{-1}), for the reactions of Q and GA with hydroxyl radical, superoxide radical anion, and peroxyl radical in water

	Water $\varepsilon = 78.35$				
	HAT	**SET-PT**		**SPLET**	
Water $\varepsilon = 78.35$	**ΔH_{BDE}**	**ΔH_{IP}**	**ΔH_{PDE}**	**ΔH_{PA}**	**ΔH_{ETE}**
		29			
Q-OH–3 + $^{\bullet}$OH	−158		−187	−72	−86
Q-OH–3′ + $^{\bullet}$OH	−147		−176	−78	−69
Q-OH–4′ + $^{\bullet}$OH	−154		−183	−87	−67
GA-OH–4 + $^{\bullet}$OH	−159	57	−216	−101	−58
		282			
Q-OH–3 + $^{\bullet}$OO^{-}	48		−234	40	7
Q-OH–3′ + $^{\bullet}$OO^{-}	58		−223	35	24
Q-OH–4′ + $^{\bullet}$OO^{-}	51		−231	26	26
GA-OH–4 + $^{\bullet}$OO^{-}	48	311	−263	9	39
		127			
Q-OH–3 + $^{\bullet}OOCH_3$	−21		−148	−34	13
Q-OH–3′ + $^{\bullet}OOCH_3$	−10		−137	−40	30
Q-OH–4′ + $^{\bullet}OOCH_3$	−17		−144	−49	31
GA-OH–4 + $^{\bullet}OOCH_3^{-}$	−18	160	−178	−64	46

be scavenged in reaction with Q, because all values for reaction enthalpies are positive, as can be seen in Table 7.1. On the other hand, exothermic reaction of Q with $^{\bullet}OH$ and $^{\bullet}OOCH_3$ implies inactivation of these two radicals via reactions of HAT and SPLET mechanisms. In the case of hydroxyl radical, low negative values of reaction enthalpies suggest HAT as more plausible antioxidative mechanism. The values of the enthalpies for reactions with the methylperoxyl radical indicate that SPLET is the most likely mechanism of an antioxidant action. Hydroxyl group in position C4′ is more reactive than others in both cases. This result is in agreement with results obtained by standard procedure. It is obvious that all three investigated radicals react by the same mechanisms with both Q and GA, in aqueous environment.

The following text presents the results obtained for the inactivation of the previously mentioned free radicals with DHBAs in solvents of different polarity. From Tables 7.2–7.4 it can be noted that DHBAs can inactivate all examined free radicals. The polarity of solvent plays an important role in determining the mechanism of antioxidant action. Unlike GA and Q, DHBAs will react with superoxide anion radical in nonpolar solvent benzene via SPLET mechanism (Table 7.3).

Regarding hydroxyl and methylperoxyl radicals, both can be inactivated via HAT or SPLET mechanism depending on the solvent polarity (Table 7.2). In the case of inactivation of hydroxyl radical, there is a competition between HAT and SPLET. The HAT is prevalent mechanism in water, while SPLET is a predominant scavenging mechanism in benzene. As for the deactivation of methylperoxyl radical, it was found that SPLET is dominant antiradical mechanism in both solvents for all examined DHBAs (Table 7.4).

8 Mechanistic approach to analysis of the antioxidant action

8.1 Reaction of quercetin via HAT mechanism

Bearing in mind the predicted reactivity of OH groups of Q, obtained on the basis of the BDE values (Marković et al., 2010), the potential antioxidant activity of Q is studied with the simulation of the reaction with hydroperoxyl radical, $Q + {}^{\bullet}OOH \rightarrow Q^{\bullet} + HOOH$ (Scheme 7.1).

The analysis of the NBO charges and orbital occupancies (Tables 7.5 and 7.6) as well as the SOMOs of the reactants, transition states, and products (Fig. 7.3) showed that the reaction of Q with the hydroperoxyl radical is

Table 7.2 Calculated reaction free energy (kJ mol^{-1}) for the reactions of DHBAs with hydroxyl radical

DHBAs	Water $\varepsilon = 78.35$					Benzene $\varepsilon = 2.27$				
	HAT	SET-PT		SPLET		HAT	SET-PT		SPLET	
	ΔH_{BDE}	ΔH_{IP}	ΔH_{PDE}	ΔH_{PA}	ΔH_{ETE}	ΔH_{BDE}	ΔH_{IP}	ΔH_{PDE}	ΔH_{PA}	ΔH_{ETE}
2.3-DHBA		93					360			
2-OH	−127		−220	−4	−123	−113		−473	−170	57
3-OH	−138		−231	7	−136	−124		−484	−159	42
2.4-DHBA		75					389			
2-OH	−97		−171	−64	−33	−70		−460	−140	70
4-OH	−104		−179	−81	−23	−99		−488	−191	93
2.5-DHBA		29					345			
2-OH	−132		−161	−57	−75	−101		−446	−131	30
5-OH	−146		−176	−57	−89	−140		−485	−148	8
2.6-DHBA		46					361			
2-OH	−114		−159	−61	−52	−87		−448	−145	57
6-OH	−125		−171	−73	−52	−109		−470	−169	60
3.4-DHBA		57					379			
3-OH	−144		−201	−80	−64	−144		−524	−192	47
4-OH	−141		−198	−91	−50	−143		−522	−210	67
3.5-DHBA		62					384			
3-OH	−122		−183	−66	−56	−115		−499	−164	49
5-OH	−122		−184	−67	−56	−116		−500	−165	49

Table 7.3 Calculated reaction free energy ($kJ\,mol^{-1}$) for the reactions of DHBAs with superoxide anion radical

DHBAs	Water $\varepsilon = 78.35$					Benzene $\varepsilon = 2.27$				
	HAT	SET-PT		SPLET		HAT	SET-PT		SPLET	
	ΔH_{BDE}	ΔH_{IP}	ΔH_{PDE}	ΔH_{PA}	ΔH_{ETE}	ΔH_{BDE}	ΔH_{IP}	ΔH_{PDE}	ΔH_{PA}	ΔH_{ETE}
2.3-DHBA		348					938			
2-OH	81		−267	106	−26	121		−817	−15	136
3-OH	70		−278	117	−39	110		−828	−4	122
2.4-DHBA		329					968			
2-OH	111		−219	47	64	164		−804	14	150
4-OH	103		−226	29	74	135		−833	−37	172
2.5-DHBA		284					426			
2-OH	76		−208	53	22	41		−384	−69	111
5-OH	61		−223	53	8	3		−423	−86	89
2.6-DHBA		300					939			
2-OH	94		−207	49	45	147		−792	10	137
6-OH	82		−218	37	45	125		−814	−15	140
3.4-DHBA		312					958			
3-OH	64		−248	31	33	90		−868	−37	127
4-OH	66		−245	19	47	91		−867	−55	146
3.5-DHBA		316					963			
3-OH	85		−231	44	41	119		−844	−9	128
5-OH	85		−231	43	42	118		−845	−11	129

Table 7.4 Calculated reaction free energy (kJ mol^{-1}) for the reactions of DHBAs with methylperoxyl radical

	Water $\varepsilon = 78.35$					Benzene $\varepsilon = 2.27$				
	HAT	SET-PT		SPLET		HAT	SET-PT		SPLET	
DHBAs	ΔH_{BDE}	ΔH_{IP}	ΔH_{PDE}	ΔH_{PA}	ΔH_{ETE}	ΔH_{BDE}	ΔH_{IP}	ΔH_{PDE}	ΔH_{PA}	ΔH_{ETE}
2.3-DHBA		197					441			
2-OH	14		−183	33	−19	30		−411	−108	138
3-OH	3		−194	44	−33	19		−422	−97	123
2.4-DHBA		178					470			
2-OH	44		−134	−26	71	73		−398	−78	151
4-OH	37		−142	−44	81	44		−426	−130	174
2.5-DHBA		133					426			
2-OH	9		−124	−20	29	41		−384	−69	111
5-OH	−6		−138	−20	14	3		−423	−86	89
2.6-DHBA		149					442			
2-OH	27		−122	−24	51	56		−386	−83	138
6-OH	16		−133	−36	52	34		−408	−107	141
3.4-DHBA		161					460			
3-OH	−3		−164	−42	40	−2		−462	−130	128
4-OH	0		−161	−54	54	0		−460	−148	148
3.5-DHBA		165					465			
3-OH	19		−146	−29	48	28		−437	−102	130
5-OH	18		−147	−30	48	27		−438	−103	130

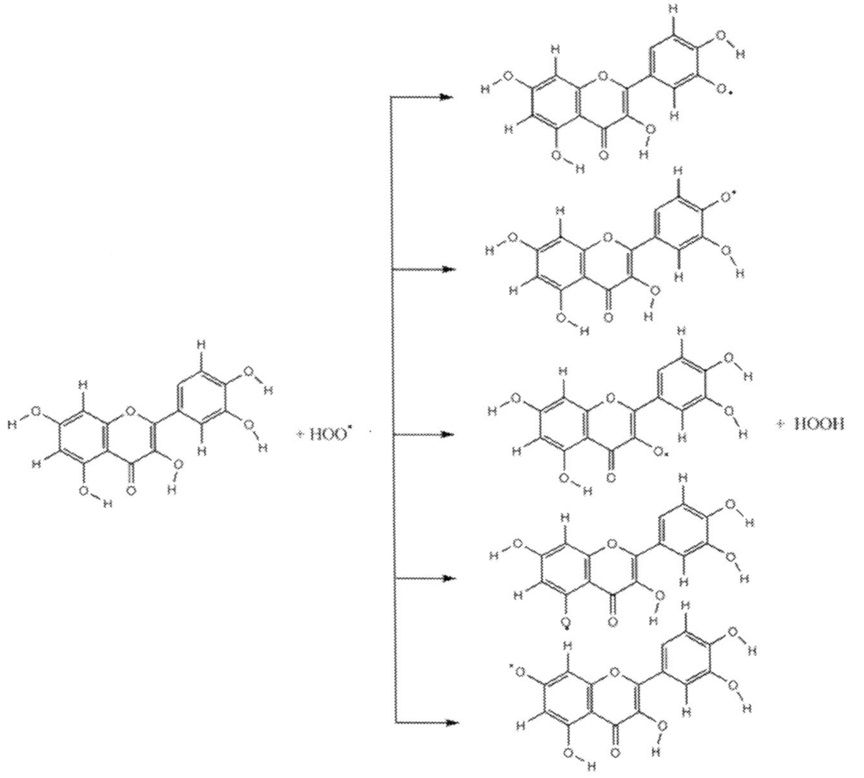

Scheme 7.1 The five reaction pathways for reaction of quercetin with hydroperoxyl radical.

governed by HAT mechanism (Marković et al., 2010; Dhaouadi et al., 2009). Namely, the analysis of NBO charges of TSs shows that the QO and HOO fragments have negative charge, while transferred H has a positive charge (amount 0.5) in all cases (Table 7.5). The spin density is also located on QO and HOO fragments. In comparison with the RC, the spin density on QO moiety increased, while spin density related with OOH fragment decreased. However, the spin density on the transferred hydrogen atom is equal to zero. These facts indicate that reactions of OH groups of quercetin with •OOH radical proceed via HAT mechanism.

From Table 7.6, it can be seen that an unpaired electron is located in *p*-orbital on the Op1 oxygen atom in •OOH radical (Scheme 7.2). There is overlapping between *p*-orbital and *s*-orbital of Hn atom from OH group of Q leading to a formation of the H-Op1 bond (occupancy is close to 1.7).

Table 7.5 Some NBO charges and spin densities for the reactants, transition states, and products

Reactants	Nat. charg.	Spin dens.	TS	Nat. charg.	Spin dens.	Products	Nat. char.	Spin dens.
QO4′H4′	0	0.00	QO4′	−0.28	0.46	QO4′	0	1
			H4′	0.52	−0.03			
OOH	0	1.00	OOH	−0.24	0.57	HOOH	0	0
			QO3′	−0.27	0.47	QO3′	0	1
			H3′	0.52	−0.03			
			OOH	−0.25	0.56	HOOH	0	0
			QO3	−0.17	0.49	QO3	0	1
			H3	0.52	−0.03			
			OOH	−0.35	0.54	HOOH	0	0
			QO5	−0.27	0.56	QO5	0	1
			H5	0.53	−0.02			
			OOH	−0.26	0.46	HOOH	0	0
			QO7	−0.25	0.54	QO7	0	1
			H7	0.52	−0.03			
			OOH	−0.27	0.49	HOOH	0	0

Table 7.6 Some selected NBO orbital occupancies and orbital energies

	Quercetin		Transition		Products	
NBO	**Occ.**	**Ener**	**Occ.**	**Ener**	**Occ.**	**Ener**
BD1 C4′O4′	1.99		1.99	−1.074	1.99	−1.171
BD2 C4′O4′			1.74	−0.531	1.92	−0.479
BD1* C4′O4′			0.03	0.471	0.01	0.622
BD2* C4′O4′			0.79	−0.095	0.35	−0.010
BD H4′O4′	1.99	−0.879				
BD H4′O1			1.70	−0.570		
BD *H4′O1			0.29	0.259		
BD1 C3′O3′			1.99	−1.090	1.99	−1.160
BD2 C3′O3′			1.76	−0.494	1.90	−0.481
BD1* C3′O3′			0.01	0.506	0.01	0.607
BD2 *C3′O3′			0.70	−0.073	0.40	−0.026
BD H3′O3′	1.99	−0.853				
BD H3′O1			1.76	−0.640		
BD *H3′O1			0.23	0.335		
BD1 C3O3			1.99	−1.090	1.99	−1.174
BD2 C3O3			1.78	−0.508	1.91	−0.479
BD1* C3O3			0.01	0.503	0.01	0.634
BD2* C3O3			0.78	−0.091	0.35	−0.016
BD H3O3	1.99	−0.850				
BDH3O1			1.73	−0.601		
BD*H3O1			0.27	0.301		
BD1 C5O5	1.99	−1.048	1.99	−1.120	1.99	−1.175
BD2 C5O5			1.82	−0.484	1.89	−0.471
BD1* C5O5			0.01	0.546	0.01	0.638
BD2* C5O5			0.54	−0.029	0.27	0.015
BD H5O5	1.99	−0.860				
BD H5O1			1.83	−0.712		
BD* H5O1			0.19	0.403		
BD1 C7O7	2	−1.053	1.99	−1.093	2	−1.160
BD2 C7O7			1.75	−0.498	1.84	−0.474
BD1* C7O7			0.02	0.503	0.01	0.616
BD2* C7O7			0.63	−0.057	0.32	−0.004
BD H7O7	1.99	−0.868				
BD H7O1			1.77	−0.640		
BD* H7O1			0.23			
HOO radical						
BD1 OO1	2	−1.049				
BD2 OO1						
Lp1 O	2	−0.836				
Lp2 O	1.74	−0.508				
Lp3 O						
Lp1 O1	2	−0.836				
Lp2 O1	1.98	−0.515				
Lp3 O1	1.25	−0.496				
BD OH	2	−0.904				

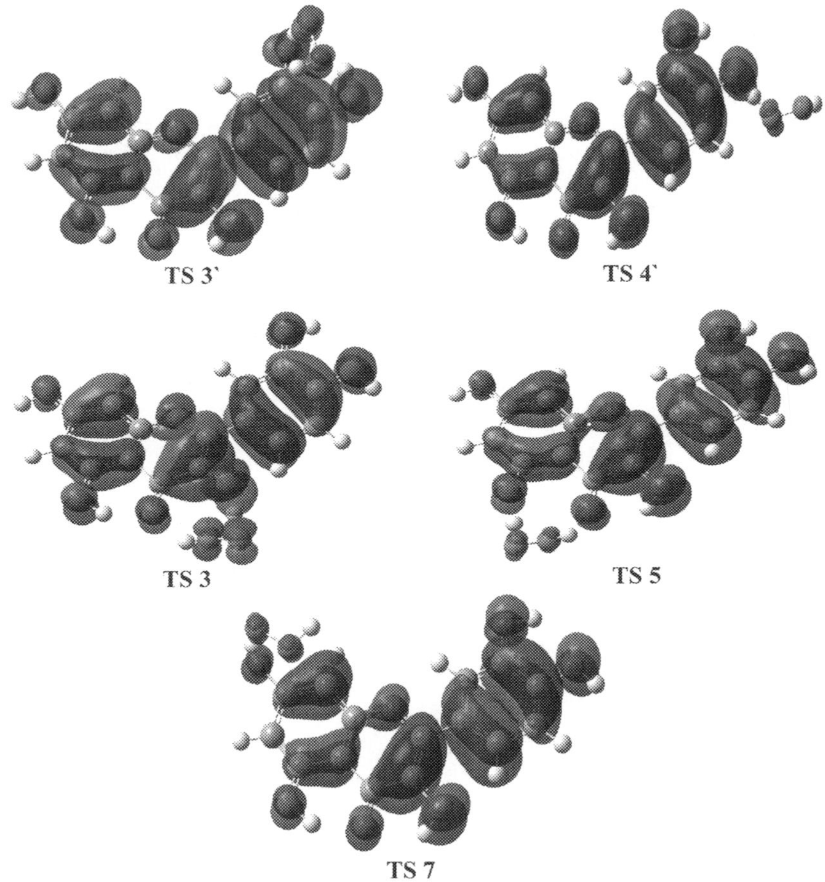

Fig. 7.3 The shapes of SOMOs in different transition states.

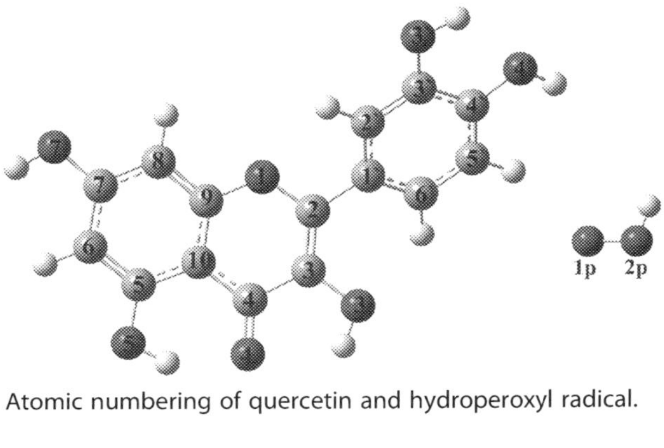

Scheme 7.2 Atomic numbering of quercetin and hydroperoxyl radical.

In addition, after cleavage of On–Hn (index n represents to the number of positions of OH groups) bond, the unpaired electron stays in the *p*-orbital of On leading to the overlapping between the *p*-orbitals of On and Cn and formation of the *p*-bonds with relatively high occupancies (about 1.7). In each TSn, the unpaired electron is mainly located in antibonding orbitals (BD(2)*CnOn) and (BD(1)*HnOp1) that are relatively highly occupied.

The SOMO of TSs (Fig. 7.3) shows that unpaired electron is mainly delocalized among ring B, C3 atom of the ring C, and corresponding oxygen atoms that are deprotonated. In addition, the delocalization of SOMOs is significantly weaker in TS5 and TS7 than in other three TSs. These differences can be explained in the higher activation energies for the reactions in position 5 and 7.

The most reactive sites are 3′–OH and 4′–OH. The reaction in the 3′–OH position is faster than one in the 4′–OH position. This result is slightly different from the BDE results, since the 4′–OH radical form has lower BDE value. It means that the reaction with the $^{\bullet}$OH radical is kinetically controlled. Also, the differences between HAT and PCET mechanisms, examining geometry orientations of peroxyl radical toward Q in transition states (Fig. 7.3), were considered. It was found that PCET is energetically unfavorable (e.g., for Q3′ and Q4′, the activation energies are higher by 3.6 and 2.6 kcal mol^{-1}) compared with HAT. These results are in good agreement with the results obtained by the examination of the reactions of flavonoids with peroxyl and DPPH radicals in nonpolar solvent (Marković et al., 2010; Musialik et al., 2009).

8.2 Reaction of quercetin via SET-PT mechanism

The reaction $Q^{\bullet+} + R^{-} \rightarrow Q^{\bullet} + RH$ (Scheme 7.3) plays a significant role in determining which mechanism of Q is dominant. The thermodynamic parameter that characterizes this mechanism is PDE. For this purpose, reactions of the radical cation of quercetin ($Q^{\bullet+}$) with different anions (OH^{-}, CH_3S^{-}, and CH_3NH_2) were examined in gas and water.

8.2.1 Mechanism $Q^{\bullet+}$ with the hydroxide anion in the gaseous and aqueous phases

The reaction with OH^{-} in gas phase (Scheme 7.3) takes place via the HAT mechanism (transition state is found), while the SET-PT mechanism is dominant in the aqueous phase (reaction is occurring without TS, Fig. 7.4). The finding that in some cases $Q^{\bullet+}$ undergoes HAT mechanism was quite unexpected. Namely, the hydroxide anion has two very high lying *p*-orbitals

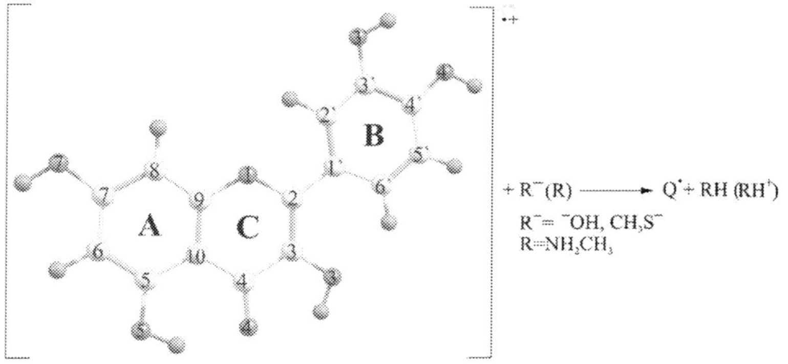

Scheme 7.3 General scheme of the investigated reactions.

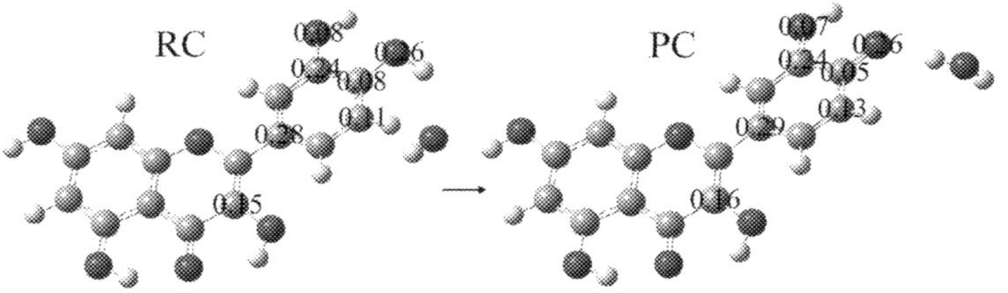

Fig. 7.4 Reaction in the position 4′ of $Q^{\bullet+}$ in the aqueous solution.

where the lone pairs are placed. Their energy (−0.02274 au) is higher than that of the SOMO of $Q^{\bullet+}$ (−0.41278 au). Thus, one of the HO^- electrons is spontaneously transferred to $Q^{\bullet+}$, yielding Q and hydroxyl radical, which further reacts according to the HAT mechanism (gas phase). On the other hand, in water, a spontaneous electron transfer from the weak base to $Q^{\bullet+}$ does not occur (due to stronger attractions between the nuclei and lone pairs of the base), because the energy of the HOMOs of the hydroxide anion (−0.31160 au) is lower than of SOMO of $Q^{\bullet+}$ is (−0.29050 au). For this reason, a spontaneous proton transfer from $Q^{\bullet+}$ to the base occurs (SET-PT) (Fig. 7.4). The reactions of $Q^{\bullet+}$ with the hydroxide anion via SET-PT in water proceed without reaction barrier, which require the reaction energy amount −63.2, −67.8, −98.4, −114.7, and −121.1 kJ mol^{-1} for transformation of RCs to the corresponding PCs.

The thermodynamic and kinetic parameters for HAT reactions, in gas phase, are calculated. These parameters are collected in Table 7.7. The presented results show that the 4′–OH position is reactive site. The optimized geometries of RC, TS, and PC, for the most favorable reaction pathway (HAT), are depicted in Fig. 7.5.

Table 7.7 The calculated values of kinetic parameters, in gas phase, for the reaction of $Q^{\bullet+}$ and HO^{-}

Site	$\Delta G^{\neq}$ (kJ mol^{-1})	k^{TST} (s^{-1})	k^{ZCT} (s^{-1})	ΔG_r (kJ mol^{-1})
7	24.5	2.88×10^{8}	7.40×10^{9}	−98.7
3′	23.2	6.01×10^{8}	5.93×10^{9}	−128.6
3	23.1	7.46×10^{8}	3.69×10^{10}	−130.7
4′	13.0	4.22×10^{10}	2.28×10^{11}	−163.4

$\Delta G^{\neq}$, k^{TST}, k^{ZCT}, and ΔG_r denote activation free energy and rate constants and reaction free energy, respectively.

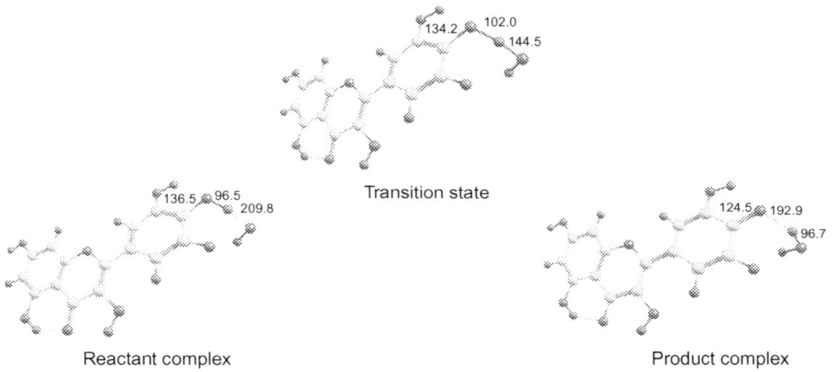

Fig. 7.5 The optimized geometries of the RC, TS, and PC for the most favorable hydrogen atom transfer reaction in position 4′–OH, with selected bond distances (pm).

The obtained results emphasize the tunneling effect as responsible for accelerating the reaction between Q and hydroxyl radical. These indicate that the reaction of hydrogen atom abstraction involves the motion of hydrogen atom that can easily tunnel through the reaction barrier (reaction of HAT mechanism). This effect is responsible for increasing the reaction rate constant (k^{ZCT}) in all positions (Table 7.7). It is observed that this effect decreases with the increase of the temperature (Fig. 7.6). In addition, the reactions at the 4′ position of Q are extremely exothermic (Table 7.7). The obtained values of activation energies are low and corresponding rate constants are high. The homolytic bond cleavage between O4′ and H4′ atoms requires the lowest activation energy (and shows the highest rate constant values). The bond breaking at position 4′ is favorable due to the involvement of O4′ in a relatively strong hydrogen bond with H3′, leading to the weakness of the O4′–H4′ bond.

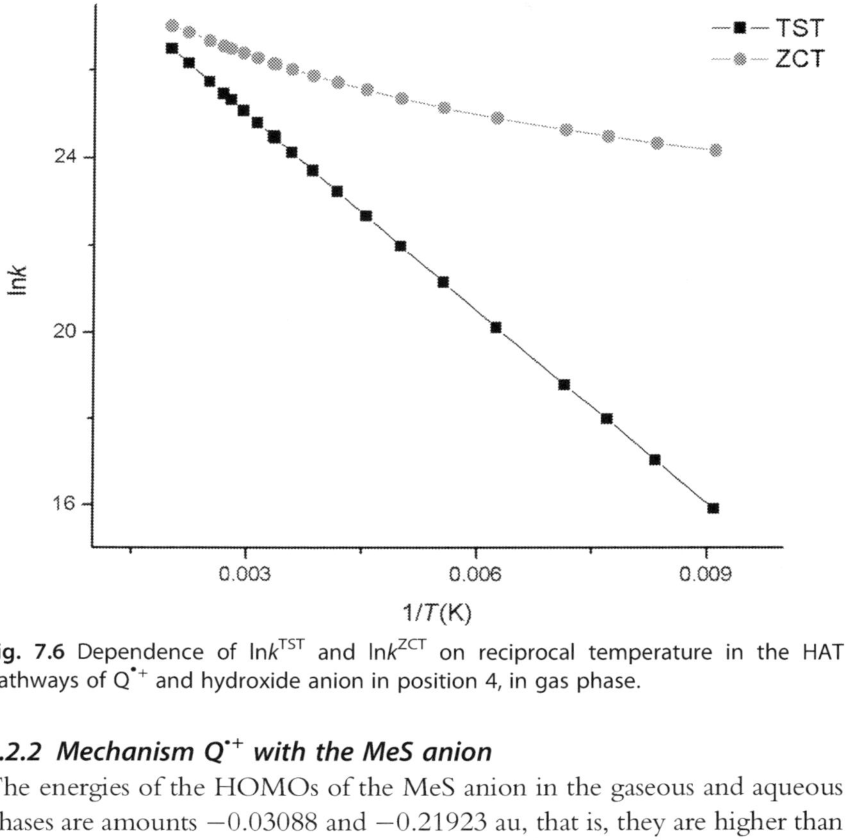

Fig. 7.6 Dependence of $\ln k^{TST}$ and $\ln k^{ZCT}$ on reciprocal temperature in the HAT pathways of $Q^{\bullet+}$ and hydroxide anion in position 4, in gas phase.

8.2.2 Mechanism $Q^{\bullet+}$ with the MeS anion

The energies of the HOMOs of the MeS anion in the gaseous and aqueous phases are amounts −0.03088 and −0.21923 au, that is, they are higher than the corresponding energies of the SOMOs of $Q^{\bullet+}$ (−0.41278 and −0.29050 au). The reaction $Q^{\bullet+} + CH_3S^- \rightarrow Q + CH_3S^\bullet$ (Fig. 7.7) is exothermic in both phases ($\Delta G_r = -562.2$ and $-146.6\,kJ\,mol^{-1}$). For these reasons, an electron from a lone pair of the MeS anion spontaneously transfers to $Q^{\bullet+}$, yielding to Q and $CH_3S^\bullet$, which further proceed via the HAT mechanism.

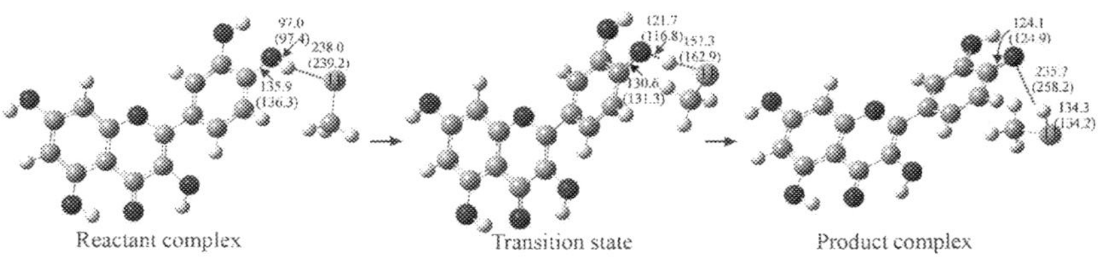

Fig. 7.7 Reaction path for the H-atom transfer from the 4′ position of Q to the $CH_3S^\bullet$ radical, with selected bond distances (pm). The values for bond distances in the aqueous solution are given in the brackets.

The reactions of Q and $CH_3S^{\bullet}$, in position 4′, proceed via the TSs that require the free activation energies of 44.2 kJ mol^{-1} (gas) and 61.0 kJ mol^{-1} (water). Corresponding reaction paths are shown in the Fig. 7.7.

8.2.3 Mechanism $Q^{\bullet+}$ with the methylamine

The reaction of $Q^{\bullet+}$ with methylamine (CH_3NH_2) takes place via SET-PT mechanism, both in gaseous and aqueous media. It should be pointed out that in this case, TSs were not located, because these reactions are taking place without reaction barriers. Namely, the HOMO energy of CH_3NH_2 in the aqueous phase is amount −0.31322 au, that is, it is lower than the corresponding energy of the SOMO of $Q^{\bullet+}$. On the other hand, HOMO energy of CH_3NH_2 (−0.30685 au) is higher than the SOMO energy of $Q^{\bullet+}$ in the gas phase. The hypothetic reaction $Q^{\bullet+} + CH_3NH_2 \rightarrow Q + {}^{\bullet}CH_3NH_2$ is endothermic ($\Delta G_r = 132.4$ kJ mol^{-1}). The energetics of this reaction is a consequence of the pronounced instability of the methylamine radical cation in comparison with the neutral parent molecule. On the basis of these facts, it can be supposed that an electron from a lone pair of methylamine will not spontaneously transfer to $Q^{\bullet+}$ and that Q can undergo the SET-PT mechanism, in both gaseous and aqueous phases.

The results for the reaction of $Q^{\bullet+}$ with methylamine in the gaseous and aqueous phases (Fig. 7.8) are mutually very similar. The reaction of the

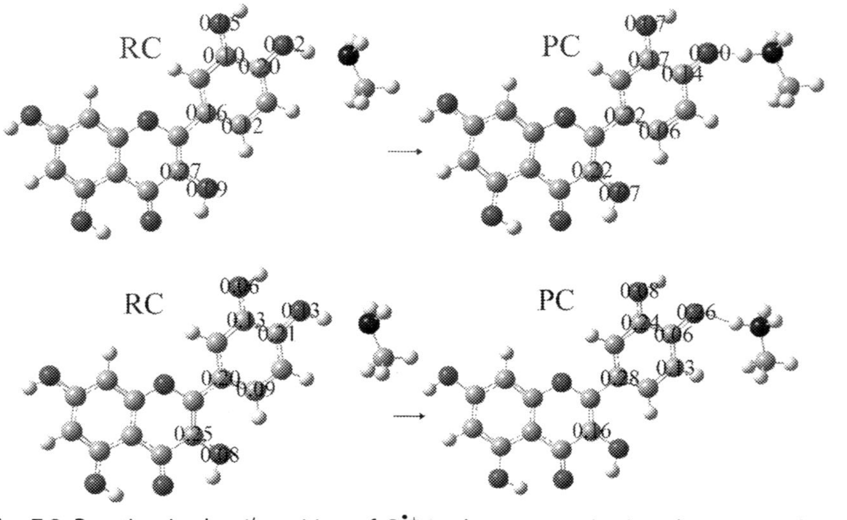

Fig. 7.8 Reaction in the 4′ position of $Q^{\bullet+}$ in the gaseous (*top*) and aqueous phase (*bottom*).

spontaneous transfer of phenolic hydrogen to N11 takes place without an activation barrier, followed by a system stabilization of −72.6 kJ/mol (gas phase) and −77.6 kJ/mol (aqueous solution). Once spontaneously formed, $Q^{\bullet +}$ donates proton to methylamine in both phases, with system stabilization. These findings show that SET-PT is an acceptable mechanism of Q with methylamine.

9 Kinetics of HAT and PCET mechanism

The examination of differences between HAT and PCET mechanisms, including different positions of the reacting species and different electronic characters (Inagaki and Yamamoto, 2014), is one of the useful methods. By examining orientations of alkyl peroxyl radicals toward GA in transition states, the differences between HAT and PCET could be considered. The optimized structures of stationary points along reaction pathways of HAT mechanism in water and PCET mechanism in benzene with methylperoxyl radical are presented in Fig. 7.9.

The calculated values of kinetic parameters for the reaction of GA and free peroxy radicals are presented in Table 7.8. The obtained results show that the barriers are systematically higher in benzene than in polar media.

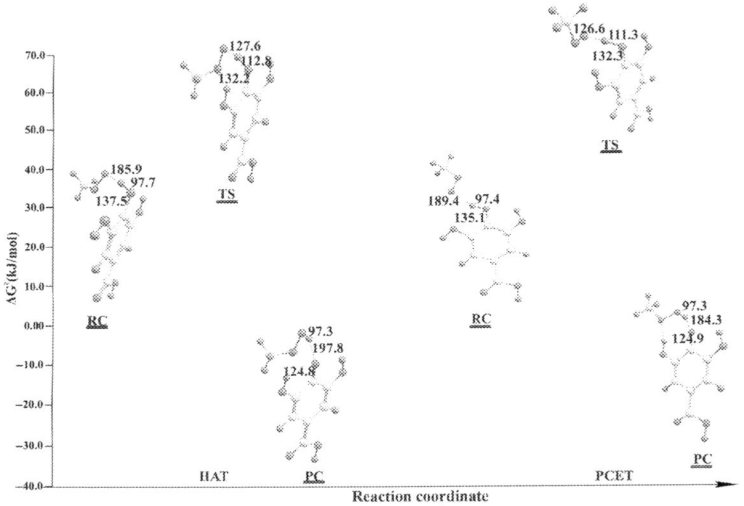

Fig. 7.9 Reaction pathway for the H-atom transfer from the C4–OH position of GA to the $CH_3OO^{\bullet}$ radical, in water and benzene. The distances between C=O, O–H, and H–O9 bonds are given in pm.

Table 7.8 The calculated values of kinetic parameters for the reaction of GA and peroxy radicals

	HAT			PCET		
	$\Delta G^{\neq}$ (kJ mol^{-1})	k^{TST} (M^{-1} s^{-1})	k^{ZCT} (M^{-1} s^{-1})	$\Delta G^{\neq}$ (kJ mol^{-1})	k^{TST} (M^{-1} s^{-1})	k^{ZCT} (M^{-1} s^{-1})
GA–3O$^{\bullet}$			***Water***			
MP	54.4	1.83×10^{3}	2.32×10^{6}	/	/	/
EP	63.3	5.01×10^{1}	1.05×10^{5}	/	/	/
iPP	65.8	1.86×10^{1}	4.07×10^{5}	/	/	/
tBP	69.3	4.00×10^{0}	9.28×10^{5}	/	/	/
GA–4O$^{\bullet}$						
MP	33.4	8.59×10^{6}	8.54×10^{8}	/	/	/
EP	34.0	6.82×10^{6}	8.19×10^{8}	/	/	/
iPP	36.9	2.16×10^{6}	6.49×10^{8}	/	/	/
tBP	39.8	6.51×10^{5}	1.23×10^{9}	/	/	/
GA–3O$^{\bullet}$			***Benzene***			
MP	59.9	1.97×10^{2}	1.15×10^{5}	73.1	9.75×10^{-1}	1.69×10^{4}
EP	61.2	1.16×10^{2}	8.29×10^{4}	74.7	5.00×10^{-1}	7.17×10^{4}
iPP	62.3	7.63×10^{1}	5.94×10^{4}	75.8	3.28×10^{-1}	2.09×10^{4}
tBP	62.7	6.37×10^{1}	1.31×10^{5}	76.0	3.06×10^{-1}	1.33×10^{4}
GA–4O$^{\bullet}$						
MP	39.7	6.79×10^{5}	5.17×10^{7}	63.9	3.99×10^{1}	2.20×10^{6}
EP	40.7	4.59×10^{5}	4.71×10^{7}	64.4	2.36×10^{1}	1.83×10^{6}
iPP	40.6	4.70×10^{5}	6.33×10^{7}	65.8	1.82×10^{1}	1.16×10^{6}
tBP	41.3	3.59×10^{5}	1.84×10^{8}	65.6	2.02×10^{1}	2.02×10^{6}

$\Delta G^{\neq}$, k^{TST}, and k^{ZCT} denote activation free energy and rate constants, respectively.

Comparison of the reactivity of GA in positions 3 and 4 in both solvents shows that the energy barrier is significantly lower for the transfer of the H atoms of the 4-OH group. It should be pointed out that there is difference of about 20 kJ mol^{-1} in the case of HAT and about 10 kJ mol^{-1} for PCET. The obtained values of kinetic parameters are given in Table 7.8 (Milenković et al., 2017). The obtained values for activation energies and k^{TST} in water and benzene (Table 7.8) are in good accordance with thermodynamic results. On the basis of obtained values for ΔG_{BDE} and for the k^{TST} rate constants, it is clear that HAT is a predominant mechanism in both solvents. Moreover, PCET mechanism is not possible in water for investigated reaction. In addition, as a consequence of the barrier sharpening, a large tunneling effect is observed for both mechanisms (Table 7.8).

The highest rate constant values can be attributed to the involvement of O4 in the relatively strong hydrogen bonds within H3 and H5, contributing to the weakening of the O4–H4 bond. The obtained results, for thermodynamic and kinetic parameters, show that the reaction of methylperoxyl radical is preferred. The values of the rate constant in benzene, for PCET mechanism, indicated that position 4 is also more probable reactivity site (Table 7.8). However, the HAT mechanism is more probable mechanism than PCET, if comparing the obtained parameters for both mechanisms (Milenković et al., 2017).

Additionally, to explain the differences between HAT and PCET mechanisms, the shapes of SOMOs of TSs are analyzed (Fig. 7.10). The obtained results show that SOMOs are not localized over O4 ⋯ H4 ⋯ O9 transition vector in TSs for PCET, while the HAT mechanism is characterized by a significant SOMO density along the donor ⋯ H ⋯ acceptor transition vector. In addition, the obtained values for HAT activation barriers are more favorable than the values for the PCET mechanism in all cases (about 24 kJ mol^{-1}). This unambiguously indicates HAT as the preferred mechanistic path under the investigated reaction conditions.

10 Radical adduct formation (RAF) mechanism

Marino and coworkers used DFT approach to examine the reaction of GA and its monoanion with $^{\bullet}$OH and $^{\bullet}$OOH radicals via RAF mechanism in water and benzene, as solvents (Marino et al., 2014). Comparison of the reactivity toward $^{\bullet}$OH and $^{\bullet}$OOH indicate that investigated compounds possess the higher reactivity toward $^{\bullet}$OH. It was found that the reactions that involve $^{\bullet}$OH are exergonic, while the reactions with $^{\bullet}$OOH are

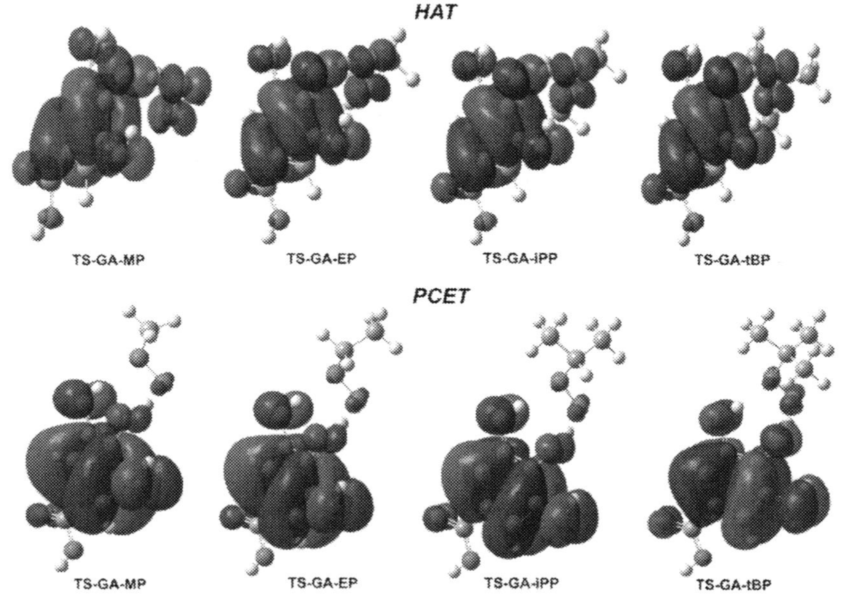

Fig. 7.10 The shape of SOMOs in different transition states of HAT and PCET mechanisms in position 4, in benzene.

thermodynamically unfavorable (endogenic). Based on these facts, the exergonic reactions are selected for further mechanistic investigation.

To calculate the rate constants for RAF mechanistic pathway with hydroxyl radical, transition states were found in water and benzene. The obtained values for activation energies and rate constants, in both solvents, are in excellent agreement with thermodynamic results (Marino et al., 2014), namely, kinetic data for the investigated compound indicating position C1 as the most reactive one.

When the total rate constants are calculated for the reaction with $^{\bullet}OH$, the results show that the reactions of GA with this radical are diffusion limited. It is worth mentioning that the reactivity of GA toward $^{\bullet}OH$ is more heterogeneous for different mechanisms and reaction positions. In addition, all mechanical pathways play a significant role in total reactivity, n nonpolar media. There is only one exception, in the case where the reaction takes place by the RAF mechanism in position 4. Based on a separate analysis of the HAT and RAF mechanisms, it can be concluded that the contribution of the HAT mechanism is greater than the contribution of the RAF mechanism. On the other hand, taking into account all reaction pathways for

Table 7.9 [a]Results related to electron transfer from the 3, 3′, and 4′ positions of Q^- to $HO^•$ in water, at the M062X/6-311++G(d,p) level of theory

Position	ΔG°_{ET} (kJ mol^{-1})	λ (kJ mol^{-1})	k_{ET} (M^{-1} s^{-1})	k_{app} (M^{-1} s^{-1})
3	−80.8	78.6	6.17×10^{12}	4.14×10^{9}
3′	−69.3	65.7	6.09×10^{12}	4.10×10^{9}
4′	−66.9	62.8	6.05×10^{12}	5.87×10^{9}

[a]ΔG°_{ET} and λ denote free energy of reaction and reorganization energy, whereas k_{ET} and k_{app} stand for rate constant arising from TST, rate constant for an irreversible bimolecular and apparent rate constant, respectively.

HAT and RAF mechanism, it is evident that both of them have important role in the scavenging activity of •OH with GA. The SPLET mechanism becomes the major reaction path, at physiological pH, in aqueous solution.

10.1 Electron-transfer reaction of quercetin

According to Marcus' approach, the important parameters for the investigation of the electron-transfer reactions are the changes of the reaction in Gibbs free energy, reorganizational energy, and diffusion rate constant. The free radical-scavenging activities of Q were also studied via the second step of SPLET mechanism, electron transfer. For this research study, molecular simulation was performed in water, as polar solvent. The calculated values of the electron-transfer reactions in the second step of SPLET are presented in Table 7.9.

It is evident from Table 7.9 that reaction Gibbs energy change is negative. When the apparent rate constant is corrected for diffusion, the values of the electron-transfer rate constants are of the order of magnitude 10^9. The 3-OH group of Q has the lowest value of free energy change for the reaction with hydroxyl radical in water. This result implies 3-OH group as the most favorable position for the reaction via electron-transfer mechanism (ET). In addition, thermochemical and kinetic analyses are in agreement and also show good compatibility with our previous results.

10.2 Electron-transfer mechanism of GA

Kinetic parameters of free radical-scavenging activity of GA were calculated by using three different radicals: $^•OH$, $CH_3OO^•$, and $^•OOH$ (Fig. 7.11). The thermochemical parameters of the reactions between selected radicals and GA were analyzed in terms of their Gibbs free energies.

The free radical-scavenging activities of gallic acid were studied via electron transfer, the second step of SPLET mechanism (Table 7.10).

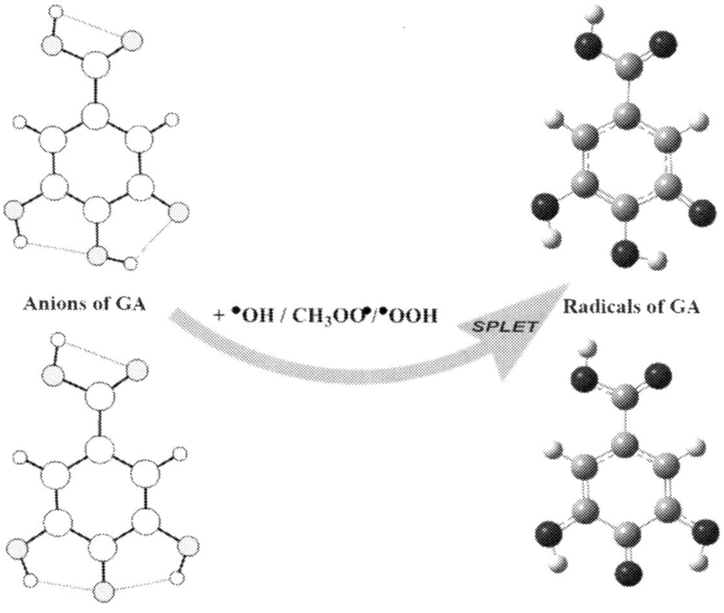

Fig. 7.11 General scheme of the examined reactions.

The molecular simulation was performed in water and pentyl ethanoate as solvents. Some thermochemical and kinetic data for reactions of gallic acid molecule with three different free radicals are calculated (Đorović et al., 2015). In water, the most favorable reaction is with hydroxyl radical. In the second place is the reaction with hydroperoxyl radical, and the least possible is the reaction with methylperoxyl radical, which is lipid peroxyl radical. Value of the activation energy for reaction with $^{\bullet}$OH radical is slightly lower for position 3 of GA molecule, while values of the activation energies for reaction with other two radicals are somewhat lower for position 4. Free energies of reactions are in agreement with activation energies and are comparable with rate constants of the examined reactions. When the values of activation energies are lower, the corresponding values of rate constants are higher, and possible reaction is faster. Similar results are obtained in pentyl ethanoate with difference in the lowest values of activation energy in reaction with hydroxyl radical. The pentyl ethanoate is used to mimic lipid environment (Pérez-González et al., 2014). In general, thermochemical and kinetic analyses are in agreement and also show good compatibility with our previous results (Table 7.10).

Table 7.10 DFT calculations of rate constants related to SPLET mechanism

Gallic acid	$\Delta G^{\ddagger}_{SPLET}$ (kJ mol^{-1})	ΔG^{0}_{SPLET} (kJ mol^{-1})	λ (kJ mol^{-1})	k_d (M^{-1} s^{-1})	k_{app} (M^{-1} s^{-1})
Water $\varepsilon = 78.35$					
$3+{}^{\bullet}OH$	21.5	−60.6	19.6	8.06×10^{9}	9.6×10^{8}
$3+CH_3OO^{\bullet}$	73.4	46.0	190.7	6.68×10^{9}	8.4×10^{-1}
$3+{}^{\bullet}OOH$	63.1	37.6	169.0	7.21×10^{9}	5.4×10^{1}
$4+{}^{\bullet}OH$	22.6	−62.9	20.2	8.22×10^{9}	6.2×10^{8}
$4+CH_3OO^{\bullet}$	71.6	43.4	189.8	6.84×10^{9}	1.7
$4+{}^{\bullet}OOH$	61.6	35.3	168.3	7.37×10^{9}	1.0×10^{2}
Pentyl ethanoate $\varepsilon = 4.73$					
$3+{}^{\bullet}OH$	29.4	23.7	61.1	8.44×10^{9}	4.4×10^{7}
$3+CH_3OO^{\bullet}$	137.5	118.6	258.3	7.13×10^{9}	4.4×10^{-12}
$3+{}^{\bullet}OOH$	128.3	111.4	238.3	7.79×10^{9}	2.1×10^{-10}
$4+{}^{\bullet}OH$	17.9	24.7	57.2	8.12×10^{9}	2.8×10^{8}
$4+CH_3OO^{\bullet}$	112.8	132.5	254.4	7.08×10^{9}	3.8×10^{-11}
$4+{}^{\bullet}OOH$	105.6	123.3	234.4	7.74×10^{9}	1.5×10^{-9}

11 Conclusion

OS is a disbalance of the oxidoreduction processes in the organism and is caused with the excessive production of free radicals. Antioxidants can prevent cell damage induced by oxidants. Antioxidative protection is a very complex process and can be investigated using both experimental (in vivo or in vitro) and theoretical (in silico) approaches; albeit combined studies are probably the best choice. The most common computational strategies used for that purpose include those based on reactivity, thermochemical, and kinetic data. The antiradical properties of antioxidants are based on their ability to donate hydrogen atom to a free radical species. In these reactions, newly formed radical is generated from antioxidant molecule, and that new formed species is more stable and less reactive than the initial free radical. There are numerous mechanisms of the antioxidant actions of free radicals, and the most studied are hydrogen atom transfer (HAT), single-electron transfer followed by proton transfer (SET-PT), and sequential proton loss electron transfer (SPLET). Antioxidant activity can be determined on the basis of the calculated thermodynamic properties of the parent molecules and the corresponding radicals, radical cations, and anions. BDE, PA, and IP values serve to confirm thermodynamically most favorable reaction pathway. The HAT mechanism is the most favorable reaction pathway for antioxidative action in nonpolar solvents. On the other hand, the SPLET mechanism is the most preferable reaction pathway for antioxidative action in water, as the polar solution. The IP values for all investigated molecules are notably high. For this reason, it has been generally accepted that SET-PT is not a plausible mechanism for investigated compounds. These facts are confirmed in the case of all examined compounds presented in this chapter. The antiradical mechanisms of the inactivation of free radical species have also been estimated on the bases of the values of reaction enthalpies related to HAT, SPLET, and SET-PT mechanisms. The influence of free radical species is also taken into account. This approach is used to verify results achieved with the standard thermodynamic approach previously mentioned. Good consistency has been achieved with results obtained with the standard procedure. At the end, antioxidative action can be examined using the kinetics of reactions. The mechanistic approach is applied as affirmation of results obtained with previously mentioned thermodynamical approaches. In general, thermodynamic and kinetic studies are in agreement.

References

Alberty, R.A., Hammes, G.G., 1958. Application of the theory of diffusion-controlled reactions to enzyme kinetics. J. Phys. Chem. 62 (2), 154–159.

Anouar, E., Calliste, C.A., Kosinova, P., Di Meo, F., Duroux, J.L., Champavier, Y., Marakchi, K., Trouillas, P., 2009. Free radical scavenging properties of guaiacol oligomers: a combined experimental and quantum study of the guaiacyl-moiety role. J. Phys. Chem. A 113 (50), 13881–13891.

Augustyniak, A., Bartosz, G., Čipak, A., Duburs, G., Horáková, L.U., Łuczaj, W., Majekova, M., Odysseos, A.D., Rackova, L., Skrzydlewska, E., Stefek, M., 2010. Natural and synthetic antioxidants: an updated overview. Free Radic. Res. 44 (10), 1216–1262.

Barzegar, A., 2012. The role of electron-transfer and H-atom donation on the superb antioxidant activity and free radical reaction of curcumin. Food Chem. 135 (3), 1369–1376.

Benavente-García, O., Castillo, J., Marin, F.R., Ortuño, A., Del Río, J.A., 1997. Uses and properties of citrus flavonoids. J. Agric. Food Chem. 45 (12), 4505–4515.

Bianco, M.A., Handaji, A., Savolainen, H., 1998. Quantitative analysis of ellagic acid in hardwood samples. Sci. Total Environ. 222 (1-2), 123–126.

Block, G., Patterson, B., Subar, A., 1992. Fruit, vegetables, and cancer prevention: a review of the epidemiological evidence. Nutr. Cancer 18 (1), 1–29.

Bors, W., Heller, W., Michel, C., Saran, M., 1990. Flavonoids as antioxidants: determination of radical-scavenging efficiencies. In: Methods in Enzymology. vol. 186. Academic Press, pp. 343–355.

Bugg, T.D., 2003. Dioxygenase enzymes: catalytic mechanisms and chemical models. Tetrahedron 59 (36), 7075–7101.

Burton, G.W., Ingold, K.U., 1984. Beta-carotene: an unusual type of lipid antioxidant. Science 224 (4649), 569–573.

Cai, Y.Z., Sun, M., Xing, J., Luo, Q., Corke, H., 2006. Structure–radical scavenging activity relationships of phenolic compounds from traditional Chinese medicinal plants. Life Sci. 78 (25), 2872–2888.

Cao, G., Sofic, E., Prior, R.L., 1997. Antioxidant and prooxidant behavior of flavonoids: structure-activity relationships. Free Radic. Biol. Med. 22 (5), 749–760.

Carpenter, J.E., Weinhold, F., 1988. Analysis of the geometry of the hydroxymethyl radical by the "different hybrids for different spins" natural bond orbital procedure. J. Mol. Struct. Theochem. 169, 41–62.

Chao, C.Y., Yin, M.C., 2009. Antibacterial effects of roselle calyx extracts and protocatechuic acid in ground beef and apple juice. Foodborne Pathog. Dis. 6 (2), 201–206.

Chatgilialoglu, C., D'Angelantonio, M., Guerra, M., Kaloudis, P., Mulazzani, Q.G., 2009. A reevaluation of the ambident reactivity of the guanine moiety towards hydroxyl radicals. Angew. Chem. Int. Ed. 48 (12), 2214–2217.

Collins, F.C., Kimball, G.E., 1949. Diffusion-controlled reaction rates. J. Colloid Sci. 4 (4), 425–437.

Day, A.J., Cañada, F.J., Díaz, J.C., Kroon, P.A., Mclauchlan, R., Faulds, C.B., Plumb, G.W., Morgan, M.R., Williamson, G., 2000. Dietary flavonoid and isoflavone glycosides are hydrolysed by the lactase site of lactase phlorizin hydrolase. FEBS Lett. 468 (2-3), 166–170.

De Graff, W.G., Myers, L.S., Mitchell, J.B., Hahn, S.M., 2003. Protection against Adriamycin® cytotoxicity and inhibition of DNA topoisomerase II activity by 3, 4-dihydroxybenzoic acid. Int. J. Oncol. 23 (1), 159–163.

Dhaouadi, Z., Nsangou, M., Garrab, N., Anouar, E.H., Marakchi, K., Lahmar, S., 2009. DFT study of the reaction of quercetin with· O2-and· OH radicals. J. Mol. Struct. Theochem. 904 (1–3), 35–42.

DiLabio, G.A., Johnson, E.R., 2007. Lone pair–π and π–π interactions play an important role in proton-coupled electron transfer reactions. J. Am. Chem. Soc. 129 (19), 6199–6203.
Dimić, D., Milenković, D., Marković, J.D., Marković, Z., 2017. Antiradical activity of catecholamines and metabolites of dopamine: theoretical and experimental study. Phys. Chem. Chem. Phys. 19 (20), 12970–12980.
Đorović, J., Marković, J.M.D., Stepanić, V., Begović, N., Amić, D., Marković, Z., 2014. Influence of different free radicals on scavenging potency of gallic acid. J. Mol. Model. 20 (7), 2345.
Đorović, J.R., Milenković, D.A., Marković, Z.S., 2015. Study of electron transfer mechanism of gallic acid. In: 2015 IEEE 15th International Conference on Bioinformatics and Bioengineering (BIBE). IEEE, pp. 1–5.
Douki, T., Cadet, J., 1996. Peroxynitrite mediated oxidation of purine bases of nucleosides and isolated DNA. Free Radic. Res. 24 (5), 369–380.
Duncan, W.T., Bell, R.L., Truong, T.N., 1998. TheRate: Program for ab initio direct dynamics calculations of thermal and vibrational-state-selected rate constants. J. Comput. Chem. 19 (9), 1039–1052.
Eigen, M., Hammes, G.G., 1963. Elementary steps in enzyme reactions (as studied by relaxation spectrometry). Adv. Enzymol. Relat. Areas Mol. Biol. 25, 1–38.
Estévez, L., Otero, N., Mosquera, R.A., 2010. A computational study on the acidity dependence of radical-scavenging mechanisms of anthocyanidins. J. Phys. Chem. B 114 (29), 9706–9712.
Foti, M.C., 2007. Antioxidant properties of phenols. J. Pharm. Pharmacol. 59 (12), 1673–1685.
Fridovich, I., 1978. The biology of oxygen radicals. Science 201 (4359), 875–880.
Galano, A., 2011. On the direct scavenging activity of melatonin towards hydroxyl and a series of peroxyl radicals. Phys. Chem. Chem. Phys. 13 (15), 7178–7188.
Galano, A., 2015. Free radicals induced oxidative stress at a molecular level: the current status, challenges and perspectives of computational chemistry based protocols. J. Mex. Chem. Soc. 59 (4), 231–262.
Galano, A., Alvarez-Idaboy, J.R., 2009. Guanosine + OH radical reaction in aqueous solution: a reinterpretation of the UV–vis data based on thermodynamic and kinetic calculations. Org. Lett. 11 (22), 5114–5117.
Galano, A., Alvarez-Idaboy, J.R., 2013. A computational methodology for accurate predictions of rate constants in solution: application to the assessment of primary antioxidant activity. J. Comput. Chem. 34 (28), 2430–2445.
Galano, A., Francisco Marquez, M., Pérez-González, A., 2014. Ellagic acid: an unusually versatile protector against oxidative stress. Chem. Res. Toxicol. 27 (5), 904–918.
Galano, A., Francisco-Marquez, M., 2009. Reactions of OOH radical with β-carotene, lycopene, and torulene: hydrogen atom transfer and adduct formation mechanisms. J. Phys. Chem. B 113 (32), 11338–11345.
Galano, A., Mazzone, G., Alvarez-Diduk, R., Marino, T., Alvarez-Idaboy, J.R., Russo, N., 2016. Food antioxidants: chemical insights at the molecular level. Annu. Rev. Food Sci. Technol. 7, 335–352.
Galano, A., Vargas, R., Martínez, A., 2010. Carotenoids can act as antioxidants by oxidizing the superoxide radical anion. Phys. Chem. Chem. Phys. 12 (1), 193–200.
Halliwell, B., 2001. Free Radicals and Other Reactive Species in Disease. Els., pp. 1–9.
Halliwell, B., Aeschbach, R., Löliger, J., Aruoma, O.I., 1995. The characterization of antioxidants. Food Chem. Toxicol. 33 (7), 601–617.
Halliwell, B., Gutteridge, J.M., 2015. Free Radicals in Biology and Medicine. Oxford University Press, USA.
Havsteen, B., 1983. Flavonoids, a class of natural products of high pharmacological potency. Biochem. Pharmacol. 32 (7), 1141–1148.

Hertog, M.G., Hollman, P.C., 1996. Potential health effects of the dietary flavonol quercetin. Eur. J. Chem. Nutr. 50, 63–71.

Hill, T.J., Land, E.J., McGarvey, D.J., Schalch, W., Tinkler, J.H., Truscott, T.G., 1995. Interactions between carotenoids and the CCl3O2.bul.Radical. J. Am. Chem. Soc. 117 (32), 8322–8326.

Hollman, P.C., 2004. Absorption, bioavailability, and metabolism of flavonoids. Pharm. Boil. 42 (1), 74–83.

Huang, M.T., Ferraro, T., 1992. Phenolic compounds in food and cancer prevention. In: Phenolic Compounds in Food and Their Effects on Health. II. Antioxidants and Cancer PreVention. American Chemical Society, Washington, DC, pp. 8–34.

Inagaki, T., Yamamoto, T., 2014. Critical role of deep hydrogen tunneling to accelerate the antioxidant reaction of ubiquinol and vitamin E. J. Phys. Chem. B 118 (4), 937–950.

Ivanović, N., Jovanović, L., Marković, Z., Marković, V., Joksović, M.D., Milenković, D., Djurdjević, P.T., Ćirić, A., Joksović, L., 2016. Potent 1, 2, 4-triazole-3-thione radical scavengers derived from phenolic acids: synthesis, electrochemistry, and theoretical study. ChemistrySelect 1 (13), 3870–3878.

Jacob, R.A., Burri, B.J., 1996. Oxidative damage and defense. Am. J. Clin. Nutr. 63 (6), 985S–990S.

Jeremić, S., Filipović, N., Peulić, A., Marković, Z., 2014. Thermodynamical aspect of radical scavenging activity of alizarin and alizarin red S. Theoretical comparative study. Comput. Theor. Chem. 1047, 15–21.

Jeremić, S.R., Šehović, S.F., Manojlović, N.T., Marković, Z.S., 2012. Antioxidant and free radical scavenging activity of purpurin. Monatsh. Chem. 143 (3), 427–435.

Kaul, T.N., Middleton, E., Ogra, P.L., 1985. Antiviral effect of flavonoids on human viruses. J. Med. Virol. 15 (1), 71–79.

Kim, D.O., Lee, C.Y., 2004. Comprehensive study on vitamin C equivalent antioxidant capacity (VCEAC) of various polyphenolics in scavenging a free radical and its structural relationship. Crit. Rev. Food Sci. Nutr. 44 (4), 253–273.

Klein, E., Lukeš, V., Ilčin, M., 2007. DFT/B3LYP study of tocopherols and chromans antioxidant action energetics. Chem. Phys. 336 (1), 51–57.

Koga, T., Moro, K., Nakamori, K., Yamakoshi, J., Hosoyama, H., Kataoka, S., Ariga, T., 1999. Increase of antioxidative potential of rat plasma by oral administration of proanthocyanidin-rich extract from grape seeds. J. Agric. Food Chem. 47 (5), 1892–1897.

Kwon, Y.I.I., Vattem, D.A., Shetty, K., 2006. Evaluation of clonal herbs of Lamiaceae species for management of diabetes and hypertension. Asia Pac. J. Clin. Nutr. 15 (1), 107–118.

León-Carmona, J.R., Alvarez-Idaboy, J.R., Galano, A., 2012. On the peroxyl scavenging activity of hydroxycinnamic acid derivatives: mechanisms, kinetics, and importance of the acid–base equilibrium. Phys. Chem. Chem. Phys. 14 (36), 12534–12543.

Lin, C.Y., Huang, C.S., Huang, C.Y., Yin, M.C., 2009. Anticoagulatory, antiinflammatory, and antioxidative effects of protocatechuic acid in diabetic mice. J. Agric. Food Chem. 57 (15), 6661–6667.

Litwinienko, G., Ingold, K.U., 2007. Solvent effects on the rates and mechanisms of reaction of phenols with free radicals. Acc. Chem. Res. 40 (3), 222–230.

Marcus, R.A., 1964. Chemical and electrochemical electron-transfer theory. Annu. Rev. Phys. Chem. 15 (1), 155–196.

Marcus, R.A., 1997. Transfer reactions in chemistry. Theory and experiment. Pure Appl. Chem. 69 (1), 13–30.

Marenich, A.V., Cramer, C.J., Truhlar, D.G., 2009. Performance of SM6, SM8, and SMD on the SAMPL1 test set for the prediction of small-molecule solvation free energies. J. Phys. Chem. B 113 (14), 4538–4543.

Marino, T., Galano, A., Russo, N., 2014. Radical scavenging ability of gallic acid toward OH and OOH radicals. Reaction mechanism and rate constants from the density functional theory. J. Phys. Chem. B 118 (35), 10380–10389.
Marković, Z., Amić, D., Milenković, D., Dimitrić-Marković, J.M., Marković, S., 2013c. Examination of the chemical behavior of the quercetin radical cation towards some bases. Phys. Chem. Chem. Phys. 15 (19), 7370–7378.
Marković, Z., Đorović, J., Dekić, M., Radulović, M., Marković, S., Ilić, M., 2013b. DFT study of free radical scavenging activity of erodiol. Chem. Pap. 67 (11), 1453–1461.
Marković, Z., Đorović, J., Marković, J.M.D., Živić, M., Amić, D., 2014c. Investigation of the radical scavenging potency of hydroxybenzoic acids and their carboxylate anions. Monatsh. Chem. 145 (6), 953–962.
Marković, Z.S., Marković, S., Dimitrić Marković, J.M., Milenković, D., 2012a. Structure and reactivity of baicalein radical cation. Int. J. Quantum Chem. 112 (8), 2009–2017.
Marković, Z.S., Marković, J.M.D., Doličanin, Ć.B., 2010. Mechanistic pathways for the reaction of quercetin with hydroperoxy radical. Theor. Chem. Accounts 127 (1-2), 69–80.
Marković, J.M.D., Milenković, D., Amić, D., Mojović, M., Pašti, I., Marković, Z.S., 2014b. The preferred radical scavenging mechanisms of fisetin and baicalein towards oxygen-centred radicals in polar protic and polar aprotic solvents. RSC Adv. 4 (61), 32228–32236.
Marković, J.M.D., Milenković, D., Amić, D., Popović-Bijelić, A., Mojović, M., Pašti, I.A., Marković, Z.S., 2014a. Energy requirements of the reactions of kaempferol and selected radical species in different media: towards the prediction of the possible radical scavenging mechanisms. Struct. Chem. 25 (6), 1795–1804.
Marković, Z., Milenković, D., Đorović, J., Jeremić, S., 2013a. Solvation enthalpies of the proton and electron in polar and non-polar solvents. JSSCM 7 (2), 1–9.
Marković, Z., Milenković, D., Đorović, J., Marković, J.M.D., Stepanić, V., Lučić, B., Amić, D., 2012b. PM6 and DFT study of free radical scavenging activity of morin. Food Chem. 134 (4), 1754–1760.
Marković, Z., Milenković, D., Đorović, J., Marković, J.M.D., Stepanić, V., Lučić, B., Amić, D., 2012c. Free radical scavenging activity of morin 2′-O − phenoxide anion. Food Chem. 135 (3), 2070–2077.
Marković, Z., Tošović, J., Milenković, D., Marković, S., 2016. Revisiting the solvation enthalpies and free energies of the proton and electron in various solvents. Comput. Theor. Chem. 1077, 11–17.
Martínez, A., Galano, A., Vargas, R., 2011. Free radical scavenger properties of α-mangostin: thermodynamics and kinetics of HAT and RAF mechanisms. J. Phys. Chem. B 115 (43), 12591–12598.
Martínez, A., Hernández-Marin, E., Galano, A., 2012. Xanthones as antioxidants: a theoretical study on the thermodynamics and kinetics of the single electron transfer mechanism. Food Funct. 3 (4), 442–450.
Mayer, J.M., 2004. Proton-coupled electron transfer: a reaction chemist's view. Annu. Rev. Phys. Chem. 55, 363–390.
Mayer, J.M., Hrovat, D.A., Thomas, J.L., Borden, W.T., 2002. Proton-coupled electron transfer versus hydrogen atom transfer in benzyl/toluene, methoxyl/methanol, and phenoxyl/phenol self-exchange reactions. J. Am. Chem. Soc. 124 (37), 11142–11147.
Mazzone, G., Galano, A., Alvarez-Idaboy, J.R., Russo, N., 2016. Coumarin–Chalcone hybrids as peroxyl radical scavengers: kinetics and mechanisms. J. Chem. Inf. Model. 56 (4), 662–670.
Medina, M.E., Galano, A., Alvarez-Idaboy, J.R., 2014. Theoretical study on the peroxyl radicals scavenging activity of esculetin and its regeneration in aqueous solution. Phys. Chem. Chem. Phys. 16 (3), 1197–1207.

Medina, M.E., Iuga, C., Alvarez-Idaboy, J.R., 2013. Antioxidant activity of propyl gallate in aqueous and lipid media: a theoretical study. Phys. Chem. Chem. Phys. 15 (31), 13137–13146.

Milenković, D., Đorović, J., Jeremić, S., Dimitrić Marković, J.M., Avdović, E.H., Marković, Z., 2017. Free radical scavenging potency of dihydroxybenzoic acids. J. Chem. 2017.

Min, D.B., Boff, J.M., 2002. Chemistry and reaction of singlet oxygen in foods. Compr. Rev. Food Sci. Food Saf. 1 (2), 58–72.

Mortensen, A., Skibsted, L.H., Sampson, J., Rice-Evans, C., Everett, S.A., 1997. Comparative mechanisms and rates of free radical scavenging by carotenoid antioxidants. FEBS Lett. 418 (1-2), 91–97.

Musialik, M., Kuzmicz, R., Pawłowski, T.S., Litwinienko, G., 2009. Acidity of hydroxyl groups: an overlooked influence on antiradical properties of flavonoids. J. Org. Chem. 74 (7), 2699–2709.

Nakanishi, I., Ohkubo, K., Miyazaki, K., Hakamata, W., Urano, S., Ozawa, T., Okuda, H., Fukuzumi, S., Ikota, N., Fukuhara, K., 2004. A planar catechin analogue having a more negative oxidation potential than (+)-catechin as an electron transfer antioxidant against a peroxyl radical. Chem. Res. Toxicol. 17 (1), 26–31.

Nakanishi, I., Shimada, T., Ohkubo, K., Manda, S., Shimizu, T., Urano, S., Okuda, H., Miyata, N., Ozawa, T., Anzai, K., Fukuzumi, S., 2007. Involvement of electron transfer in the radical-scavenging reaction of resveratrol. Chem. Lett. 36 (10), 1276–1277.

Nelsen, S.F., Blackstock, S.C., Kim, Y., 1987. Estimation of inner shell Marcus terms for amino nitrogen compounds by molecular orbital calculations. J. Am. Chem. Soc. 109 (3), 677–682.

Ness, A.R., Powles, J.W., 1997. Fruit and vegetables, and cardiovascular disease: a review. Int. J. Epidemiol. 26 (1), 1–13.

Neta, P., Huie, R.E., Mosseri, S., Shastri, L.V., Mittal, J.P., Maruthamuthu, P., Steenken, S., 1989. Rate constants for reduction of substituted methylperoxyl radicals by ascorbate ions and N, N, N', N'-tetramethyl-p-phenylenediamine. J. Phys. Chem. 93 (10), 4099–4104.

Pérez-González, A., Galano, A., 2012. On the outstanding antioxidant capacity of edaravone derivatives through single electron transfer reactions. J. Phys. Chem. B 116 (3), 1180–1188.

Pérez-González, A., Galano, A., Alvarez-Idaboy, J.R., 2014. Dihydroxybenzoic acids as free radical scavengers: mechanisms, kinetics, and trends in activity. New J. Chem. 38 (6), 2639–2652.

Petrović, Z.D., Đorović, J., Simijonović, D., Petrović, V.P., Marković, Z., 2015. Experimental and theoretical study of antioxidative properties of some salicylaldehyde and vanillic Schiff bases. RSC Adv. 5 (31), 24094–24100.

Pettersen, R.C., Ward, J.C., Lawrence, A.H., 1993. Detection of northern red oak wetwood by fast heating and ion mobility spectrometric analysis. Holzforschung 47 (6), 513–522.

Prasad, K., Laxdal, V.A., 1994. Hydroxyl radical-scavenging property of indomethacin. Mol. Cell. Biochem. 136 (2), 139–144.

Prütz, W.A., Mönig, H., Butler, J., Land, E.J., 1985. Reactions of nitrogen dioxide in aqueous model systems: oxidation of tyrosine units in peptides and proteins. Arch. Biochem. Biophys. 243 (1), 125–134.

Pryor, W.A., 1986. Oxy-radicals and related species: their formation, lifetimes, and reactions. Annu. Rev. Physiol. 48 (1), 657–667.

Pryor, W.A., 1988. Why is the hydroxyl radical the only radical that commonly adds to DNA? Hypothesis: it has a rare combination of high electrophilicity, high thermochemical reactivity, and a mode of production that can occur near DNA. Free Radic. Biol. Med. 4 (4), 219–223.

Radi, R., Peluffo, G., Alvarez, M.N., Naviliat, M., Cayota, A., 2005. Unraveling peroxynitrite formation in biological systems. Free Radic. Biol. Med. 39 (10), 1286.
Rice-Evans, C.A., Miller, N.J., 1996. Antioxidant activities of flavonoids as bioactive components of food. Biochem. Soc. Trans. 24, 790–795.
Rice-Evans, C.A., Miller, N.J., Paganga, G., 1996. Structure-antioxidant activity relationships of flavonoids and phenolic acids. Free Radic. Biol. Med. 20 (7), 933–956.
Rose, R.C., Bode, A.M., 1993. Biology of free radical scavengers: an evaluation of ascorbate. FASEB J. 7 (12), 1135–1142.
Sakagami, H., Yokote, Y., Akahane, K., 2001. Changes in amino acid pool and utilization during apoptosis in HL-60 cells induced by epigallocatechin gallate or gallic acid. Anticancer Res. 21 (4A), 2441–2447.
Sies, H., 1997. Oxidative stress: oxidants and antioxidants. Exp. Physiol. 82 (2), 291–295.
Smoluchowski, M.V., 1918. Versuch einer mathematischen Theorie der Koagulationskinetik kolloider Lösungen. Z. Phys. Chem. 92 (1), 129–168.
Squadrito, G.L., Pryor, W.A., 1998. Oxidative chemistry of nitric oxide: the roles of superoxide, peroxynitrite, and carbon dioxide. Free Radic. Biol. Med. 25 (4-5), 392–403.
Stagos, D., Kazantzoglou, G., Theofanidou, D., Kakalopoulou, G., Magiatis, P., Mitaku, S., Kouretas, D., 2006. Activity of grape extracts from Greek varieties of Vitis vinifera against mutagenicity induced by bleomycin and hydrogen peroxide in Salmonella typhimurium strain TA102. Mutat. Res. 609 (2), 165–175.
Stokes, S.G.G., 1903. Mathematical and Physical Papers. vol. 3. Cambridge University Press, Cambridge. esp. Sect. IV, 55.
Sueishi, Y., Hori, M., Kita, M., Kotake, Y., 2011. Nitric oxide (NO) scavenging capacity of natural antioxidants. Food Chem. 129 (3), 866–870.
Thomas, C., 1998. Oxygen Radicals and the Disease Process. CRC Press.
Tošović, J., Marković, S., Marković, J.M.D., Mojović, M., Milenković, D., 2017. Antioxidative mechanisms in chlorogenic acid. Food Chem. 237, 390–398.
Van Dorsten, F.A., Peters, S., Gross, G., Gomez-Roldan, V., Klinkenberg, M., De Vos, R.C., Vaughan, E.E., Van Duynhoven, J.P., Possemiers, S., Van de Wiele, T., Jacobs, D.M., 2012. Gut microbial metabolism of polyphenols from black tea and red wine/grape juice is source-specific and colon-region dependent. J. Agric. Food Chem. 60 (45), 11331–11342.
Vijayalaxmi, Reiter, R.J., Tan, D.X., Herman, T.S., Thomas, C.R., 2004. Melatonin as a radioprotective agent: a review. Int. J. Radiat. Oncol. Biol. Phys. 59 (3), 639–653.
Xie, J., Schaich, K.M., 2014. Re-evaluation of the 2, 2-diphenyl-1-picrylhydrazyl free radical (DPPH) assay for antioxidant activity. J. Agric. Food Chem. 62 (19), 4251–4260.
Yan, X.T., Lee, S.H., Li, W., Sun, Y.N., Yang, S.Y., Jang, H.D., Kim, Y.H., 2014. Evaluation of the antioxidant and anti-osteoporosis activities of chemical constituents of the fruits of Prunus mume. Food Chem. 156, 408–415.

Further reading

Glendening, E.D., Badenhoop, J.K., Reed, A.E., Carpenter, J.E., Bohmann, J.A., Morales, C.M., Weinhold, F., 2004. NBO 5.0; Theoretical Chemistry Institute, University of Wisconsin: Madison, WI, 2001.
Milenković, D., Đorović, J., Petrović, V., Avdović, E., Marković, Z., 2018. Hydrogen atom transfer versus proton coupled electron transfer mechanism of gallic acid with different peroxy radicals. React. Kinet. Mech. Catal. 123 (1), 215–230.

CHAPTER 8

Computational modeling of dry-powder inhalers for pulmonary drug delivery

Contents

1 Theoretical background

1.1 Introduction

Drug delivery to the respiratory tract is a field of research that has shown great potential and has been rapidly growing, especially because of its application in patients with asthma (Patton et al., 2004; Mossaad, 2014). It is one of the noninvasive systemic routes of drug administration to the respiratory tract. However, opposed to other routes of drug administration, the pulmonary route is accompanied by several unique challenges out of which the generation of aerosol particles in a physical form suitable for inhalation is the major one (Mossaad, 2014). Improving the quality of life of patients with

Computational Modeling in Bioengineering and Bioinformatics
https://doi.org/10.1016/B978-0-12-819583-3.00008-4

asthma and other obstructive lung disorders is done mainly via respiratory treatment. Asthma and COPD are treated mainly not only through the control of symptoms, prevention of exacerbations, keeping the normal lung function and maintaining the normal activity levels, including exercise, but also through the targeted pulmonary drug delivery. To achieve this, in the inhalation process, the drug is mixed with a powder or carrier and aerosolized before it is being inhaled by the patient (Milenkovic, 2015).

Medical aerosol (aerosolized drug) is defined as a suspension of liquid or solid particles in a carrier gas produced by an aerosol generator (Dean, 2008). Our respiratory system evolved to have filtration and elimination systems that must be overcome or bypassed in the process of providing local delivery of medications to the lung (Gardenhire, 2013).

Synthesis and manipulation of micro- and nanoparticles has allowed drug delivery to be applied not only as local but also as a systemic therapy. Additionally, the drug encapsulation in micro- and nanoparticles has the advantage of reduced drug toxicity and prolongation of the biological half-life. Drug delivery systems to the respiratory tract can be classified into two broad categories according to Mossaad (2014):

1. Immediate release systems (pure drug is in a physical form, which is adequate for dry-powder inhaler administration),
2. Controlled release systems (micro- and nanoparticles based on polymeric matrix).

By modifying the drug bioavailability, which is a function of enhancement in the drug absorption and reduction in the drug metabolism, therapeutic index of the drug can be improved.

Generally speaking, it was stated that the accepted rule is that aerosol particles of 1–5 μm are required for deposition at the site for systemic absorption, namely, the pulmonary alveoli (Mossaad, 2014). Numerous conventional and newly developed techniques are available to synthesize carrier particles for inhaling, such as micronization and blending, spray-drying (SD), spray-freeze-drying (SFD), freeze drying (FD), supercritical fluid (SCF), supercritical antisolvent (SAS), ultrasound-assisted antisolvent (UAS), and some of the most advance techniques like electrospinning. In that sense, simulation of the electrospinning process could be of use, such as research presented by Šušteršič et al. (2018) who simulated electrospinning process from the needle to the collector, using commercial ANSYS and their in-house software PAK. They showed that computational approaches such as FEM and FVM could be used to implicitly determine the homogeneity of the obtained electrospun fibers based on jet shape during electrospinning

(Šušteršič et al., 2018; Ferouka et al., 2018). However, in this chapter, we will concentrate on two traditional main techniques for production of fine particle aerosols that have been mostly used—jet milling and spray drying.

Jet milling, as a successful tool for producing very fine particles, provides a broad size particle distribution (Mossaad, 2014). The main drawback of jet milling is that the fluidized particles might suffer from a considerable degree of breakage due to intense interparticle collisions (Taylor et al., 1999). Additionally, particle size, shape, and morphology could be hardly controlled by jet milling, and control of the surface properties and electrostatic charges is limited (Bentham et al., 2004). Due to mechanical forces applied for micronization, the formed powders are not naturally grown, which leads to the breakage of the crystals at the cleavage plane with the lowest attachment energy (Roberts et al., 1994). The production of disordered structures in the therapeutic substance also affects the processing properties, powder flow and cohesion. Jet milling also leads to reduction in therapeutic bioavailability and decreased powder dispersibility, due to the association between the active sites of the carriers and the micronized drug substance (Taylor et al., 1999; Niven, 2002).

Spray drying is another conventional method for the production of uniform microparticles. However, since stability issue is associated with the production of spray-dried particles due to thermodynamically active amorphous particles, these particles tend to recrystallize (Mossaad, 2014). This property leads to alteration of drug characteristics. As an improvement for spray drying, a procedure called spray-freeze-drying technique is favorable for thermolabile drugs. It combines both the atomization step from the spray-drying technique and freezing step involved in the freeze-drying. Based on the location of the atomization nozzle, different particle characteristics can be obtained. As a result, spray-freeze-dried product can be formulated not only in the size range less than 5 μm (Yu et al., 2002; Zijlstra et al., 2004) but also on a nanoscale (Hu et al., 2003; Leach et al., 2005). Diameter of particles can be manipulated by controlling the mass flow rate of the liquid feed; an increase in mass flow ratio leads to a decrease in particle size, while the addition of excipients (e.g., trehalose and ammonium sulfate) may cause an increase in particle size (Costantino et al., 2002).

Methods for generating aerosols, formulating drugs, and administering medications effectively to the desired site of action include small-volume nebulizer (SVN), the pressurized metered-dose inhaler (pMDI), or the dry-powder inhaler (DPI) (Gardenhire, 2013). Aerosols generated by dry-powder inhalers (DPIs) are among the most promising forms to deliver therapeutic agents by inhalation to the lungs, as stated by Gardenhire (2013), especially offering advantages such as fast delivery of particles to the lungs and higher deposition of aerosol particles within the lungs (Clark, 2004; Islam and Gladki, 2008).

1.2 Dry-powder inhalers (DPIs)

In order to eliminate drawbacks of other drug administration types and improve delivery performance, dry-powder inhaler devices (DPIs) are introduced (Milenkovic, 2015). The drug delivery to the respiratory tract using DPIs, as a noninvasive systemic route of administration, has attracted significant attention (Todo et al., 2004). Dry-powder inhalers are generally defined as devices, which either store the medication as fine particle aggregates, pure drug substance, or encapsulated in a nano- or microparticulate formulation (Mossaad, 2014). Generally speaking, the drug aerosol lies inside the powder container, and once the process of inhalation begins, fine particles are dispersed by turbulent airflow, aiming for the lungs and airways (Concessio et al., 1999; Milenkovic, 2015). Furthermore, the drug can be separately packed in individual storage compartments in some multidose inhalers (Mossaad, 2014).

DPIs have captured attention and have become popular passive devices for pulmonary drug administration (Velaga et al., 2018). The main reason for that is that they are devices that are easy to work with and patients report on higher stability of the drug (because it is solid) and better compliance in comparison with the metered-dose inhalers (MDIs). The main application of the DPIs is in local treatments such as local inflammation or infections in the lungs (i.e., asthma, COPD, or cystic fibrosis infections) (according to Velaga et al., 2018). In order to produce an effective and safe inhalation therapy, a test DPI must be able to reproduce the adequate delivery of so-called fine particle dose (FPD) to the targeted site (infection, receptor, and absorption site) in the respiratory system (Demoly et al., 2014). Main determinants to meet these requirements are not only DPI design and powder formulation but also the correct use of the inhaler and adherence to therapy. General rules for API powder is that its aerodynamic particle size less than 3 μm shows high FPF and peripheral lung deposition (Corradi et al., 2014). Aerodynamic size of formulated particles will have the most dominant effect on the deposition of these particles. The size is dependent on a combination of factors, drug-carrier agglomerate size, shape, and density characteristics (Riley et al., 2012). Therefore, great attention has been paid to the methods for dissolution to appropriately characterize the in vitro behavior of particles from DPI in the last decades (Velaga et al., 2018). It can be seen that the local delivery of drug to the lung, by using a passive DPI, depends on many factors that can be categorized as device and formulation properties, as well as patient inspiratory flow (Shur et al., 2012). Design of the DPI device and physicochemical properties of the powder dose affect the control the fluidization, deaggregation, and aerodynamic particle size of the drug (Shur et al., 2012). Implicitly, this complex relationship influences the efficacy of a DPI and regional deposition of drug particles in the lung. Therefore, in the production of a test DPI that will have the in vitro performances that match the reference DPI, it is important to pay attention to the fluidization and drug dose deaggregation behavior in the test and reference DPI.

When using DPIs, at the beginning of inhalation, the particles are initially in the form of loose powder, which is broken up and dispersed as particle aggregates under the action of airflow, and then further broken up into fine particles (Fig. 8.1). Particles in DPIs can be also distinguished as porous or nonporous. Flow-enhancing filters are included in the powder formulation to improve the flowability of nonspherical powder particles. In that sense, most promising transfer of the drug is via carrier particles (e.g., lactose) (Milenkovic, 2015).

Aerosolization performance (powder dispersion, deposition, and the breakage of particle aggregates) of inhaled particles is affected by the particle morphology (cohesion, electrostatic charge, size, and size distribution),

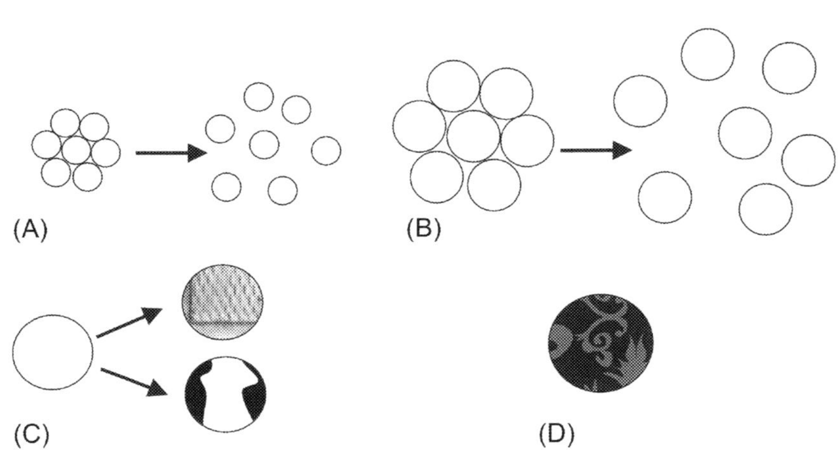

Fig. 8.1 Particle types: (A) fine particles, (B) large carrier particles, (C) macro porous particles, and (D) heterogeneous mixture.

density, and their aggregation profile (Mossaad, 2014; Milenkovic, 2015). Furthermore, several studies demonstrated the importance of particle size on the deposition and clinical efficacy of inhaled dry-powder therapeutics (Mossaad, 2014). Therefore, it is commonly accepted to propose the aerodynamic diameter as a parameter to describe the diameter of particles moving in an air stream (Mossaad, 2014). Also, drugs with a molecular weight below 30 kDa can easily penetrate the alveolar membrane to the blood circulation, which enables effective targeting of the respiratory epithelium (Mossaad, 2014).

Selection of the appropriate inhalation delivery system is a pivotal decision in pharmaceutical manufacturing that is dependent on the clinical objective (acute or chronic treatment) and target patient features (infant, elderly, or ambulatory) (Dailey et al., 2003). The patient inspiration is the main force that initiates the inhalation process. Recent DPI technology is making attempts to reduce the dependence of a patient's inspiratory effort by adding a battery-driven propeller that helps with the dispersion of the powder or by using compressed air to aerosolize the powder and convert it into a standing cloud in a holding chamber. Independent of these efforts, DPIs are primarily constructed to be breadth actuated and patient's PIFR of 30–130 L min − 1 is necessary to achieve an aerosol within the respirable range (Milenkovic, 2015). Although the inspiratory airflow rate is a critical factor that influences the delivery of medication, some dry-powder inhaler devices appear to be relatively independent on the patient's inspiratory rate. Therefore, in comparison with metered-dose inhalers, the need for good coordination between the patient's inspiration and inhaler device actuation is eliminated.

Different dry-powder inhalation devices are available in the market, yet no single inhaler device possesses all the properties of an ideal inhaler (Dailey et al., 2003). Fractional deposition of drug and its depth of penetration have to be accurately assessed in order to evaluate inhalation drug delivery systems. Most marketed DPIs are passive inhalers, which range from single-dose devices loaded by the patient (e.g., Aerolizer and Rotahaler) to multidose devices provided in a blister pack (e.g., Diskhaler); multiple unit doses sealed in blisters on a strip, which moves through the inhaler (e.g., Diskus); or reservoir-type (bulk powder) systems (e.g., Turbuhaler) (Milenkovic, 2015). They employ the patient's inspiratory effort to generate the necessary airflow, fluidize the powder bed into a respirable aerosol, and overcome the cohesive nature of the respirable active pharmaceutical ingredient (API) (Wong et al., 2012).

Problem with DPIs is that 10%–30% of the drug is retained within the device (Pedersen, 1996; Dolovich, 1999; Newman et al., 1989; according to Milenkovic, 2015), while between 10% and 40% of the emitted dose reaches the lungs, depending on the type of inhaler used (Islam and Gladki, 2008). Many factors (e.g., inefficient deaggregation of the fine drug particles, humidity, temperature changes, charge, low PIFR, and angle of inhaler when distributing the drug) may cause drug deposition within the devices, which leads to reduce the efficiency of pulmonary drug delivery (Milenkovic, 2015). With most DPIs, drug delivery to the lungs is increased by fast inhalation, which is not the case with MDIs, that require slow inhalation and breath holding to enhance lung deposition of the drug. For example, Borgstrom et al. (1994) demonstrated that an increase of the PIFR from 35 to 60 $\mathrm{L\,min^{-1}}$ through the Turbuhaler leads to the increase in total lung dose of terbutaline (from 14.8% of nominal dose to 27.7%) (Milenkovic, 2015).

1.2.1 Aerolizer

This chapter focuses on analyzing dry-powder inhaler Aerolizer because it is a readily available commercial device. Aerolizer, as one of the DPI, consists of a cap to protect the mouthpiece and base that allows the proper release of drug from the capsule. Smaller parts within the base are a circulation chamber, a capsule chamber, a button with "winglets" (projecting side pieces) and pins on each side, two air inlet channels, and grid structure (Fig. 8.2).

Operating with Aerolizer includes removing protecting cap and the inhaler opens by twisting the mouthpiece. Capsule filled with drug is placed in a capsule chamber. After closing, the capsule is pierced by pressing buttons with "winglets." During inhalation, air enters the inhaler through tangential air inlets, rotating the capsule and ejecting the drug powder into the circulation chamber. Deagglomerating forces, provided by the device flow field,

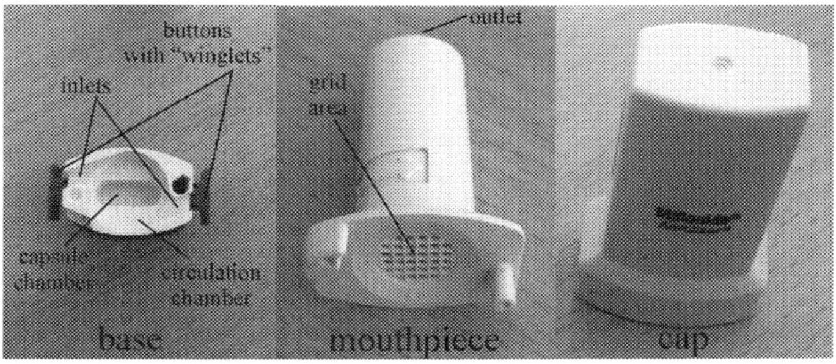

Fig. 8.2 Parts of dry-powder inhaler Aerolizer.

disperse the drug agglomerates to produce a fine respirable aerosol cloud. It should be also emphasized that an existing grid area at the bottom of the mouthpiece prevents the capsule from exiting the inhaler (Gardenhire, 2013; Coates et al., 2006).

1.3 Clinical efficacy of inhalation dry powders

Delivering aerosolized drugs directly into the lungs has the advantages of higher drug concentration delivered more effectively to the airways and reduced systemic adverse effects (Dean, 2008). Other advantages include simplification of the inhalation technique and help in fluctuation of the inhaled dose reduction as well as local and systemic side effects (Crompton, 1982; Jackson, 1995; Brocklebank et al., 2001; Terzano, 2008).

The findings presented by Mossaad (2014) suggest that the deposition profile of inhalation by dry powders and therefore their efficiency (dose entrainment, fine particle fraction (FPF), and lung deposition) is affected by two major independent factors:

1. patient-related factors (anatomical and physiological aspects of the respiratory system and the inhalation airflow rate and time),
2. physical properties of dry powders (subdivided into properties of pure drug and design of dry-powder inhaler device) (Table 8.1).

The main advantage of DPIs is the ease of use and cost-effectiveness, because of which they became one of the principle means of delivering pharmaceuticals (Milenkovic et al., 2014a). Some of the advantages DPIs offer are that they do not require propellants for their administration, which makes them more respectful of the environment, and that many of them have an indicator of the doses remaining.

The main drawback that the researches showed is the loss of powder/drug due to deposition within the device (Coates et al., 2004). Donovan et al. found that in Aerolizer, in comparison with the HandiHaler, aerosolization performance was more dependent on the carrier particle size, which appeared to be related to a greater number of carrier particle-inhaler

Table 8.1 Independent factors influencing patient inhalation performance.

Patient-related factors	Physical properties of DPIs
• Instruction • Clinical parameters • Age, gender, training • Smoker/nonsmoker	• DPI design • Dose system • Powder formulation • Deagglomeration principle

Table 8.2 Advantages and disadvantages of DPIs.

Advantages	Disadvantages
• Compact • Portable • Breath actuated • Easy to use • No requirements for hand-mouth coordination	• Humidity may cause powders to aggregate and capsules to soften • Dose lost if patient inadvertently exhales into the DPI • Most DPIs contain lactose • High oral deposition

collisions (Donovan et al., 2012). One of the disadvantages of DPIs is also an increased airflow resistance. If the resistance of the device is high, an inspiratory flow necessary for achieving the maximum dose from the inhaler is more difficult to generate (Milenkovic, 2015). Furthermore, according to authors cited in Milenkovic (2015), it is implied that deposition in the lung is increased when using high-resistance inhalers.

Main advantages and drawbacks of DPIs are given in Table 8.2.

Additionally, local delivery of the drugs is said to be recommended for patients with cystic fibrosis, asthma (Mossaad, 2014), chronic obstructive pulmonary disease (Ryan et al., 2011), and lung cancer (Newhouse and Corkery, 2001). This way of distribution is highly beneficial in these cases due to significant reduction of systemic side effects, in addition to the localized concentration of medication at the site of drug action. Therefore, hormones and toxic chemotherapeutic agents seem promising drug candidates for local pulmonary administration and in treatment of lung cancer (Mossaad, 2014).

1.4 Forces during inhalation

As already mentioned, the drug release efficiency inside the DPI is dependent not only on the properties of the drug itself but also on the construction and material of the DPI. All these properties would affect adhesion and cohesion forces and influence the flow field inside the DPI (Milenkovic, 2015). Interactions such as particle/aggregate interactions with other particles/aggregates or with the DPI walls, control breakage of aggregates into smaller aggregates and fine particles (by fine particles are considered particles less than 5 μm in size). These fine particles are most adequate in targeting upper respiratory tract. Cohesion and adhesion forces, investigated by Kroeger et al. (2010) and Wong et al. (2010a, b), affect particle/particle and particle/wall interactions. They performed particle aerosolization in an entrainment tube apparatus experimentally and using CFD simulations, further evaluating effects of grid geometry, turbulence, and aggregate impaction with variable impact angles.

Particle trajectories and deposition are generally determined using Eulerian fluid/Lagrangian particle approach, solving the force balance equation for each particle considering an unperturbed airflow solution (Milenkovic, 2015).

$$m_p \frac{du_p}{dt} = F_D + F_g - F_B + F_s \tag{8.1}$$

where m_p is the particle mass, F_D is the drag force, F_g is the gravitational force, F_B is the buoyant force, and F_s represents additional forces. This assumption is valid for dilute systems, meaning with dispersed phase volume fractions less than 10%.

In Eq. (8.1) the drag force F_D exerted by the fluid on the solid acts opposite to the direction of flow and is given by

$$F_D = C_D \frac{\pi}{8} \rho D^2 V^2 \tag{8.2}$$

where V is the relative velocity between the gas and the spherical particle and D is the diameter of the particle, while C_D is the drag coefficient.

Gravitational force F_g is defined as

$$F_g = \frac{\pi D^3}{6} \rho_p g \tag{8.3}$$

where ρ_p is the particle density and g is the gravitational acceleration.

Buoyant force F_B is a result of displacement of fluid by the particle given by

$$F_B = \frac{\pi D^3}{6} \rho g \tag{8.4}$$

where ρ is the gas density.

As already said, F_s represents additional forces that can be included in the particle force balance equation, when special circumstances are taken into account. These include "virtual mass" force, which is required to accelerate the fluid that surrounds the particle and the Saffman's lift force (lift due to shear).

During particle elastic deformation, the forces will tend to return the particle to its spherical shape when it rebounds. These collisions occur at normal velocities on the order of 1 m/s, and most particle sizes are on the order of

2 μm (Milenkovic, 2015). In the particle capture by the wall, van der Waals forces develop and the capillary forces over the contact line of the meniscus. It is reported that the process of collision is too short and then gravitational and inertial drag forces do not have any effects. Usually, during the investigation of particle deposition inside DPIs, humidity is not investigated. Electrostatic effects are said to be nonexistent for the materials used in DPIs, according to some experimental studies; if static electricity was a contributing factor, DPIs would have more than 90% deposition, which is not the real case. However, excessive humidity could be a problem and increase the aggregation and deposition (Milenkovic, 2015).

Critical normal velocity v_c above which particles rebound is connected to the capture (sticking) efficiency σ. This has been defines as

$$\sigma = 1 - H\frac{v_n}{v_c} \tag{8.5}$$

where H is the Heaviside step function and v_n is the normal component of the particle velocity at the point when the particle collides with the DPI wall. This equation is used to determine when the particles will rebound ($\sigma = 0$, $v_n > v_c$), losing some of tangential and normal momentum, and when they will deposit ($\sigma = 1$, $v_n < v_c$). Aforementioned momentum losses are dependent on the particle/wall material properties, angle of collision, collision velocity, surface roughness, etc. (Milenkovic, 2015).

To summarize this part, in properly formulated powders, particle-particle cohesion forces are dominant during the initial powder release and breakage of particle aggregates. After this, capture efficiency and particle-wall collision frequency (which is controlled by adhesion forces) will determine the deposition rate, while particle cohesion forces are in control of the rate of collision-induced breakage.

Further along the lung pathways, factors that influence the particle deposition in the lung, beside the anatomy of the airways and mucosal layer, are physical mechanisms of inertial impaction, sedimentation, diffusion, interception, and electrostatic precipitation (Boshhiha, 2011; Gonda, 1992). In this process, starting from the mouth and all the way toward the alveoli, the flow changes its direction. As a result of the inertial impaction, particles with the largest diameters are deposited in the throat and in largest airway pathways (more specifically bifurcations in large airways). This happens due to the particle inertia that is higher than the inertia of the gas phase in these parts of the airway system (Milenkovic, 2015). Other forces, such as gravitational, gain more importance downwards in the lung and affect

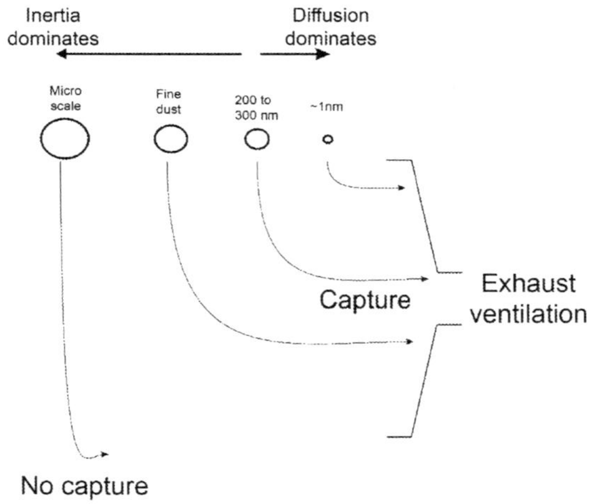

Fig. 8.3 Main impacts on the particles depending on their diameter.

the remaining smaller particles. The smallest particles can reach the alveoli because of the Brownian motion (diffusion) (Fig. 8.3). Additionally, it should be said that a charged particle may deposit in the respiratory system due to electrostatic forces, van der Waals, capillary-viscous forces, etc. (de Koning et al., 2001; Approaches to Safe Nanotechnology, 2016).

2 Literature review: Particle engineering strategies for pulmonary drug delivery

Many researchers have examined different dry-powder inhalers in order to have better understanding of their overall performance. Computational fluid dynamics (CFD) has emerged as a computational tool used in describing dispersed phase flows (Milenkovic, 2015). However, not many CFD studies have been performed, mainly not only due to the complex and transient flow structures that are observed in most commercial DPIs but also because of the dynamic nature of powder breakup and dispersion in a DPI (Coates et al., 2004). In order to gain knowledge and better understanding of particle dispersion within the Aerolizer inhaler, computational fluid dynamics (CFD) can be used for modeling both laminar and turbulent flow (Wong et al., 2012). However, CFD can only simulate airflow without considering particle interaction. For this purpose, the field of CFD analysis is coupled with discrete phase modeling or discrete element method (DEM) to allow for aerosol transport and deposition to be calculated in realistic three-dimensional (3D) models of the inhaler (Longest and Holbrook, 2012;

Tong et al., 2010; Calvert et al., 2011). DEM simulations are computationally expensive and only a limited number of particles can be described, while DPM can only be used in limited cases, when flow rates are small and normal impact velocities are smaller than critical velocity (Milenkovic, 2015). Both approaches provide valuable information on aggregate stability and powder dispersion. The information obtained in this way can help maximize the dispersion performance of a dry-powder inhaler (Coates et al., 2006).

Wong et al. (2010a, b) examined particle aerosolization in an entrainment tube apparatus, performing experiments and CFD simulations, which lead to evaluation of the effect of grid geometry, turbulence, and aggregate impaction with variable impact angles. Milenkovic et al. (2014a, b) examined commercial dry-powder inhaler turbohaler and considered different flow models combined with particle motion and deposition. Coates et al. (2004, 2005, 2006) investigated the effects of grid structure, mouthpiece length, inlet size, and airflow on Aerolizer DPI performance. They have showed that the structure of the inhaler grid has a significant effect on the amount of powder retained in the device without affecting the inhaler dispersion performance. Since the size of inlet controls the air velocity that enters the device, the changes in dimensions will influence both the levels of turbulence generated in the device and the intensity of particle collisions with the device walls (Coates et al., 2006). Some experimental investigations have further examined particle deposition and the fine particle fraction and particle size distribution of particles that reach the outlet (de Koning et al., 2001; Hoe et al., 2009). Tong et al. (2012) used combined CFD and DPM method to better understand dispersion mechanisms in commercial DPIs, while Zhou et al. (2013) analyzed how device design effects aerosolization and Yang et al. (2014) investigated airflow-induced detachment of particles. The study by Shur et al. (2012) examined the effects of different design modifications on the performance of DPI Cyclohaler. They preformed CFD modeling and in vitro characterization (using multistage cascade impactor [CI]) in order to investigate key device modification attributes that affect the aerosolization performance for unit-dose and capsule-based DPI device. The conclusion is that in the design and development of a test DPI, the attention should be payed to the pressure drops for different flow rates and aerosolization performance should be manipulated so it matches the performance of a reference DPI. Another research by Donovan et al. (2012) has employed CFD to examine aerosolization performance of carrier particles with different physical properties in two DPIs—Aerolizer and HandiHaler. They concluded that the aerosolization performance of the Aerolizer depended more on the carrier particle size in comparison with the Aerolizer, what they explained to be a consequence of greater number of carrier particle-inhaler collisions. Although there are those studies that use CFD modeling to define optimal device design parameters, this kind of research is still limited, as in vitro comparability in the presence of multiple flow rates has to be preformed (Donovan et al., 2012). It should be also emphasized that many authors have used different CFD methods to examine lung behavior in animals (mostly mice, as the lungs of mice have similar structural composition of acinus) and drug deposition in human lungs, but we did not focus on those models in this chapter. More information about these models could be found, for example, in Filipovic et al. (2014), Tsuda et al. (2008), and Koullapis et al. (2018a, b).

De Boer et al. (2012) examined the performance of the Twincer DPI using CFD and total particle deposition as well as the spatial distribution of deposited particles in the device. They showed that the effect of particle/aggregate interactions with other particles/aggregates or with the DPI walls is of key importance as these interactions control breakage of aggregates into smaller aggregates and fine particles. The effects of particle/particle and particle/wall interactions were investigated by Kroeger et al. (2010) and are shown to be controlled by cohesion and adhesion forces, respectively. Additionally, Cui et al. (2014) investigated Lagrangian detachment models and combined with multiscale approach to allow the prediction of inhaler performance and efficiency. Using OpenFOAM, they investigated flow field inside the Cyclohaler and later combined it with Lattice-Boltzmann method (LBM) to examine fixed carrier particle covered with hundreds of drug particles (Cui et al., 2014). The results of determined drug particle detachment possibility by lifting off, sliding, or rolling are verified using measurements by atomic force microscope (AFM).

To explain further the connection of the mentioned models, it should be said that Koullapis et al. (2018a) took a step toward large-scale simulations of the human lung by introducing a numerical methodology in order to predict particle deposition in a simplified model of the deep lung during a full breathing cycle. More specifically, they investigated generations 10–19 of the conducting zone and a heterogeneous acinar model to estimate deposition in the deep lung (Koullapis et al., 2018a).

It should be emphasized that an outstanding work has been done within the SimInhale COST Action in the field of pharmaceutical technologies and simulation for patient inhaled medicines (SimInhale Website, 2016). During the duration of the project, a benchmark case has been created with intention to use it for the validation of computational tools for regional deposition studies in the upper airways. Complex geometry has been created to preform in vitro deposition measurements using various inhalation flow rates. Several simulation approaches have been used in numerical testing of the benchmark case. In the paper, authors investigated experimental data used for quality assurance of Computational Fluid Particle Dynamics (CFPD) studies in the upper airways, performed a review of the existing modeling approaches, and provided practical recommendations, where it was possible. The authors also investigated the flow and aerosol deposition and discussed the efficiency of different simulation methods (Koullapis et al., 2018a, b).

3 Simulations performed on dry-powder inhaler Aerolizer

The motivation for analyzing Aerolizer inhaler lies also in its construction, which can be further optimized as shown by Coates et al. (2004, 2005, 2006). Simulation of the flow of particles carried by air during inhalation can help in the determination of the weak points in construction of the inhaler with the possibility of shape and dimensions change in order to optimize and maximize efficiency of drug delivery to the lungs.

3.1 Geometry

The first step in simulating powder dispersion was to create 3D models of original and modified inhalers using commercial software for computer-aided design (e.g., CATIA V5R21). We firstly modeled the solid area of dry-powder inhaler Aerolizer (Fig. 8.4), while only the fluid area was taken into account for simulations (Fig. 8.5).

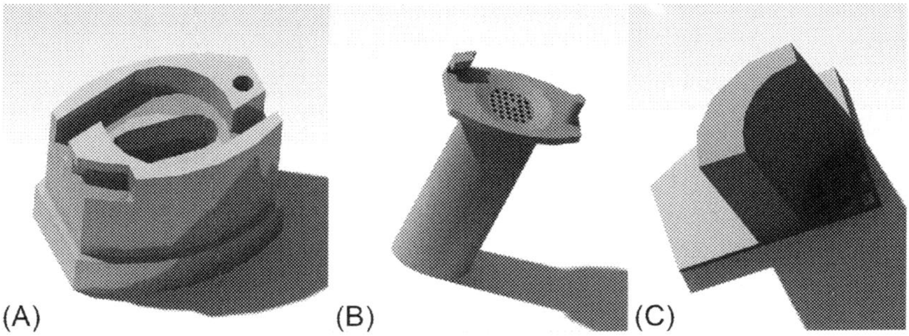

Fig. 8.4 Geometrical model of solid domain of Aerolizer: (A) base, (B) mouthpiece, and (C) one-quarter cross section of the mouthpiece.

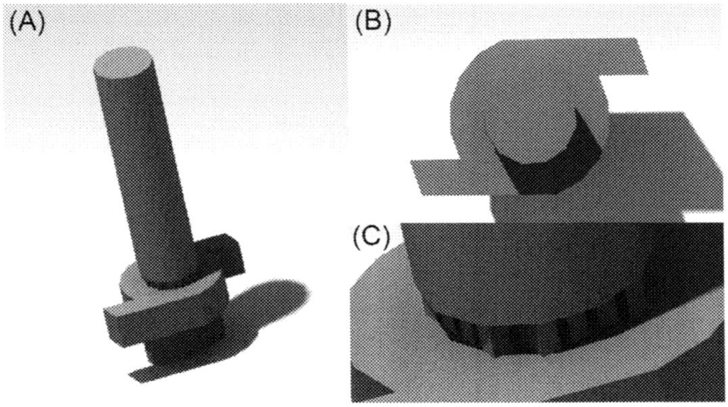

Fig. 8.5 Fluid domain of Aerolizer that was used for aerosol powder dispersion simulations: (A) isometric view, (B) view from above, and (C) detail of the grid area.

Geometry and dimensions for commercial Aerolizer were adopted by taking detailed measurements from the aforementioned device using a micrometer. Some simplifications on the model included chamfered and curved edges in areas where those modifications would not account for great effect on the process of inhalation.

Only few studies have been found to be connected to the determination of the weak points in construction of the inhaler. Since circulation chamber has not yet been thoroughly investigated, some modifications in dimensions of the standard, original inhaler have been made in order investigate the efficiency of drug delivery to the lungs connected to different heights of the circulation chamber. A standard device (SD) Aerolizer and four modified devices (MD1–MD4) were adopted with the heights of the circulation chamber of the modified inhalers as follows: MD1, 2 times greater; MD2, 1.5 times greater; MD3, 1.5 times smaller; and MD4, 2 times smaller than the original standard device circulation chamber height (Fig. 8.6).

Mentioned fluid domains were exported as .stp file for further processing (meshing, setting boundary conditions, and performing simulations).

3.2 Meshing

Unstructured tetrahedral mesh of the fluid domain was made using commercial software ANSYS 16.0. After conversion to the polyhedral cells, with purpose to reduce the skewness of the overall cell count, the model grid of standard inhaler Aerolizer consisted of 281,782 polyhedral cells and 1,402,608 nodes. The grid was refined in the areas where large gradients

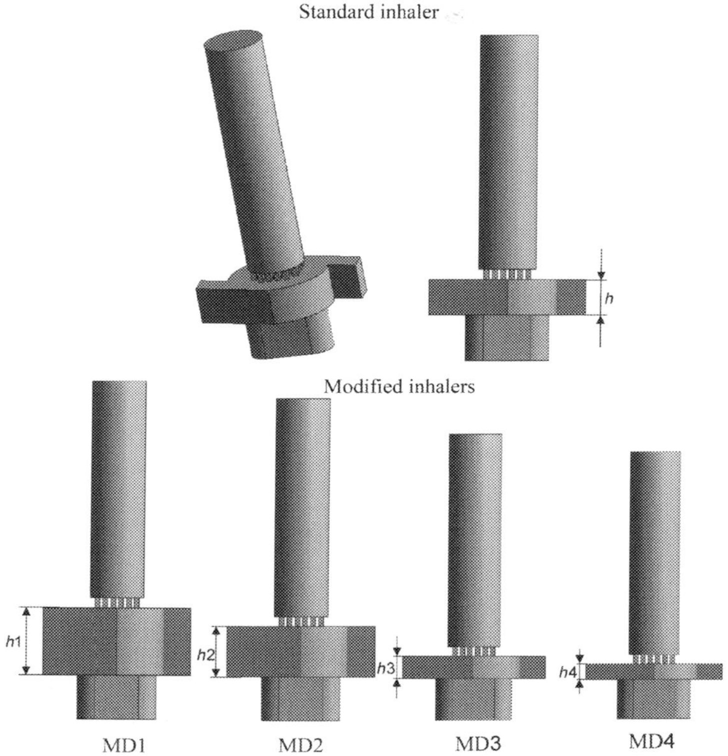

Fig. 8.6 Modifications MD1–MD4 in comparison with the standard (SD) Aerolizer circulation chamber dimensions.

of flow were expected and then additionally refined near the wall based on the initial solutions (Fig. 8.7).

Element quality distribution is examined in order to determine if the mesh was appropriate (Fig. 8.8). It was shown that almost all the elements were in the range from 0.5428 to 0.99992 for the element quality.

A more detailed explanation of the created model could be found in Šušteršič et al. (2017a, b) and Vulović et al. (2017).

3.3 Simulation assumptions and boundary conditions

The commercial CFD software ANSYS Fluent (version 6.3, ANSYS, the United States) was used to simulate fluid flow that is governed by the Navier-Stokes equations. The boundary conditions of air inlets, outlet of mouthpiece, and other walls in the computational models were set as

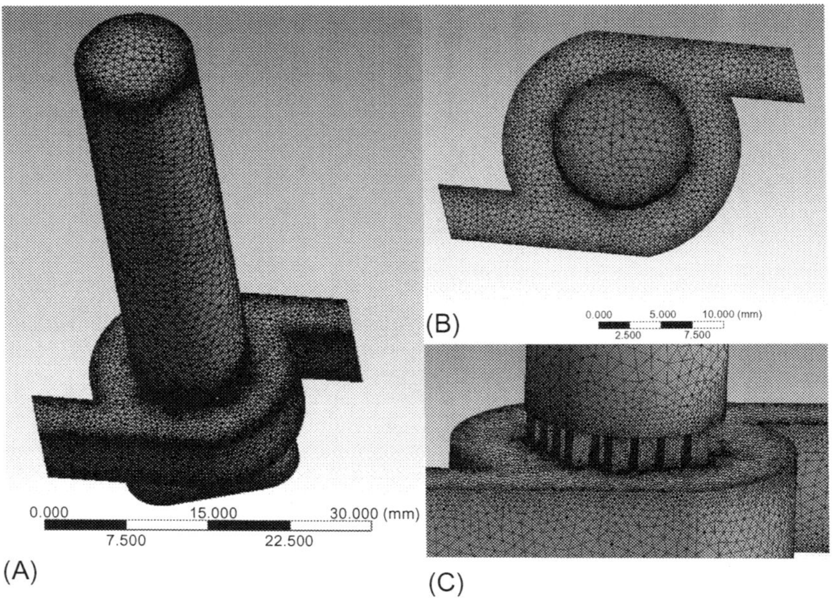

Fig. 8.7 Fluid domain mesh: (A) isometric view, (B) view from above, and (C) detail of the grid area.

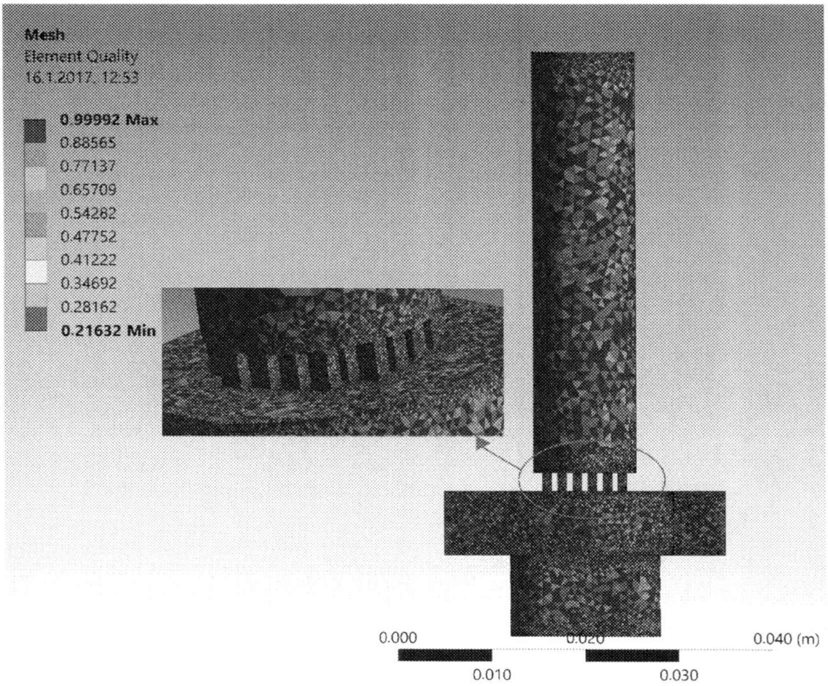

Fig. 8.8 General mesh quality with extracted detail of the grid area.

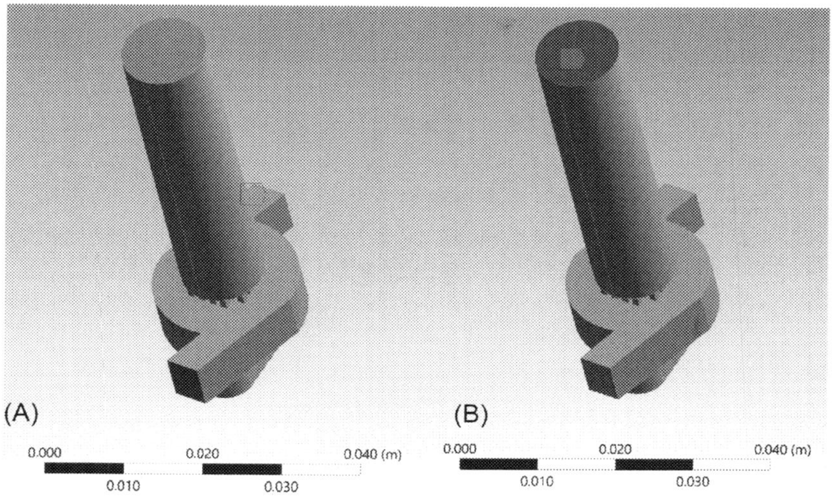

Fig. 8.9 Boundary condition surfaces: (A) inlet surface and (B) outlet surface.

velocity inlet, pressure outlet, and wall, respectively. Fig. 8.9 shows inlet and outlet surfaces colored in red.

Some simplifications are assumed, such as that air in the model was incompressible, flow in the inlets was steady and normal to the surface area of the inlet, and the friction heat was negligible in this process. Steady flow can also be assumed as instantaneous flow rate is approximately equal to the highest value during inhalation (during forced inhalation the instantaneous volumetric flow rate rapidly increases and reaches its highest value) according to Milenkovic et al. (2013). The gravitational acceleration was not included throughout the simulations as it was shown that the gravity has the insignificant impact due to small aerosol diameter sizes and short residence times in the inhaler itself (Milenkovic et al., 2013).

As far as the turbulence model is concerned, standard k-ω model with low Reynolds number (LRN) was used for airflow simulation. The k-ω turbulent model is a two-transport-equation model for turbulence prediction based on two partial difference equations for two variables (kinetic energy k and specific turbulent dissipation rate or turbulent frequency ω). This model has shown good accuracy when combined with high efficiency in comparison with complex methods, such as large eddy simulation (LES) (Longest and Holbrook, 2012). The LRN k-ω model can be used for accurate prediction of pressure drop, velocity profiles, and shear stress for transitional and turbulent flows (Ghalichi et al., 1998). All transport equations were discretized to be at least second-order accuracy in space (Shur et al., 2012). A second-order upwind scheme was used to discretize the pressure equations, and a quadratic upwind interpolation for Convection Kinetics scheme was used to discretize the momentum and turbulence equations.

Rosin-Rammler logarithmic diameter distribution of inert particles injected using face normal direction was used as input for Fluent's discrete phase model (DPM). The assumption was that the behavior of particles in contact with the wall is that they stay "trapped" and the particles that reach the outlet surface "escape" the device domain. This assumption is valid due to the fact that normal component of the particle velocity at the impact moment with the inhaler wall (υ_n) is smaller than the critical normal velocity throughout the simulation (υ_c) (Milenkovic et al., 2013) for the given flow rate. In that way, it could be calculated how many particles were present at the start of the calculation and how many particles escaped the device. The initial assumption was also that powder instantaneously, at the beginning, breaks up into a population of particles and aggregates identical to that of the free-flowing powder after which no more breakage or aggregation occurs. Particle size is considered constant throughout simulation. A more detailed explanation of the simulation boundary conditions Šušteršič et al. (2017a, b) and Vulović et al. (2017). Normally, particle cohesion forces affect the initial powder release and dispersion dynamics as well as aggregate deposition and breakage, which has been observed and explained in Milenkovic et al. (2014b). In order to consider collision breakage or particle-particle collision, additional tools such as discrete element method (DEM) are necessary. DPM method neglects interparticle collisions (Kloss et al., 2009) but was chosen in comparison with the DEM method because of the less CPU effort that DPM requires (Kloss et al., 2009). Due to the number of different simulations performed, it was not possible to use DEM, and this approach will be left for future work.

Boundary conditions for this simulation included velocity at the two inlets and pressure outlet (underpressure to simulate suction effect, which comes from the inhalation). Convergence criteria to ensure full convergence included applied normalized Reynolds stress residuals in the range of $10e^{-4}$. Inlet velocity was calculated based on the flow rate 28.3 L/min (Djokic et al., 2014) and inlet surface area. Particle mass at the beginning of simulation was 20 mg. Input materialistic characteristics for particles were also used based on Djokic et al. (2014) using jet-milled particles (JM1, JM2, JM3, and JM4) and spray-dried particles (SD).

3.4 Results and discussion

Results for the velocity magnitudes for the airflow show the distribution (Fig. 8.10) that is consistent with literature. Highest velocity magnitude can be seen in grid area and is 47.543 m/s, while the lower values are in the mouthpiece. These results can be compared with other published papers (Tong et al., 2012), which show around 45 m/s as the maximum velocity. Outlet velocity vectors (Fig. 8.10, upper right) are helpful in determining particle distribution at the beginning of simulations of particle deposition in the oral cavity.

It can be implied that the grid suppresses the turbulence in the barrel and also reduces swirl flow that is generated in the capsule chamber. Higher air inlet velocity leads to wider impact velocity and higher

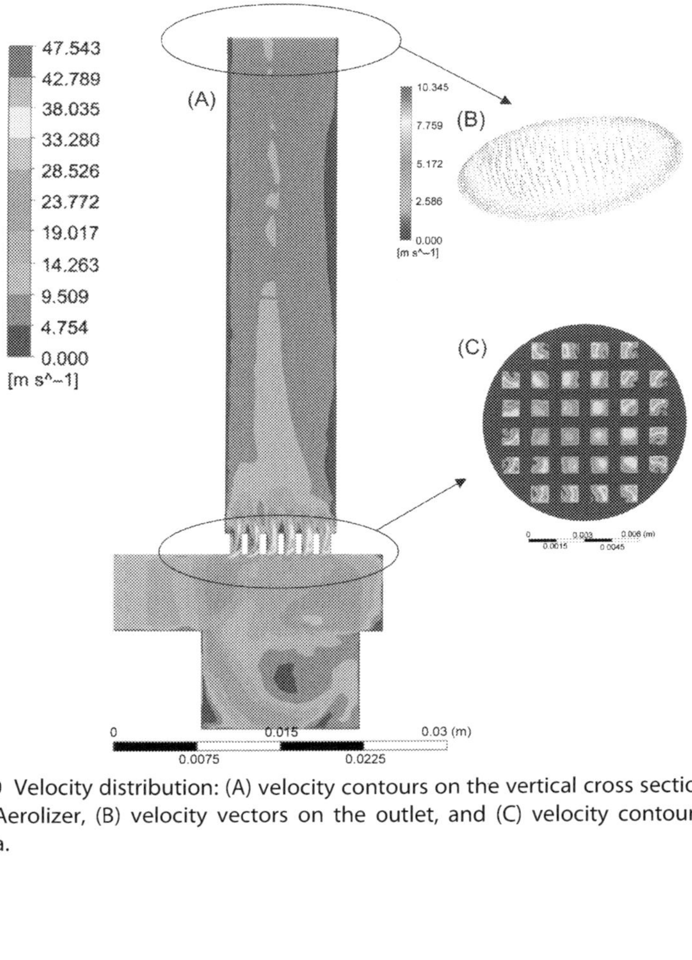

Fig. 8.10 Velocity distribution: (A) velocity contours on the vertical cross section of the inhaler Aerolizer, (B) velocity vectors on the outlet, and (C) velocity contours in the grid area.

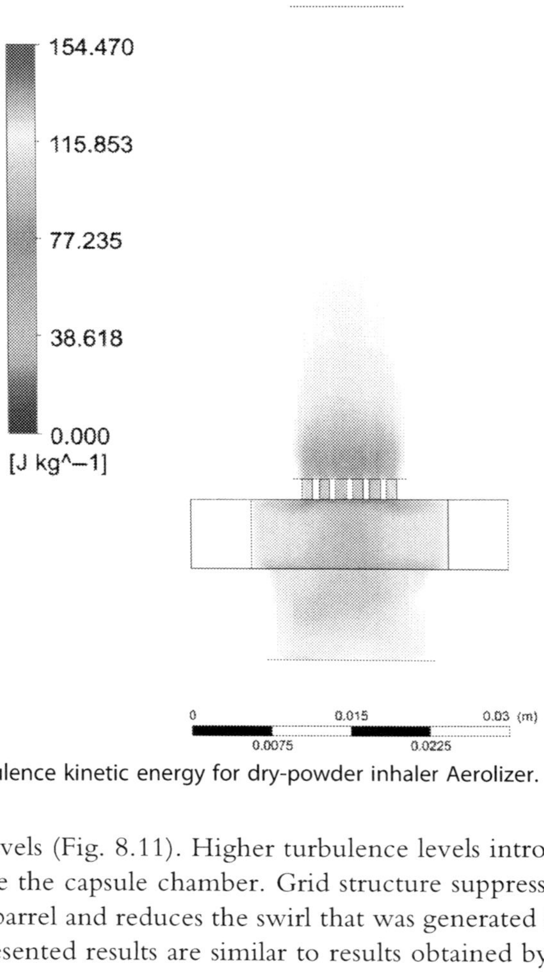

Fig. 8.11 Turbulence kinetic energy for dry-powder inhaler Aerolizer.

turbulence levels (Fig. 8.11). Higher turbulence levels introduce swirling motion inside the capsule chamber. Grid structure suppresses this turbulence in the barrel and reduces the swirl that was generated in the capsule chamber. Presented results are similar to results obtained by Coates et al. (2006).

Based on the results, the fluid flow inside the Aerolizer leads to particle collisions with the inhaler wall, which can be especially seen in the area of capsule and circulation chamber below grid structure. In general, particle-particle collisions generally have to be taken into account. Due to the small-volume fraction (10^{-4}) in this work, the effect of particle-particle collisions were not considered or taken into account, which is the case when the particle volume fraction is less than about 10^{-3} (Sommerfeld et al., 2008; Milenkovic et al., 2014a).

Particles mostly get "trapped" inside the carrier area because of the before mentioned swirl flow. This is due to the very short residency times

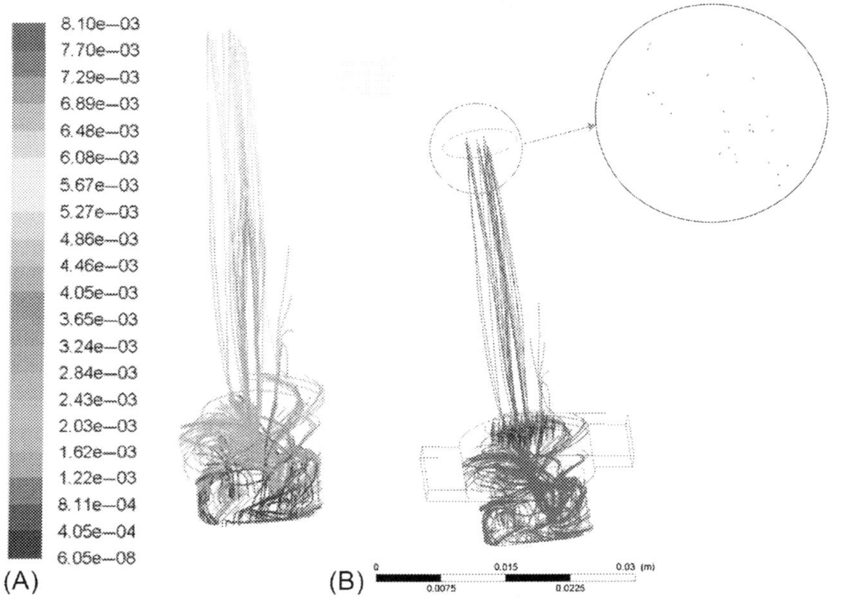

Fig. 8.12 Particle residency times (*left*) and particle tracks with separate zoom in view of the outlet (*right*).

in these areas (Fig. 8.12). These results are similar to the results obtained by Donovan et al. (2012).

Simulations of the dispersion of particles were repeated for different number of particles in order to determine if the particle impaction velocities obtained were independent of the number of particles simulated. Since the velocities proved independent of the number of particles, smaller number of particles was tracked, since the simulation time and necessary computer memory drastically increases with particle number.

Particle-device collisions and therefore deposition are more common than carrier-carrier collisions and mostly occur in the capsule and circulation chamber (Fig. 8.13).

Main reason that explains why carrier-carrier collisions rarely occur is not only the very small particle volume fraction (particle-particle collision rate depends on the second power of the local particle number density function) but also the turbulence levels in the mouthpiece. The flow in the mouthpiece is less turbulent than the flow in the chamber (tangential air that introduces a swirl component does not have such great influence upper from the grid structure), and particles enter the barrel at the different times, which reduces the chances for collision.

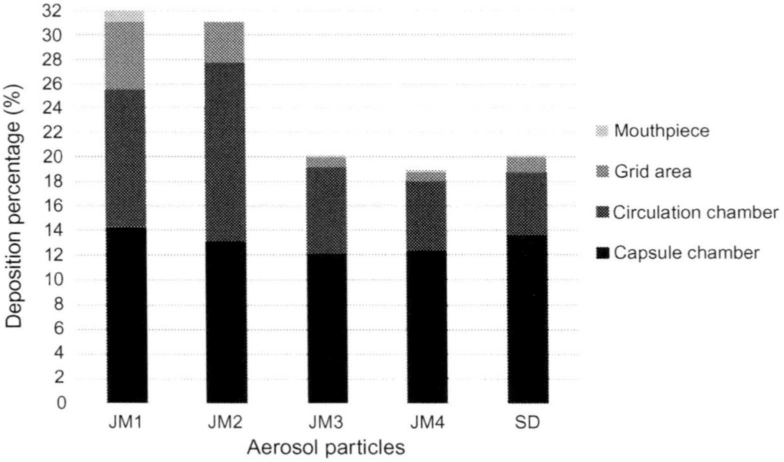

Fig. 8.13 Deposition sites with corresponding percentages for different particle types.

Finally, looking at the inhaler performance, emitted doses for each of the investigated aerosol types are shown in Fig. 8.14. The results are compared with the experimental results in Djokic et al. (2014). Best match between in silico and simulation results was obtained for JM1. Generally, efficiency of the inhaler was lowest with 69% for JM1. The results obtained are

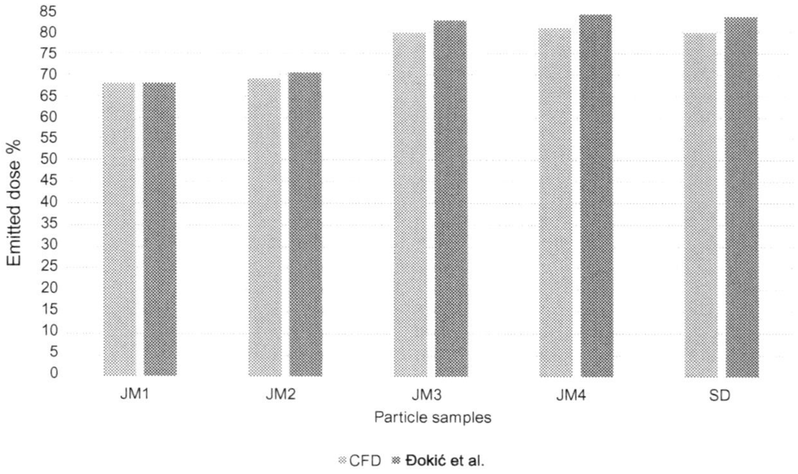

Fig. 8.14 Emitted doses of aerosol on the outlet (comparison between experimental and simulation results).

consistent with the results obtained by Djokic et al. (2014) where they showed that the emitted dose for JM1 particles was $68.15 \pm 1.58\%$.

Differences in other jet-milled samples were also not greater than 4%, whereas for spray-dried sample, the obtained difference was about 4.5%. These results demonstrate that the generated DPI inhaler model adequately describes aerosolization of five model formulations.

Further simulations included the comparison of the standard with modified inhalers (modifications were made in circulation chamber height) that showed that the velocity profiles change with modifications in circulation chamber height (Fig. 8.15).

Maximal velocities are generally smaller in cases of smaller height of the circulation chamber. As already mentioned, the highest velocity of air was 44.811 m/s for standard inhaler dimensions (tested with JM2 aerosol particles), while the highest velocity magnitude was for modification MD1 with 89.974 m/s and smallest for modification MD4 with 27.958 m/s.

Different heights of the circulation chambers also influence turbulence levels, resulting in different percentages in aforementioned regional deposition sites (Fig. 8.16). Total deposition for standard dimensions (tested for

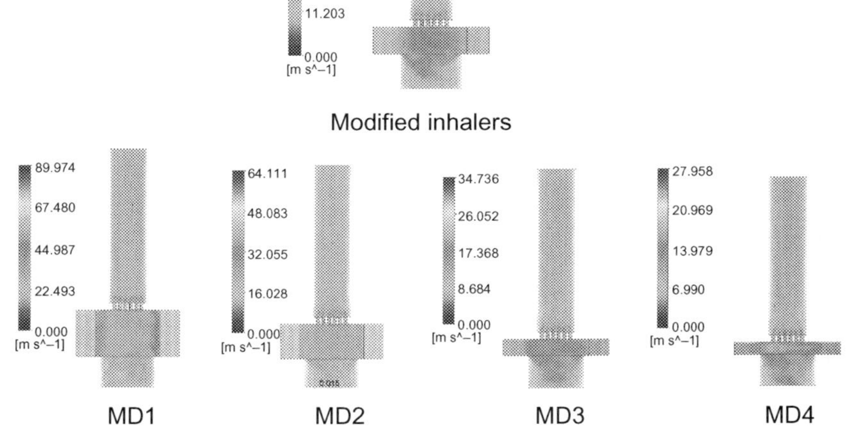

Fig. 8.15 Velocity volume rendering for standard and modified Aerolizer inhalers.

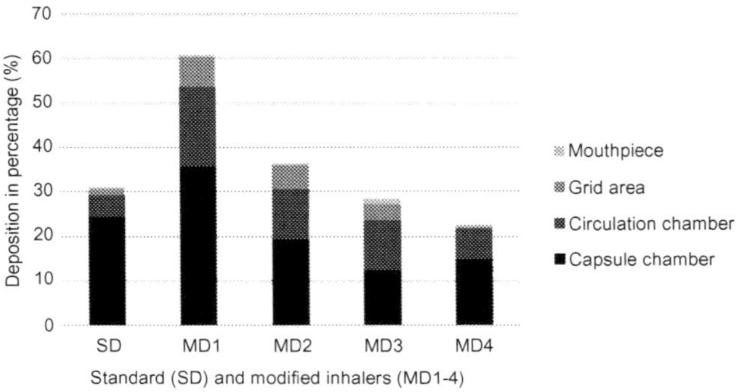

Fig. 8.16 Regional deposition sites for standard (SD) and modified inhaler (MD1–4).

JM2 aerosol particles) was around 31%, out of which capsule chamber retained the greatest percentage of particles—around 24% and circulation chamber in the second place—around 5%. In comparison with the efficiency in SD inhaler, performance of the modified inhalers MD1 and MD2 (which had an increase in circulation chamber height) decreased with total deposition around 61% and 37%, respectively. To explain in detail, most of the deposited particles were captured in capsule and circulation chamber. Looking at modified devices MD3 and MD4, total deposition was around 29% and 23%, respectively. It should be emphasized that for the modified inhaler MD3, deposition percentages in capsule and circulation chambers were of similar values (12% and 11%, respectively). Additional explanations of the results could be found in Šušteršič et al. (2017b).

The results show that computational modeling presented in this chapter could be used in further analysis of the dry-powder inhaler Aerolizer and as a first step in optimization of Aerolizer dimensions. The same analysis could be applied to other dry-powder inhalers in order to have a complete insight into deposition sites of the inhalers and possible approaches to their performance optimization.

4 Conclusions and recommendations

This chapter introduces computational modeling of dry-powder inhalers for pulmonary drug delivery. Firstly, short theoretical background is given in the form of introduction to the pulmonary drug delivery and dry-powder inhalers, their clinical efficiency, while special attention is paid to the dry-powder inhaler Aerolizer. Literature review of particle engineering strategies for pulmonary drug delivery is also analyzed after which practical part of this chapter is dedicated to the simulations performed on dry-powder inhaler Aerolizer.

The geometry is made based on detailed measurements, and during meshing, special attention was paid to the near wall and grid area based on initial solution. A part of this chapter is also dedicated to making the modifications in the geometry of a standard inhaler, specifically circulation chamber height, in order to analyze the effect of dimension change on inhaler performance. The particle dispersion in Aerolizer was simulated by a combined CFD-DPM model. The analysis of the flow field using CFD analysis has provided information about air behavior during the inhalation process. Based on this airflow field, by using DPM, we were able to determine how aerosol drug particles are progressing through the Aerolizer inhaler (from the capsule chamber until the end of the mouthpiece) and the number of particles that reach the outlet and enter patient's throat. Investigations on particle deposition in different modifications of standard inhaler showed that greater turbulence levels were present when circulation chamber height was greater, resulting in higher levels of deposition. Efficiency was the highest (around 77%) for the modification MD4 where the height of circulation chamber was two times smaller than in the original standard device.

The study presented in this chapter had several simplifications such as that aerosol particles with its final dimensions (diameter was of micrometer order of magnitude) were set at the bottom of the chamber area at the beginning of the simulation. This means that we did not account for the effect that capsule itself has on the deposition, as this would have to be done with dynamic meshing and would increase computational complexity. In that case, we would lose some of the benefits of the CFD and DPM simulations. Future work will include dynamic meshing to account for the capsule effect and new simulations of particle progressing inside patient's mouth, throat, and toward the lungs. As it was mentioned previously, there is also a possibility of applying the same analysis to different dry-powder inhalers in order to understand better how the particles are moving and where they are mostly depositing.

The results of the work presented in this chapter are important to understand the mechanisms in different DPIs, with the emphasis on Aerolizer inhaler, in order to help in different alterations and optimization of devices. Other new DPIs could be designed and tested using CFD to predict DPI outflow properties (e.g., air velocity profile and emitted particle properties) before constructing it. Ultimately, the results will contribute to understand better the connection between DPI design and operation parameters, as well as formulation properties to DPI outflow characteristics (e.g., emitted dose and FPF). All this is done in order to increase efficacy of the inhalation therapy.

Acknowledgments

This study is based upon work from COST Action MP1404 SimInhale "Simulation and pharmaceutical technologies for advanced patient-tailored inhaled medicines," supported by European Cooperation in Science and Technology (COST) www.cost.eu. In addition, this work was supported by the Ministry of Education, Science and Technological Development, Republic of Serbia with projects III41007 and OI174028.

References

Approaches to Safe Nanotechnology. 2016. Managing the Health and Safety Concerns Associated with Engineered Nanomaterials. Department of Health and Human Services Centers for Disease Control and Prevention National Institute for Occupational Safety and Health. Website: www.cdc.gov/niosh (Accessed 20 August 2016).

Bentham, A.C., Kwan, C.C., Boerefijn, R., Ghadiri, M., 2004. Fluidised-bed jet milling of pharmaceutical powders. Powder Technol. 141 (3), 233–238.

Boshhiha, A. 2011. The Influence of the Carrter Surface Characteristics on Dry Powder Inhalation (Doctoral dissertation), University of Dusseldorf.

Brocklebank, D., Ram, F., Wright, J., Barry, P., Cates, C., Davies, L., et al., 2001. Comparison of effectiveness of inhaler devices in asthma and chronic obstructive airway disease: a systematic review of the literature. Health Technol. Assess. 5 (26), 1–149.

Borgstrom, L., Bondesson, E., Moren, F., Trofast, E., Newman, S.P., 1994. Lung deposition of budesonide inhaled via Turbohaler: a comparison with terbutaline sulphate in normal subjects. Eur. Respir. J. 7, 69–73.

Calvert, G., Hassanpour, A., Ghadiri, M., 2011. Mechanistic analysis and computer simulation of the aerodynamic dispersion of loose aggregates. Chem. Eng. Res. Des. 89, 519–525.

Clark, A.R., 2004. Pulmonary delivery technology: recent advances and potential for the new millennium. In: Hickey, A.J. (Ed.), Drugs and the Pharmaceutical Sciences. Marcel Dekker Inc, New York, pp. 571–591.

Coates, M.S., Chan, H.K., Fletcher, D.F., Raper, J.A., 2005. Influence of air flow on the performance of a dry powder inhaler using computational and experimental analyses. Pharm. Res. 22 (9), 923–932.

Coates, M.S., Chan, H.K., Fletcher, D.F., Raper, J.A., 2006. Effect of design on the performance of a dry powder inhaler using computational fluid dynamics. Part 2: air inlet size. J. Pharm. Sci. 95 (6), 1382–1392.

Coates, M.S., Fletcher, D.F., Chan, H.K., Raper, J.A., 2004. Effect of design on the performance of a dry powder inhaler using computational fluid dynamics. Part 1: grid structure and mouthpiece length. J. Pharm. Sci. 93 (11), 2863–2876.

Concessio, N.M., VanOort, M.M., Knowles, M.R., Hickey, A.J., 1999. Pharmaceutical dry powder aerosols: correlation of powder properties with dose delivery and implications for pharmacodynamic effect. Pharm. Res. 16, 828–834.

Corradi, M., Chrystyn, H., Cosio, B.G., Pirozynski, M., Loukides, S., Louis, R., … Usmani, O.S., 2014. NEXThaler, an innovative dry powder inhaler delivering an extrafine fixed combination of beclometasone and formoterol to treat large and small airways in asthma. Expert Opin. Drug Deliv. 11 (9), 1497–1506.

Costantino, H.R., Firouzabadian, L., Wu, C., Carrasquillo, K.G., Griebenow, K., Zale, S.E., Tracy, M.A., 2002. Protein spray-freeze drying. 2. Effect of formulation variables on particle size and stability. J. Pharm. Sci. 91 (2), 388–395.

Crompton, G.K., 1982. Problems patients have using pressurized aerosol inhalers. Eur. J. Respir. Dis. Suppl. 119, 101–104.

Cui, Y., Schmalfuß, S., Zellnitz, S., Sommerfeld, M., Urbanetz, N., 2014. Towards the optimisation and adaptation of dry powder inhalers. Int. J. Pharm. 470 (1-2), 120–132.

Dailey, L.A., Schmehl, T., Gessler, T., Wittmar, M., Grimminger, F., Seeger, W., Kissel, T., 2003. Nebulization of biodegradable nanoparticles: impact of nebulizer technology and nanoparticle characteristics on aerosol features. J. Control. Release 86 (1), 131–144.

De Boer, A.H., Hagedoorn, P., Woolhouse, R., Wynn, E., 2012. Computational fluid dynamics (CFD) assisted performance evaluation of the Twincer™ disposable high-dose dry powder inhaler. J. Pharm. Pharmacol. 64 (9), 1316–1325.

Dean, H.R., 2008. Aerosol delivery devices in the treatment of asthma. Respir. Care 53 (6), 699–723.

Demoly, P., Hagedoorn, P., de Boer, A.H., Frijlink, H.W., 2014. The clinical relevance of dry powder inhaler performance for drug delivery. Respir. Med. 108 (8), 1195–1203.

Djokic, M., Kachrimanis, K., Solomun, L., Djuris, J., Vasiljevic, D., Ibric, S., 2014. A study of jet-milling and spray-drying process for the physicochemical and aerodynamic dispersion properties of amiloride HCl. Powder Technol. 262, 170–176.

Dolovich, M., 1999. New propellant-free technologies under investigation. J Aerosol Med. 12 (1), 9–17.

Donovan, M.J., Kim, S.H., Raman, V., Smyth, H.D., 2012. Dry powder inhaler device influence on carrier particle performance. J. Pharm. Sci. 101 (3), 1097–1107.

Ferouka, I., Sustersic, T., Zivanovic, M., Filipovic, N., 2018. Mathematical modelling of polymer trajectory during electrospinning. J. Serbian Soc. Comput. Mech. 12 (2), 17–38.

Filipovic, N., Gibney, B.C., Nikolic, D., Konerding, M.A., Mentzer, S.J., Tsuda, A., 2014. Computational analysis of lung deformation after murine pneumonectomy. Comp. Methods Biomech. Biomed. Eng. 17 (8), 838–844.

Gardenhire, D.S., 2013. A Guide to Aerosol Delivery Devices for Respiratory Therapists, third ed. American Association for Respiratory Care.

Ghalichi, F., Deng, X., Champlain, A.D., Douville, Y., King, M., Guidoin, R., 1998. Low Reynolds number turbulence modelling of blood flow in arterial stenoses. Biorheology 35, 281–294.

Gonda, I., 1992. Targeting by deposition. In: Pharmaceutical Inhalation Aerosol Technology. pp. 61–82.

Hoe, S., Traini, D., Chan, H.K., Young, P.M., 2009. Measuring charge and mass distributions in dry powder inhalers using the electrical Next Generation Impactor (eNGI). Eur. J. Pharm. Sci. 38, 88–94.

Hu, J., Johnston, K.P., Williams, R.O., 3rd., 2003. Spray freezing into liquid (SFL) particle engineering technology to enhance dissolution of poorly water-soluble drugs: organic solvent versus organic/aqueous co-solvent systems. Eur. J. Pharm. Sci. 20 (3), 295–303.

Islam, N., Gladki, E., 2008. Dry powder inhalers (DPIs)—a review of device reliability and innovation. Int. J. Pharm. 360 (1–2), 1–11.

Jackson, W.F., 1995. Inhalers in Asthma. The New Perspective. Clinical Vision Ltd, Harwell, Oxfordshire, pp. 1–56.

Kloss, C., Goniva, C., Aichinger, G., Pirker, S., 2009. Comprehensive DEM-DPM-CFD simulations—model synthesis, experimental validation and scalability. In: Seventh International Conference on CFD in the Minerals and Process Industries, CSIRO, Melbourne, Australia, December 9–11.

de Koning, J.P., Visser, M.R., Oelen, G.A., de Boer, A.H., van der Mark, T.W., Coenegracht, P.M.J., et al., 2001. Effect of peak inspiratory flow and flow increase rate on in vitro drug deposition from four dry powder inhaler devices. In: Dry Powder Inhalation. Technical and Physiological Aspects, Prescribing and Use, pp. 83–94. Thesis, Rijksuniversiteit Groningen, Ch. 6.

Koullapis, P.G., Hofemeier, P., Sznitman, J., Kassinos, S.C., 2018a. An efficient computational fluid-particle dynamics method to predict deposition in a simplified approximation of the deep lung. Eur. J. Pharm. Sci. 113, 132–144.

Koullapis, P., Kassinos, S.C., Muela, J., Perez-Segarra, C., Rigola, J., Lehmkuhl, O., Cui, Y., Sommerfeld, M., Elcner, J., Jicha, M., Saveljic, I., Filipovic, N., Lizal, F., Nicolaou, L., 2018b. Regional aerosol deposition in the human airways: the SimInhale benchmark case and a critical assessment of in silico methods. Eur. J. Pharm. Sci. 113, 77–94.

Kroeger, R., Becker, M., Wynn, E., 2010. Developments in techniques for simulation of particles in and from inhalers using CFD. Respir. Drug Deliv. 2009 (2), 231–234.

Leach, W.T., Simpson, D.T., Val, T.N., Anuta, E.C., Yu, Z., Williams, R.O., 3rd., 2005. Uniform encapsulation of stable protein nanoparticles produced by spray freezing for the reduction of burst release. J. Pharm. Sci. 94 (1), 56–69.

Longest, P.W., Holbrook, L.T., 2012. In silico models of aerosol delivery to the respiratory tract—development and applications. Adv. Drug Deliv. Rev. 64 (4), 296–311.

Milenkovic, J., 2015. Airflow and Particle Deposition in a Dry Powder Inhaler: A CFD and Particle Computational Approach (Doctoral dissertation, Αριστοτέλειο Πανεπιστήμιο Θεσσαλονίκης (ΑΠΘ). Σχολή Πολυτεχνική. Τμήμα Χημικών Μηχανικών. Τομέας Ανάλυσης, Σχεδιασμού και Ρύθμισης Χημικών Διεργασιών και Εγκαταστάσεων).

Milenkovic, J., Alexopoulos, A.H., Kiparissides, C., 2013. Flow and particle deposition in the Turbuhaler: a CFD simulation. Int. J. Pharm. 448 (1), 205–213.

Milenkovic, J., Alexopoulos, A.H., Kiparissides, C., 2014a. Airflow and particle deposition in a dry powder inhaler: an integrated CFD approach. In: Simulation and Modeling Methodologies, Technologies and Applications. Springer International Publishing, pp. 127–140.

Milenkovic, J., Alexopoulos, A.H., Kiparissides, C., 2014b. Deposition and fine particle production during dynamic flow in a dry powder inhaler: a CFD approach. Int. J. Pharm. 461 (1–2), 129–136.

Mossaad, D.M.R., 2014. Drug Delivery to the Respiratory Tract Using Dry Powder Inhalers. Electronic Thesis and Dissertation Repository. Paper 2036.

Newhouse, M.T., Corkery, K.J., 2001. Aerosols for systemic delivery of macromolecules. Respir. Care Clin. N. Am. 7 (2), 261–275.

Newman, S.P., Moren, F., Trofast, E., Talaee, N., Clarke, S.W., 1989. Deposition and clinical efficacy of terbutaline sulphate from Turbuhaler. A new multi-dose inhaler. Eur. Respir. J. 2, 247–252.

Niven, R.W., 2002. Powders and Processing: Deagglomerating a Dose of Patents and Publications. Davis Horwood International Publishing, pp. 257–266.

Patton, J.S., Fishburn, C.S., Weers, J.G., 2004. The lungs as a portal of entry for systemic drug delivery. Proc. Am. Thorac. Soc. 1 (4), 338–344.

Pedersen, S., 1996. Inhalers and nebulizers: which to choose and why. Resp. Med. 90, 69–77.

Riley, T., Christopher, D., Arp, J., Casazza, A., Colombani, A., Cooper, A., … Sigari, N., 2012. Challenges with developing in vitro dissolution tests for orally inhaled products (OIPs). AAPS PharmSciTech. 13 (3), 978–989.

Roberts, R.J., Rowe, R.C., York, P., 1994. The relationship between indentation hardness of organic solids and their molecular structure. J. Mater. Sci. 29 (9), 2289–2296.

Ryan, G., Singh, M., Dwan, K., 2011. Inhaled antibiotics for long-term therapy in cystic fibrosis. Cochrane Database Syst. Rev. 3:Cd001021.

Shur, J., Lee, S., Adams, W., Lionberger, R., Tibbatts, J., Price, R., 2012. Effect of device design on the in vitro performance and comparability for capsule-based dry powder inhalers. AAPS J. 14 (4).
SimInhale COST Action MP1404 Website, 2016. URL: http://www.siminhale-cost.eu/ (Accessed 15 December 2018).
Sommerfeld, M., van Wachem, B., Oliemans, R., 2008. Best Practice Guidelines for Computational Fluid Dynamics of Dispersed Multiphase Flows. ERCOFTAC European Research Community on Flow, Turbulence and Combustion.
Šušteršič, T., Liverani, L., Boccaccini, A.R., Savić, S., Janićijević, A., Filipović, N., 2018. Numerical simulation of electrospinning process in commercial and in-house software PAK. Mater. Res. Exp. 6 (2), 025305.
Šušteršič, T., Vulović, A., Cvijić, S., Ibrić, S., Filipović, N., 2017a. Aerosol particle deposition in dry powder inhaler Aerolizer®. J. IPSI BgD Trans. Adv. Res. 13 (2), 42–47. ISSN 1820–4511.
Šušteršič, T., Vulović, A., Cvijić, S., Ibrić, S., Filipović, N., 2017b. Effect of circulation chamber dimensions on aerosol delivery efficiency of a commercial dry powder inhaler Aerolizer®. In: 2017 IEEE 17th International Conference on Bioinformatics and Bioengineering (BIBE). IEEE, pp. 555–558.
Taylor, K.M.G., Pancholi, K., Wong, D.Y.T., 1999. In-vitro evaluation of dry powder inhaler, formulations of micronized and milled nedocromil sodium. Pharm. Pharmacol. Commun. 5 (4), 255–257.
Terzano, C., 2008. Dry powder inhaler and the risk of error. Respiration 75, 14–15.
Todo, H., Okamoto, H., Iida, K., Danjo, K., 2004. Improvement of stability and absorbability of dry insulin powder for inhalation by powder-combination technique. Int. J. Pharm. 271 (1–2), 41–52.
Tong, Z.B., Yang, R.Y., Chu, K.W., Yu, A.B., Adi, S., Chan, H.-K., 2010. Numerical study of the effects of particle size and polydispersity on the agglomerate dispersion in a cyclonic flow. Chem. Eng. J. 164, 432–441.
Tong, Z.B., Zheng, B., Yang, R.Y., Yu, A.B., Chan, H.K., 2012. CFD-DEM investigation of the dispersion mechanisms in commercial dry powder inhalers. Powder Technol. 240, 19–24.
Tsuda, A., Filipovic, N., Haberthur, D., Dickie, R., Matsui, Y., Stampanoni, M., Schittny, J.C., 2008. Finite element 3D reconstruction of the pulmonary acinus imaged by synchrotron X-ray tomography. J. Appl. Physiol. 105, 964–976.
Velaga, S.P., Djuris, J., Cvijic, S., Rozou, S., Russo, P., Colombo, G., Rossi, A., 2018. Dry powder inhalers: an overview of the in vitro dissolution methodologies and their correlation with the biopharmaceutical aspects of the drug products. Eur. J. Pharm. Sci. 113, 18–28.
Vulović, A., Šušteršič, T., Cvijić, S., Ibrić, S., Filipović, N., 2017. Coupled in silico platform: computational fluid dynamics (CFD) and physiologically-based pharmacokinetic (PBPK) modelling. Eur. J. Pharm. Sci. 113, 171–184.
Wong, W., Adi, H., Traini, D., Chan, H.-K., Fletcher, D.F., Crapper, J., Young, P.M., 2010b. Use of rapid prototyping and CFD in the design of DPI devices. Respir. Drug Deliv. 3, 879–882.
Wong, W., Fletcher, D.F., Traini, D., Chan, H.-K., Crapper, J., Young, P.M., 2010a. Particle aerosolisation and break-up in dry powder inhalers 1: evaluation and modelling of Venturi effects for agglomerated systems. Pharm. Res. 27, 1367–1376.
Wong, W., Fletcher, D.F., Traini, D., Chan, H.K., Young, P.M., 2012. The use of computational approaches in inhaler development. Adv. Drug Deliv. Rev. 64, 312–322.
Yang, J., Wu, C.Y., Adams, M., 2014. A three-dimensional DEM–CFD analysis of air-flow-induced detachment of api particles from carrier particles in dry powder inhalers. Acta Pharm. Sin. B 4, 52–59.

Yu, Z., Rogers, T.L., Hu, J., Johnston, K.P., Williams, R.O., 3rd., 2002. Preparation and characterization of microparticles containing peptide produced by a novel process: spray freezing into liquid. Eur. J. Pharm. Biopharm. 54 (2), 221–228.
Zhou, Q.T., Tong, Z., Tang, P., Citterio, M., Yang, R., Chan, H.K., 2013. Effect of device design on the aerosolization of a carrier-based dry powder inhaler—a case study on Aerolizer® Foradile®. AAPS J. 15 (2), 511–522.
Zijlstra, G.S., Hinrichs, W.L., de Boer, A.H., Frijlink, H.W., 2004. The role of particle engineering in relation to formulation and de-agglomeration principle in the development of a dry powder formulation for inhalation of cetrorelix. Eur. J. Pharm. Sci. 23 (2), 139–149.

CHAPTER 9

Computer modeling of cochlear mechanics

Contents

1 Introduction

The human hearing system is very complex, as many other organs or senses in the human body. There are several different physical processes that take place in the hearing system (Pickles, 1988). Sound formed somewhere in the environment travels through the air as a sound wave. Such a wave can come to us and enter into our ear channel. The mechanism that sound enters to reach the central nervous system consists of several different processes. First process is the impact of sound wave to the ear drum (tympanic membrane). Second process is mechanical vibrations of the ossicles in the middle ear, caused by vibrations of ear drum. Third process is traveling wave in cochlear

Computational Modeling in Bioengineering and Bioinformatics
https://doi.org/10.1016/B978-0-12-819583-3.00009-6

chambers. Fourth process is mechanotransduction in hair cells placed in the organ of Corti, between basilar membrane and tectorial membrane. Fifth process is transmitting of electrical signal produced in the organ of Corti by vestibulocochlear nerve to the brain. Each of the aforementioned processes has its own complexity. Modeling of the whole process would be very hard, but some parts of hearing system and interaction between them can be modeled. For example, processes in the middle and inner ear can be modeled as mechanical or mechanical-acoustical mechanisms that can vibrate under the influence of an external acoustical or even mechanical excitation.

In this chapter, we present a way how the human hearing system can be modeled. The hearing system consists of three major parts: the outer ear, middle ear, and inner ear. In terms of modeling, two parts are significant: the middle ear and inner ear. The role of the outer ear is mainly to direct the sound wave to the elements of the middle ear. The middle ear is a part with only mechanical elements such as small bones (ossicles), which are the incus, malleus, and stapes; joints; and ligaments. The inner ear is more complex part in which mechanical vibrations of solid elements cause sound wave in the fluid. Hereafter, the fluid, driven by mechanical vibrations, acts on other structures of the inner ear, like the basilar membrane, tectorial membrane, and organ of Corti. The moving of those structures generates electrical signal that is sent to the brain by an auditory nerve.

All described processes can be modeled separately. Our objective is to develop algorithm for modeling of solid and fluid elements of the middle and inner ear and to couple them to make a full mechanical model of the middle and inner ear. Solid domain of ear structures can be modeled using standard Newtonian dynamic equation, derived from basic equilibrium equations for solid and also principle of virtual work. Fluid domain can be modeled in two different ways: The first approach is modeling of fluid as Newtonian fluid with Navier-Stokes partial differential equations and continuity equation. The second approach is modeling of fluid using acoustic wave equation. Coupling of those two different fields is achieved through the boundary surface, by transposing boundary conditions from one field to another and vice versa. In the second approach, solid is modeled the same as previous, but fluid is modeled by using acoustic wave equation. The coupling of those equations is obtained by the equalization of the most dominant forces on boundary surfaces: pressure gradient from fluid and normal acceleration from the solid.

The main difference between those two approaches is the order of solution of different physical fields. In the first approach, solid domain and fluid domain are solved separately: solid with boundary conditions obtained from fluid domain and fluid with boundary conditions obtained from solid domain. Both approaches have advantages and drawbacks. This will be discussed later, in this chapter.

Basic anatomy of the human ear: The outer, middle, and inner ear

The human hearing system, or sense of hearing, is a paired organ. That means there are two organs on each side of the head. The ear consists of three parts: the outer ear, middle ear, and inner ear. Each of those parts has specific role in transmitting sounds from the environment to the human brain.

In Fig. 9.1 is given general overview of parts or organs that take place in the human ear.

Looking at the Fig. 9.1, three main parts of hearing system can be noticed: the outer ear, middle ear, and inner ear. Each of these parts consists of several smaller parts or, more precisely, organs.

Going from the exterior to the interior, first part is the outer ear. It consists of earlobe (Latin: pinna) and external auditory canal. The role of earlobe is to collect sounds from the environment and to direct them to the external auditory canal. By this canal, sound from environment can come to the next part of the hearing system—the middle ear. Hence, the process that happens in the outer ear is only the motion of sound waves.

Second part of the hearing system is the middle ear. The border between the outer ear and middle ear is the tympanic membrane, which lies at the end of the external auditory canal. The tympanic membrane is a thin elastic

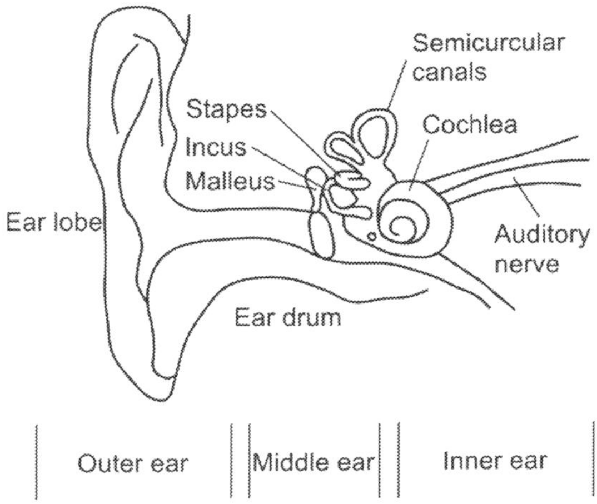

Fig. 9.1 Anatomy of the human hearing system: The outer ear, middle ear, and inner ear.

structure that can vibrate under the influence of sound waves. Those mechanical vibrations represent the beginning of the process that happens in the middle ear.

Third part of the hearing system is the inner ear. The inner ear is responsible for the sound detection and transmission by the peripheral nervous system to the brain and for the balance. The part responsible for sound detection is cochlea.

The cochlea is a small snail shell-like cavity in the temporal bone, which has two openings, the oval window and the round window. The cavity consists of two fluid chambers, the scala vestibuli and scala tympani. Those chambers, filled with fluid, are sealed by two elastic membranes that cover the windows (oval and round). External sounds set the ear drum in motion, which is conveyed to the inner ear by the ossicles—three small bones of the middle ear: the malleus, incus, and stapes. The pistonlike motion of the stapes against the oval window displaces the fluid of the cochlea, so generating traveling waves that propagate along the basilar membrane. Macromechanical system of the cochlea can be described by the Navier-Stokes equations of incompressible fluid mechanics or acoustic wave equation, coupled with equations modeling the elastic properties of the basilar membrane and the membranes of the oval and the round windows. The displacements of the basilar membrane are extremely small (on a nanometer scale), which means that the system works in a linear regime.

The cochlea model can be presented for simplicity as a single membrane with properties similar to that of the basilar membrane and two fluid chambers on both sides of membrane. Those chambers are coupled at the end of basilar membrane that is called helicotrema. The properties of the basilar membrane vary along its length, but across its length, it is assumed that they are very much similar. The cochlea is spiral-shaped but for modeling purposes can be approximated as straight (uncoiled) structure. The length of uncoiled human cochlea is about 30–35 mm (Fig. 9.2).

Between two fluid chambers, there are several small organs (Fig. 9.3). There are Reissner's membrane, tectorial membrane, organ of Corti with hair cells, auditory nerve, spiral ligament, cochlear duct, etc.

The mechanism of transforming sound wave traveling through cochlear chambers is the following: Sound waves traveling through the air reach the tympanic membrane in middle ear. Three ossicles in the middle ear transmit vibrations of the tympanic membrane to the oval window. The motion of oval window causes traveling wave in fluid placed in cochlear chambers. That traveling wave causes oscillatory motion of basilar membrane. Along with

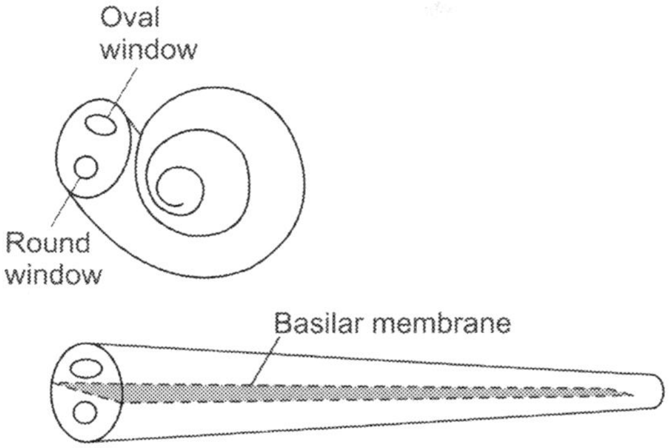

Fig. 9.2 Cochlea—real shape and uncoiled approximation.

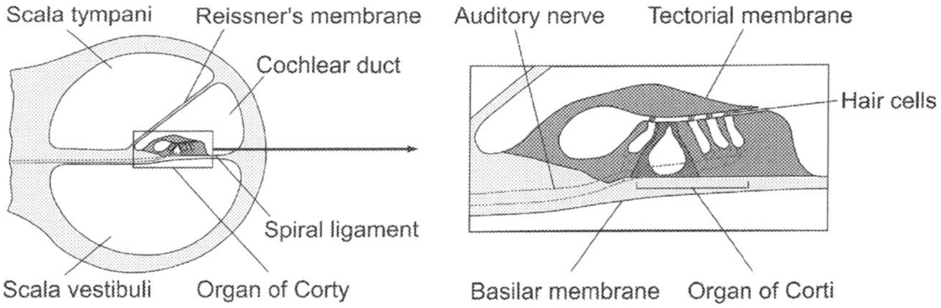

Fig. 9.3 Cochlea cross section.

basilar membrane, it starts to oscillate the organ of Corti and make shear motion relative to the tectorial membrane. That motion generates impulse in hair cells that is transferred through the auditory nerve to the brain.

Mechanical model of human hearing system, without the process of converting mechanical motion in electrical signal, will be presented in this chapter. The model will include elements of middle ear and inner ear.

2 Concepts of modeling

There are many situations in engineering and also in bioengineering where the fluid and solid interact to each other. One simple example in engineering is fluid flow through the pipe. Second example is ship floating on water. Further, bullet fired from some weapon flying through the air, tennis, golf, or soccer ball flying through the air. In all mentioned cases, fluid acts with its pressure (and shear stress also) to the walls of pipe. Also, there is an effect in opposite direction. Solid acts on the fluid by changing the position of boundary nodes that lie on the fluid side.

There are similar processes in human or animal bodies also. Let's give some examples: blood flow through vascular system, air flow through the respiratory system, interaction of blood cells and other cells with blood plasma, urine flow in the kidneys, filling or emptying of urinary bladder, and movement of the contents of the stomach.

Very similar processes appear in the outer and inner ear. In the outer ear, acoustic wave (which can be modeled in two ways) travels through the external auditory canal and interacts with the tympanic membrane. Also, in the inner ear, acoustic wave travels through the cochlear chambers. Another example of fluid-solid interaction is motion of fluid in semicircular channels, the part of the human's vestibular system, responsible for balance.

To model some process with fluid-solid interaction, we need appropriate coupling algorithm. As mentioned in the introduction, there are two different approaches in modeling mechanical processes that appear as a consequence of interaction of fluid and solid. Both of them use similar differential equations, but they are coupled in different ways.

The first concept is the so-called loose coupling algorithm, which implies independent solving of solid and fluid domains and transfer of boundary conditions between different domains. That concept is the most accepted in computational modeling of solid-fluid interaction problems. One of the advantages of that concept is a separated solving of different physical fields. This allows the system of fluid equations and the system of solid equations to be solved successively, with the transfer of boundary conditions on fluid-solid boundary: from fluid to solid and vice versa. In this case, the nodes lying on the boundary between fluid and solid domains must be placed at the same place, regardless of whether they are related to the fluid or to the solid. Geometrical position must be the same and has to be updated for each time step.

The second concept is the so-called strong coupling algorithm. In this concept, different equations that are used for modeling of fluid and solid domains are coupled and solved in the same system of equations. The coupling is achieved by using of some specific boundary conditions derived from equality of forces generated by fluid and solid. That boundary condition provides connection between different equations, corresponding to appropriate field, that have to be solved in nodes on the fluid-solid boundary. Nodes on the boundary are common for solid and fluid domains and have degrees of freedom from both domains. In other words, it means that we don't have two different nodes at the same geometrical position, as is the case with the previously described loose coupling concept, but only one node.

More details about concrete partial differential equations used for different models and specific boundary conditions applied on those equations that allow their coupling will be given later in this chapter.

3 Solid model

Equations of solid deformation and motion are starting from principle of virtual work. Starting from the equilibrium equations and using the boundary conditions, we can derive principle of virtual work:

$$\delta W_{int} = \delta W_{ext} \tag{9.1}$$

The virtual work of internal forces on virtual deformations and virtual work of external forces on virtual displacements are equal (Kojic et al., 2008).

The virtual works of internal and external forces are as follows:

$$\delta W_{int} = \int_V \sigma_{ij} \delta e_{ij} dV = \int_V \delta \boldsymbol{e}^T \boldsymbol{\sigma} dV \tag{9.2}$$

and

$$\delta W_{ext} = \int_V F_k^V \delta u_k dV + \int_S f_k^S \delta u_k^S dS + \sum_i F_k^{(i)} \delta u_k^{(i)} \quad \text{sum on } k, k = 1,2,3 \tag{9.3}$$

Here,

- F_k^V and δu_k are the volumetric forces and virtual displacements,
- f_k^S and δu_k^S are the distributed surface forces and virtual displacements at the surface,
- $F_k^{(i)}$ and $\delta u_k^{(i)}$ are the concentrated forces and virtual displacements at some point.

By discretization of Eq. (9.1), with taking into account the inertial forces,

$$dF^{in} = -\ddot{\boldsymbol{u}} dm = -\rho \ddot{\boldsymbol{u}} dV \tag{9.4}$$

where $\ddot{\boldsymbol{u}} = \frac{d^2 y}{dt^2}$ is acceleration, ρ is material density, and dV is elementary volume, we can derive basic dynamic equation:

$$\boldsymbol{M}\ddot{\boldsymbol{u}} + \boldsymbol{C}\dot{\boldsymbol{u}} + \boldsymbol{K}\boldsymbol{u} = \boldsymbol{F}^{ext} \tag{9.5}$$

where

- $\boldsymbol{M} = \int_V \rho \mathbf{N}^T \mathbf{N} dV$ is mass matrix,
- $\boldsymbol{C} = \int_V c \mathbf{N}^T \mathbf{N} dV$ is damping matrix,

$\boldsymbol{K} = \int_V \boldsymbol{B}^T \boldsymbol{C} \boldsymbol{B} dV$ is stiffness matrix,
$\boldsymbol{F}^{ext} = \int_V b \mathbf{N}^T \mathbf{N} dV$ is damping matrix.
In previous equations, c is damping coefficient, $\boldsymbol{N}$ is matrix of interpolation functions, $\boldsymbol{B}$ is strain-displacement relation matrix, and $\boldsymbol{F}^{ext}$ is external force vector that includes external concentrated, surface and body forces.

The details about previous equations are given in literature (Kojic et al., 2008).

4 Fluid model

There are two different approaches in modeling of fluid. The first approach is modeling by using Navier-Stokes equations, and the second one is by using acoustic wave equation.

The three-dimensional flow of a viscous incompressible fluid considered here is governed by the Navier-Stokes equations and continuity equation that can be written as follows (Filipovic et al., 2012; Kojic et al., 2015):

$$\rho\left(\frac{\partial v_i}{\partial t} + v_j \frac{\partial v_i}{\partial x_j}\right) = -\frac{\partial p}{\partial x_i} + \mu\left(\frac{\partial^2 v_i}{\partial x_j \partial x_j} + \frac{\partial^2 v_j}{\partial x_j \partial x_i}\right) \tag{9.6}$$

$$\frac{\partial v_i}{\partial x_i} = 0 \tag{9.7}$$

where $\boldsymbol{v_i}$ is the blood velocity in direction $\boldsymbol{x_i}$, $\boldsymbol{\rho}$ is the fluid density, p is pressure, and μ is the dynamic viscosity, and summation is assumed on the repeated indices, $i, j = 1, 2, 3$. The Eq. (9.6) represents balance of linear momentum, while Eq. (9.7) expresses incompressibility condition.

The incremental-iterative form of the equations for a time step and equilibrium iteration "i" is

$$\begin{bmatrix} \frac{1}{\Delta t}\boldsymbol{M}_v + {}^{t+\Delta t}\boldsymbol{K}_{vv}^{(i-1)} + {}^{t+\Delta t}\boldsymbol{K}_{\mu v}^{(i-1)} + {}^{t+\Delta t}\boldsymbol{J}_{vv}^{(i-1)} & \boldsymbol{K}_{vp} \\ \boldsymbol{K}_{vp}^T & 0 \end{bmatrix} \begin{Bmatrix} \Delta \boldsymbol{v}^{(i)} \\ \Delta \boldsymbol{p}^{(i)} \end{Bmatrix} = \begin{Bmatrix} {}^{t+\Delta t}\boldsymbol{F}_v^{(i-1)} \\ {}^{t+\Delta t}\boldsymbol{F}_p^{(i-1)} \end{Bmatrix} \tag{9.8}$$

The left upper index "$t+\Delta t$" denotes that the quantities are evaluated at the end of time step. The matrix $\boldsymbol{M}_v$ is mass matrix, $\boldsymbol{K}_{vv}$ and $\boldsymbol{J}_{vv}$ are convective matrices, $\boldsymbol{K}_{\mu v}$ is the viscous matrix, $\boldsymbol{K}_{vp}$ is the pressure matrix, and $\boldsymbol{F}_v$ and $\boldsymbol{F}_p$ are vectors of forces. The pressure is eliminated at the element level through the static condensation.

For the penalty formulation, we define the incompressibility constraint in the following manner:

$$\operatorname{div} \boldsymbol{v} + \frac{p}{\lambda} = 0 \tag{9.9}$$

where λ is a relatively large positive scalar so that $\frac{p}{\lambda}$ is a small number (practically zero).

Using previous equation, the incremental-iterative equilibrium equations can be written in following form:

$$\left(\frac{1}{\Delta t}\boldsymbol{M}_v + {}^{t+\Delta t}\boldsymbol{K}_{vv}^{(i-1)} + {}^{t+\Delta t}\boldsymbol{K}_{\mu v}^{(i-1)} + {}^{t+\Delta t}\boldsymbol{J}_{vv}^{(i-1)} + \boldsymbol{K}_{\lambda v}\right)\Delta \boldsymbol{v}^{(i)} = {}^{t+\Delta t}\boldsymbol{F}_v^{(i-1)} \tag{9.10}$$

The matrices and vectors from Eq. (9.10) are given in literature (Kojic et al., 2008; Filipovic et al., 2012; Kojic et al., 2015).

The second approach that can be used for modeling of fluid is using of acoustic wave equation. Acoustic wave equation is defined as

$$\frac{\partial^2 p}{\partial x_i^2} - \frac{1}{c^2}\frac{\partial^2 p}{\partial t^2} = 0 \tag{9.11}$$

where p stands for fluid pressure inside the chambers, x_i are spatial coordinates in Cartesian coordinate system, c is the speed of sound, and t is time.

Matrix form of the acoustic wave equation, obtained by using Galerkin method (Kojic et al., 2008), can be presented in the following formulation:

$$\boldsymbol{Q}\ddot{\boldsymbol{p}} + \boldsymbol{H}\boldsymbol{p} = 0 \tag{9.12}$$

where $\boldsymbol{Q}$ is the acoustic inertia matrix and $\boldsymbol{H}$ represents the acoustic "stiffness" matrix.

5 Loose coupling algorithm

Here, basic algorithmic scheme that we use for coupling different physical fields, in terms of computational modeling, will be presented. There is the main control program that calls different solvers (CFD solver and solid mechanics solver) and manages with boundary conditions that should be exchanged between different domains, that is, between different systems of equations. This concept is shown in Fig. 9.4.

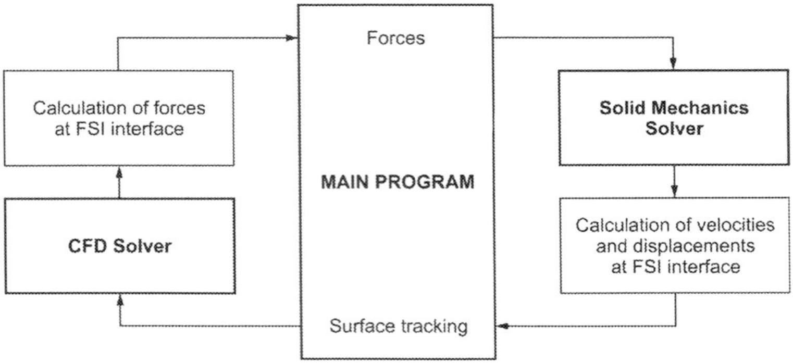

Fig. 9.4 Fluid-solid interaction: loose coupling scheme.

The algorithm consists of the following steps:

1. Solving of system of equations with prescribed boundary conditions (velocities of surface traction) for fluid domain
2. Calculation of fluid forces on the interface surface between fluid and solid, based on obtained solution for velocities and pressures from fluid domain
3. Solving of solid domain with boundary conditions obtained in fluid domain. It means the forces calculated on the fluid side of fluid-solid interface should be transferred to the system of equations for solid domain
4. Updating of geometry and velocities of the fluid domain, based on the calculated solid displacement
5. Returning to the first step

Such a set of steps are repeated over time steps. For each step, there is a verification of the convergence of the solution:

$$\left|\Delta u_{solid}^{(i)}\right| \leq \varepsilon_{disp}$$

$$\left|\Delta v_{velocity}^{(i)}\right| \leq \varepsilon_{velocity}$$

If the convergence criteria is not satisfied (in any domain, fluid or solid), program goes into the next iteration using incremental-iterative algorithm.

6 Strong coupling algorithm

The strong coupling algorithm can be achieved in two ways: The first one is to couple system of equations for solid domain with Navier-Stokes equations for fluid domain. This type of coupling is probably the best way in terms of accuracy, because it assumes minimum of approximations. But for models that we want to develop and to investigate, we can use simpler system. The acoustic wave equation can also be used for modeling of fluid motion. That equation can be coupled with solid equation by additional boundary condition, which equalizes most dominant forces that appears on common boundary: inertial force from solid and pressure gradient from fluid (Zienkievicz, 1983). Strong coupling means that the solution of solid element in the contact with fluid has impact on the solution of fluid element and vice versa. The coupling was achieved by the equalization of normal fluid pressure gradient with normal acceleration of solid domain on the fluid-solid interface, as it is shown in Eq. (9.13):

$$n \cdot \nabla p = \rho n \cdot \ddot{u} \tag{9.13}$$

where

n—normal vector on fluid-structure interface,
∇p—gradient of fluid pressure,
ρ—density of fluid,
$\ddot{u}$—acceleration of solid.

In Fig. 9.5. how the boundary between fluid and solid looks like is shown. In solid domain, there are only equations (or degrees of freedom) that are related to solid dynamics (translations in x, y, and z direction). On the other side, in fluid domain, there is only pressure as degree of freedom. But on the fluid-solid interface, all nodes have both, degrees of freedom from solid domain and from fluid domain (Fig. 9.5).

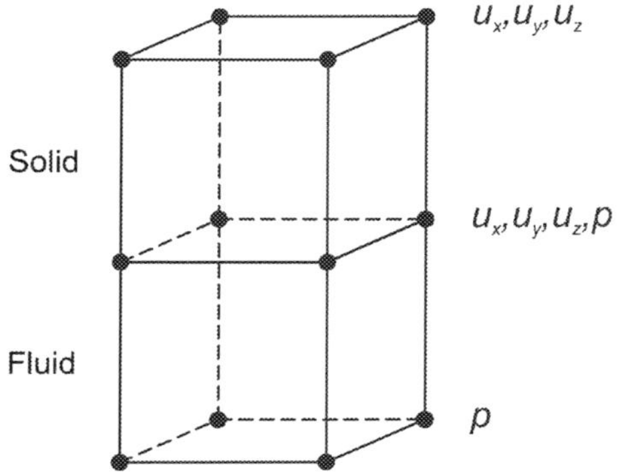

Fig. 9.5 Boundary elements between fluid and solid.

Using solid dynamic equation and acoustic wave equation and applying boundary condition (9.13), we can define system of equations that can be used for solving of fluid-solid coupled problems:

$$\begin{bmatrix} \boldsymbol{M} & \boldsymbol{0} \\ -\rho_f \boldsymbol{R} & \boldsymbol{Q} \end{bmatrix} \begin{Bmatrix} \ddot{u} \\ \ddot{p} \end{Bmatrix} + \begin{bmatrix} \boldsymbol{C} & \boldsymbol{0} \\ \boldsymbol{0} & \boldsymbol{C_p} \end{bmatrix} \begin{Bmatrix} u \\ p \end{Bmatrix} + \begin{bmatrix} \boldsymbol{K} & -\boldsymbol{S} \\ 0 & \boldsymbol{H} \end{bmatrix} \begin{Bmatrix} u \\ p \end{Bmatrix} = \begin{bmatrix} \boldsymbol{F} \\ q \end{bmatrix} \tag{9.14}$$

The details of matrices and vectors in Eq. (9.14) can be found in literature (Zienkievicz, 1983).

This kind of coupling is more suitable for analysis that we want to conduct. To make model simpler, we can analyze system only in terms of frequency response. It assumes that we analyze the response of dynamical system under periodic excitation with some prescribed frequency. This can be very helpful to clarify the nature of the processes occurring in the human ear. Using previous equations, we developed numerical software, to be able to analyze behavior of middle and inner ear elements under the excitation of sound wave that reached human ear.

7 Finite element modeling

The geometry of the human organs is pretty unstructured. Using the data from medical images, it is possible to reconstruct geometry of some organs or part of organs from the human body. The hearing system is relatively small in comparison with other organs. To get sufficiently accurate geometry, it is necessary to use medical images from micro CT scanner.

Geometry obtained from such images is very precise, but it is not completely enough for modeling. At first stage, the model must be simplified to be validated. Considering the facts that geometry of the cochlea and its moving parts is complex, as well as the properties of materials from which its parts are made, it is obvious that certain simplifications must be introduced.

The real material properties of the basilar membrane are nonlinear and anisotropic. Also, dimensions of the cross sections of the basilar membrane along the cochlea are not constant. It is known from experiments that different sounds produce different responses of the basilar membrane. Sounds with low frequency produce resonant peak near the apex and sounds with high frequency near the stapes. Distance from base to the peak is proportional to the logarithm of excitation frequency (Fig. 9.6). This feature is used for fitting of material or geometrical properties (Ni, 2012).

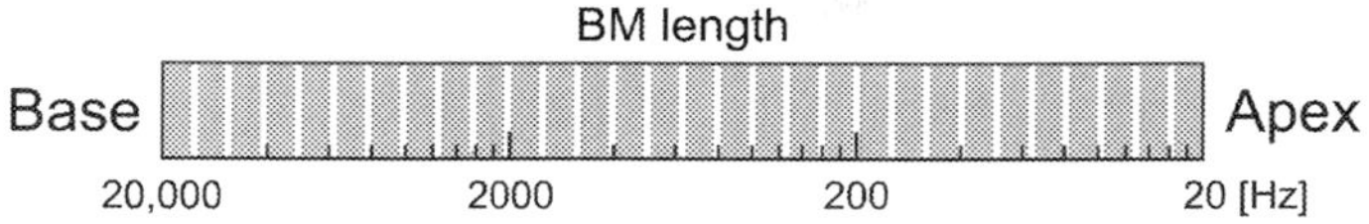

Fig. 9.6 Logarithmic distribution of frequencies along the basilar membrane. High frequencies are at the beginning (base), and low frequencies are at the end (apex) of the basilar membrane.

Material properties of basilar membrane can be approximated as orthotropic. The Young's modulus along basilar membrane is much less then in opposite (transversal) direction. There are two approximations that can be used to model this physical property (Ni, 2012; Kim et al., 2011):

1. We can approximate basilar membrane as a set of independent strips (Fig. 9.7A). Each strip has its own value of Young's modulus calculated according to the logarithmic function obtained from literature (Ni, 2012). That function is fitted by natural frequencies, because each strip should have its own natural frequency value according to the experimentally obtained frequency map.
2. Another way to approximate this feature is to make tapered model with variable cross section of the basilar membrane (Kim et al., 2011), as shown in Fig. 9.7B. In that case, we can use orthotropic material properties so the Young's modulus in longitudinal direction should be negligible according to modulus in transversal direction.

In the first case, the value of Young's modulus is defined as a function of distance from the beginning of the basilar membrane (Ni, 2012):

$$E(x) = \frac{4\pi^2 f_B^2(x) A\rho(1-\nu^2)}{\beta^4 I} \tag{9.15}$$

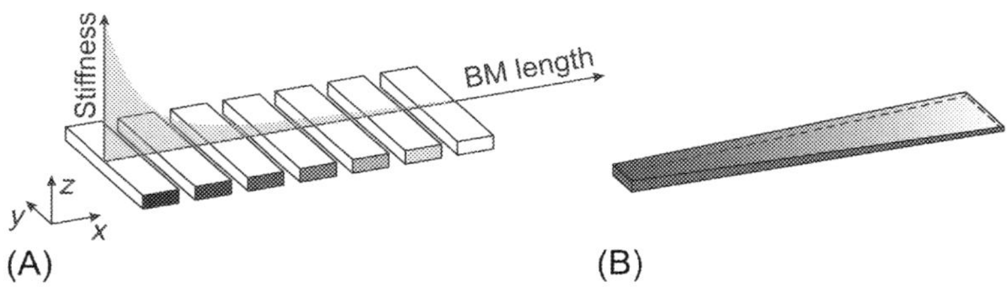

Fig. 9.7 Variable stiffness of basilar membrane: (A) variable Young's modulus along membrane, (B) variable cross section dimensions (thickness and width).

where

f_B^2—frequency distribution along the cochlear length, which has exponentially decaying character;
A—cross-sectional area of the basilar membrane;
ρ—density;
ν—Poisson's ratio;
β—coefficient that depends on the boundary conditions;
I—second moment of inertia.

In the second case, material properties are constant along basilar membrane. Orthotropic material properties used in this case are given in Table 9.1.

In the frequency analysis, damping could be included using modal damping (Ni, 2012). In that case, inside the stiffness matrix, there is an imaginary part, so Eq. (9.5) could be written in the following form:

$$\boldsymbol{M}\ddot{\boldsymbol{u}} + \boldsymbol{K}(1 + i\eta)\boldsymbol{u} = \boldsymbol{F}^{ext} \tag{9.16}$$

where η is the hysteretic damping ratio. This value is set by using exponentially increasing function as explained in literature (Ni, 2012).

The fluid-structure interaction with strong coupling was used for solving these equations.

For the mechanical model of the cochlea, we defined a system of coupled equations:

$$\begin{bmatrix} \boldsymbol{M} & 0 \\ -\rho_f \boldsymbol{R} & \mathbf{Q} \end{bmatrix} \begin{Bmatrix} \ddot{\boldsymbol{u}} \\ \ddot{\boldsymbol{p}} \end{Bmatrix} + \begin{bmatrix} \boldsymbol{K}(1 + i\eta) & -\boldsymbol{S} \\ 0 & \boldsymbol{H} \end{bmatrix} \begin{Bmatrix} \boldsymbol{u} \\ \boldsymbol{p} \end{Bmatrix} = \begin{bmatrix} \boldsymbol{F} \\ q \end{bmatrix} \tag{9.17}$$

where $\boldsymbol{R}$ and $\boldsymbol{S}$ are coupling matrices, Ni, 2012.

Table 9.1 Material properties

Symbol	Quantity	Value	Unit
E_x	Young's modulus in X direction	104	Pa
E_y	Young's modulus in Y direction	107	Pa
E_z	Young's modulus in Z direction	107	Pa
G_{xy}	Shear modulus in XY plane	2.104	Pa
G_{yz}	Shear modulus in YZ plane	106	Pa
G_{zx}	Shear modulus in ZX plane	1	Pa
ν_{xy}	Poisson's ratio in XY plane	0.005	–
ν_{yz}	Poisson's ratio in YZ plane	0.3	–
ν_{zx}	Poisson's ratio in ZX plane	0.005	–
ρ	Density of fluid	1000	kg/m^3

The solutions for displacement of the basilar membrane and pressure of fluid in the chambers were assumed in the following form:

$$\boldsymbol{u} = A_u \sin(\omega t + \alpha) \tag{9.18}$$

$$\boldsymbol{p} = A_p \sin(\omega t + \beta) \tag{9.19}$$

In Eqs. (9.18) and (9.19), A_u and A_p represent amplitudes of displacement and pressure, respectively. The circular frequency is ω, t is time, α and β are phase shift factors.

When displacement (Eq. 9.18) and pressure (Eq. 9.19) solution were substituted in the Eq. (9.17), we obtain a system of linear equations that can be solved:

$$\begin{bmatrix} \boldsymbol{K}(1+i\eta) - \omega^2 \boldsymbol{M} & -\boldsymbol{S} \\ -\rho_f \boldsymbol{R} & \boldsymbol{H} - \omega^2 \mathbf{Q} \end{bmatrix} \begin{Bmatrix} A_u \\ A_p \end{Bmatrix} = \begin{bmatrix} \boldsymbol{F} \\ \boldsymbol{q} \end{bmatrix} \tag{9.20}$$

8 Finite element models

8.1 Cochlea—Box model

To simulate mechanical processes occurred in the middle and inner ear, we have developed several different models, starting from basic models used only to validate model up to the coupled models with realistic geometry obtained from micro CT images. All models are run by in-house developed numerical software PAK (Kojic et al., 2009).

The first model developed for the validation of presented concept is straight box model of the cochlea (Fig. 9.8). As it is mentioned before, the cochlea can be modeled in that way, because response of the basilar

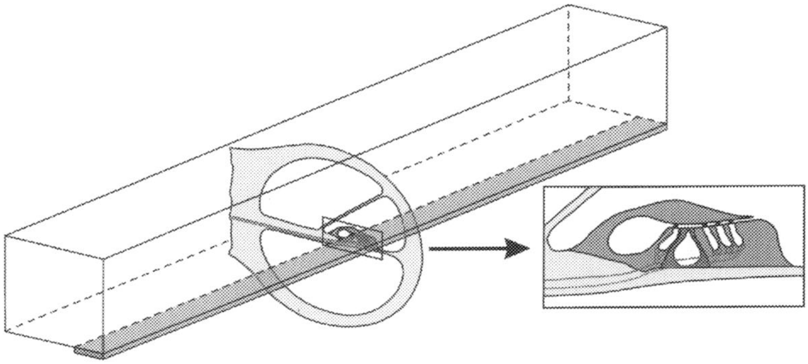

Fig. 9.8 Basic approximation—uncoiled box model.

membrane, placed inside the cochlea, does not depend on coiled cochlea geometry, (Ni, 2012).

Besides that, there are few more assumptions in this model:

- The model consists of only one fluid chamber and the basilar membrane. Other parts can be neglected in the first stage, because they do not affect much the response of the basilar membrane.
- Prescribed boundary condition is unit force applied in the fluid domain, on the side where the stapes is connected by oval window to the scala vestibuli.
- End of the chamber is opened.
- Coupling between fluid and solid is achieved through common nodes on the fluid-solid interface.
- The boundary conditions for the basilar membrane are three clamped edges and one simply supported edge. These boundary conditions correspond to the real cochlea, since clamped boundary conditions were applied on the edges where the basilar membrane is coupled with bone structure, while simply supported boundary condition was applied on the edge where there is a contact between the basilar membrane and the spiral ligament. Fig. 9.9 shows the basilar membrane with applied boundary conditions. In this model, the basilar membrane, that is, the cochlea, is uncoiled along x-axis.
- The Young's modulus is variable along the basilar membrane, and for each, finite element is calculated by function (Eq. 9.15).
- The basilar membrane is assembled of independent stripes along its length according to literature (Ni, 2012).

The first results obtained from this model are given in the Fig. 9.10. Here, response of basilar membrane for frequency of 1 kHz is presented.

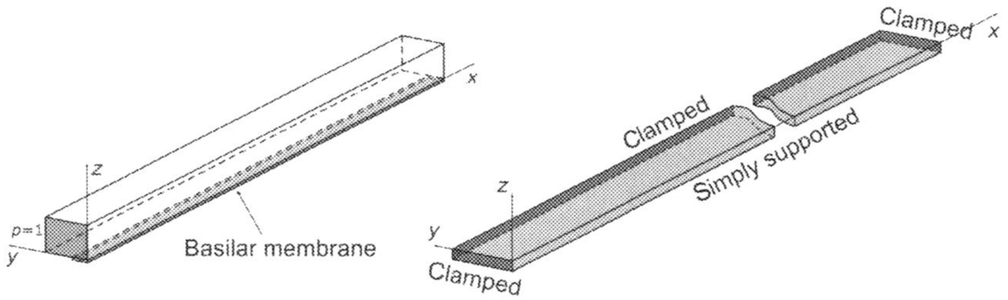

Fig. 9.9 Boundary conditions in cochlea box model.

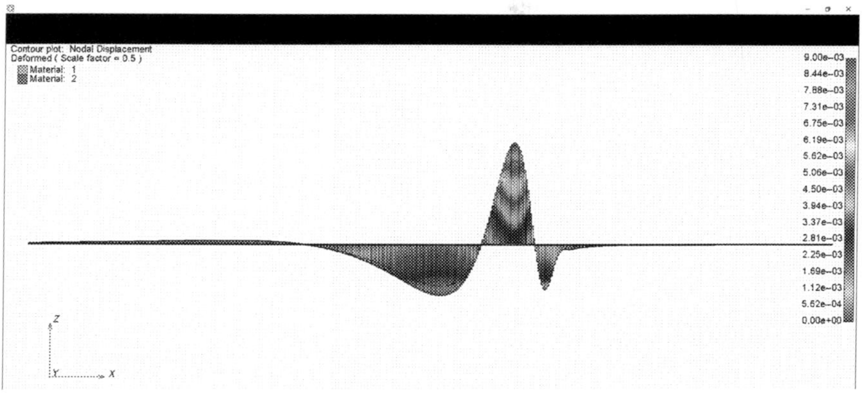

Fig. 9.10 Response of basilar membrane for frequency of 1 kHz.

This result is compared with results from literature (Ni, 2012; Steele and Taber, 1979). The distribution of amplitude of modal velocity along the basilar membrane calculated by strong coupled, full 3-D fluid-structure interaction using a finite element solver is presented in Fig. 9.11 with excitation frequency of 1 kHz. Also, we presented amplitude of modal velocity calculated by using WKB method described in literature (Steele and Taber, 1979). It can be observed that our results are in good agreement with results obtained by using WKB method.

There is one more validation example. The obtained results are compared to Greenwood function. The model shown earlier has good matching with literature data (Greenwood function) as it can be seen from the Fig. 9.12. A number of simulations were launched to compare numerical results with experimental data. Obtained results look good in whole tested range: from 100 Hz to more than 10 kHz.

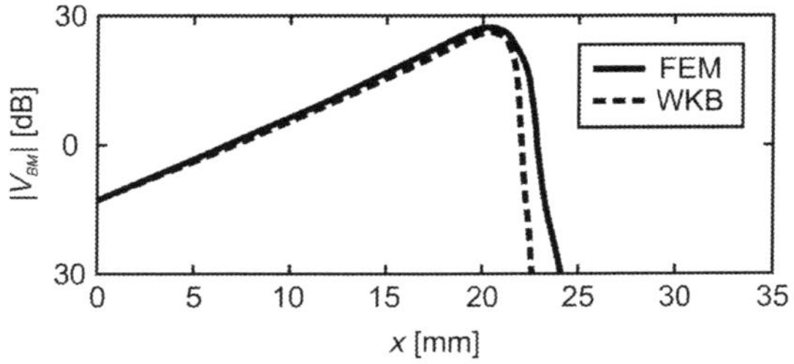

Fig. 9.11 The amplitude of modal velocity along the basilar membrane with excitation frequency of 1 kHz.

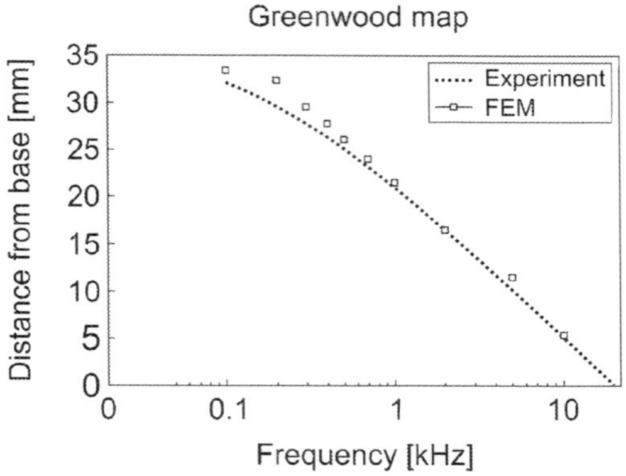

Fig. 9.12 Greenwood map—experimental versus numerical data.

8.2 Cochlea—Tapered model

The next model that we have developed is tapered model of the cochlea. In this model, orthotropic material model for modeling of basilar membrane is used. Orthotropic material properties are used to make model closer, because the real material properties have such a character (Kim et al., 2011). Model is also more precise in terms of geometry. The shape of the cochlear ducts and basilar membrane is tapered: The cochlear ducts narrow from the beginning to the end; basilar membrane spreads from beginning to the end. Those features correspond to real geometry of cochlear parts. The geometry of this model is given in Fig. 9.13.

The properties of materials used in this model are given in Table 9.2.

The boundary condition in the fluid domain is prescribed acoustic pressure at round window (Fig. 9.14), the beginning of upper fluid chamber. That excitation corresponds to reality when an audio signal comes into the hearing system. This signal is then transmitted through the elements of the the middle ear to the cochlea. The unit value is prescribed because the value is not significant. This model is used only to analyze frequency response.

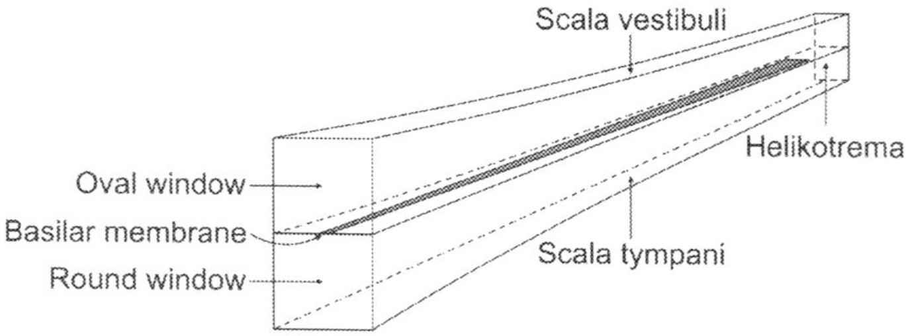

Fig. 9.13 Tapered model of the cochlea.

Table 9.2 Properties of materials used in tapered box model

Quantity	Value	Unit
Length of cochlea	35	mm
Width of fluid chamber	Tapered from 3 to 1	mm
Width of basilar membrane	Tapered from 6e − 5 to 6e − 4	mm
Width of basilar membrane	5e − 6	mm
Density of fluid	1000	kg/m^3
Speed of sound	1500	m/s
Density of solid	1000	kg/m^3
BM properties:		
E_x	1e + 4	Pa
E_y	1e + 8	Pa
E_z	1e + 8	Pa
ν_{xy}	0.005	
ν_{yz}	0.3	
ν_{xz}	0.005	
G_{xy}	2e + 4	Pa
G_{yz}	1e + 5	Pa
G_{xz}	1	Pa
Damping ratio	0.3	

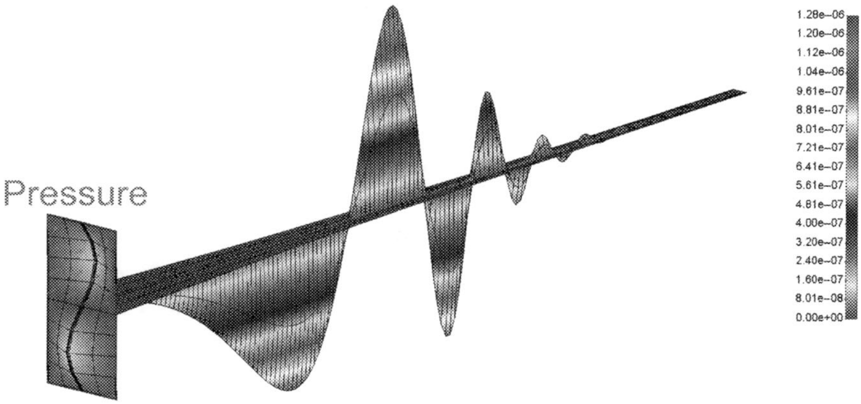

Fig. 9.14 Tapered box model: the basilar membrane response for excitation frequency of 1 kHz.

In the Fig. 9.15, results of the basilar membrane response for frequency of 1 kHz obtained with two different models are given: box model and tapered model. As can be seen from the figure, the peak in response is almost at the same place in both models.

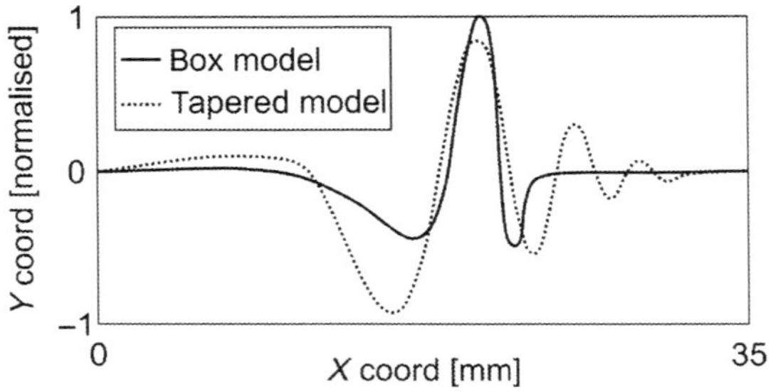

Fig. 9.15 Comparison of results obtained with box model and tapered model.

8.3 Cochlea—Coiled model

The next model that we have investigated is parametric coiled cochlea model. That model consists of two fluid chambers: the scala vestibuli and scala tympani, connected at the apex of the cochlea. The basilar membrane as a solid structure is placed between them. This model is longitudinally coupled.

The boundary conditions and prescribed values are the same as in cochlea box model. Coiled geometry of the basilar membrane in this model is similar to its real shape. The position of the peak in response of the basilar membrane and its amplitude is almost the same as in uncoiled box model. In Fig. 9.16, response of the basilar membrane for excitation frequency of 1 kHz is given.

The values on the scale are not important, because we analyze only frequency response of system.

8.4 Middle ear model

The next model that we have developed and investigated is numerical model of the middle ear (Fig. 9.17). The model consists of the tympanic membrane; three ossicles—the malleus, incus, and stapes; and supporting ligaments. In this model, we have only solid elements, without any coupling with fluid. According to that, we use only dynamic equation for solid to simulate behavior of the middle ear.

In Table 9.3, material properties of materials used for modeling of different structures in the middle ear are shown. Some of material properties are available in literature; other properties are fitted to obtain good response of whole middle ear structure.

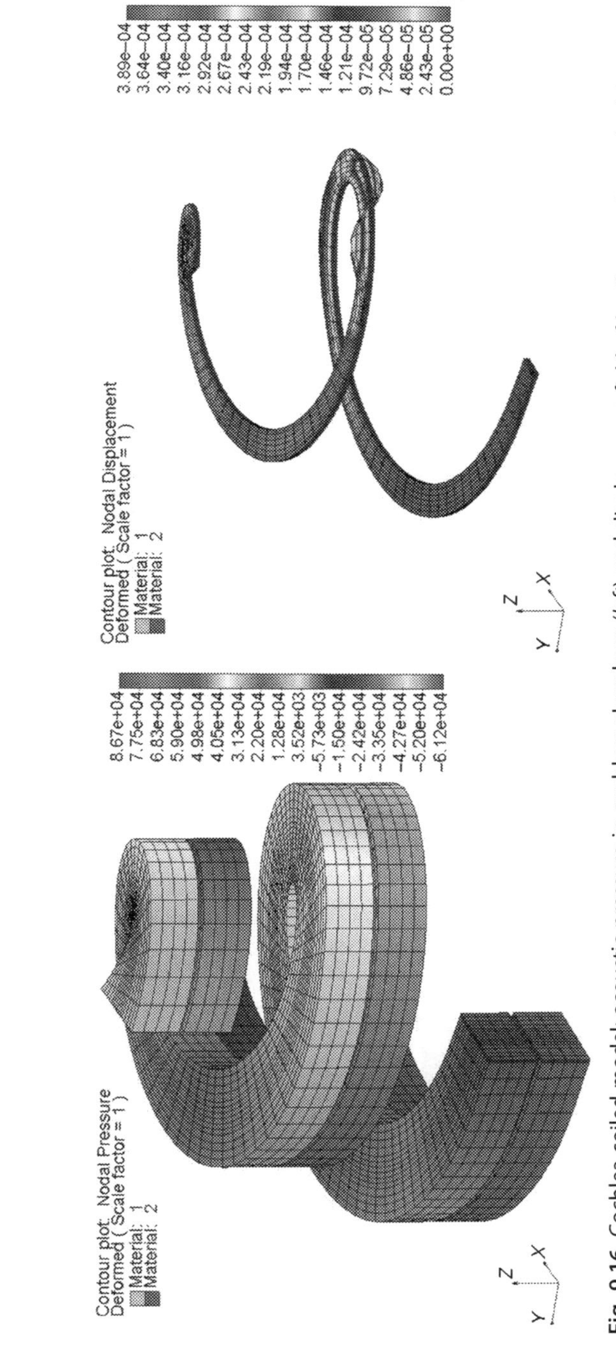

Fig. 9.16 Cochlea coiled model: acoustic pressure in cochlear chambers (*left*) and displacement field of basilar membrane (*right*).

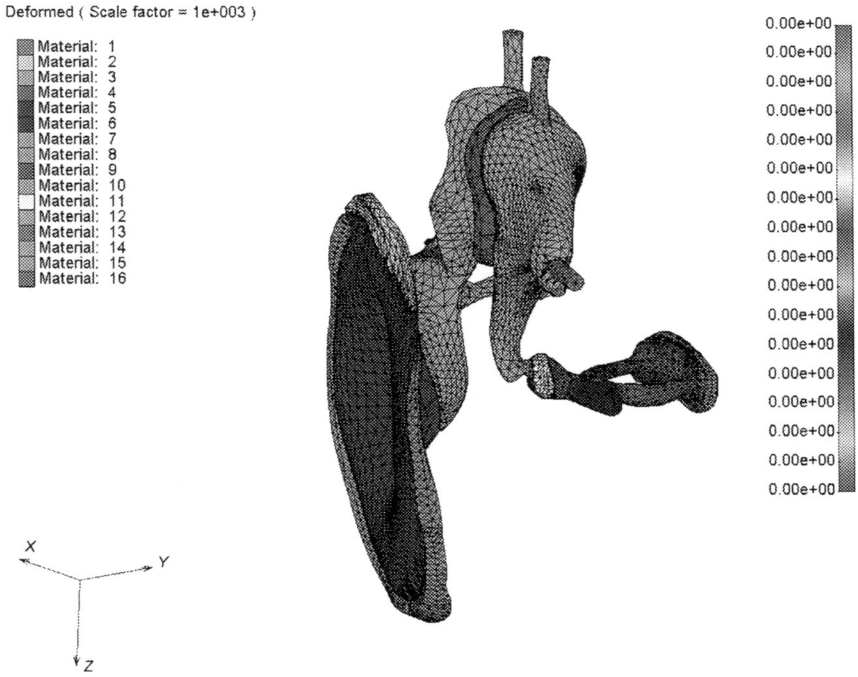

Fig. 9.17 The middle ear: pure mechanical model.

Table 9.3 Material properties used in middle ear model

Material	Young's modulus [MPa]	Poisson's ratio
1—Annulus	0.4	0.33
2—Sharpnell's membrane	7.0	0.33
3—Incus	14100.0	0.33
4—Stapes	14100.0	0.33
5—Stapedius tendon	5.0	0.33
6—Tympanic membrane	20.0	0.33
7—Tensor tympani tendon	0.33	0.33
8—Oval window	0.041	0.33
9—Incudomalleolar joint	7.0	0.33
10—Malleus	14100.0	0.33
11—Incudostapedial joint	0.6	0.33
12—Ligament	7.0	0.33
13—Ligament	7.0	0.33
14—Ligament	7.0	0.33
15—Ligament	7.0	0.33
16—Rigid coupling material	14100.0	0.33

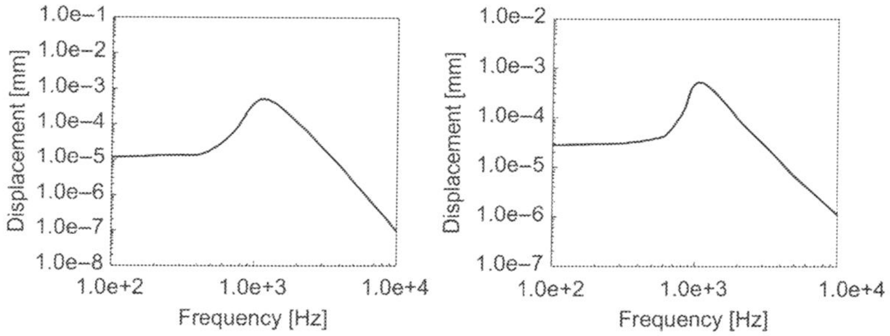

Fig. 9.18 Displacements versus frequency of characteristic points in the inner ear: umbo point (*left*) and footplate point (*right*).

The numerical model consists of 15,838 nodes and 50,116 tetrahedral finite elements. Linear elastic isotropic material model is used in this example. Frequency analysis is considered.

The boundary conditions in this model are fixed edges around the tympanic membrane, fixed ends of ligaments (that are not connected to the middle ear ossicles), and fixed edges around oval window—the structure that the stapes is connected to.

The prescribed outer load is pressure at the tympanic membrane. The value of pressure corresponds to the value of 90 dB sound pressure level (SPL) to model acoustic pressure in the external auditory canal at the frequency of 1000 Hz (Tachos et al., 2016a, b).

In Fig. 9.18, dependencies of displacement versus frequency for two characteristic points are given: umbo point at the tympanic membrane and footplate point at stapes.

The shape of umbo point displacements graph corresponds to the experimental result obtained by Gan et al. (2004).

8.5 Coupled model

The next model that we have developed is coupled model of the middle and inner ear. For the middle ear, we used the model described earlier in this chapter, and for the inner ear, we used tapered model of the cochlea. Coupling is achieved between stapes—the part of the middle ear and elastic membrane at the beginning of the upper fluid chamber of the cochlea.

The model consists of the ear drum, simply supported round structure with prescribed sound pressure level (SPL) of 90 dB; three ossicles—the incus, malleus, and stapes, supported by ligaments whose free ends are fixed; cochlea with outer bony structure; two elastic membranes, oval and round window;

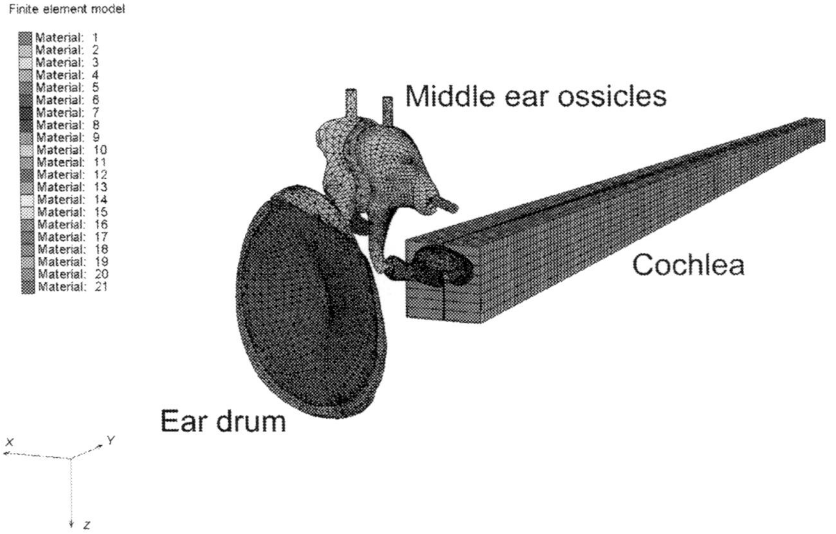

Fig. 9.19 Coupled model: the middle and inner ear.

two fluid chambers; and basilar membrane that separates fluid chambers. The frequency of external excitation is 1 kHz. The model is shown in Fig. 9.19.

The boundary conditions and material parameters in this model are taken from both models used: middle ear model and tapered cochlea model.

In Fig. 9.20, results obtained by coupled model are presented. Here, response of the basilar membrane under excitation that comes from ear drum and pressure field distribution inside cochlear chambers are given. Frequency of excitation was 1 kHz. The resonant peak is at the same place as in tapered and box model. It means that coupled model describes the processes taking place in the human ear well, and that can be used in further investigations related to other processes, such as generating of electrical signals in the organ of Corti.

9 Model of cochlea including feedforward and feedbackward forces

9.1 Feedforward and feedbackward OHC forces

The BM's vibration is in the vertical direction, with downward displacement (i.e., toward the scala tympani (ST) being positive. The BM has mass, damping, and stiffness. Its motion is directly driven by the pressure difference across it, while additional force on the BM is OHC force defined as

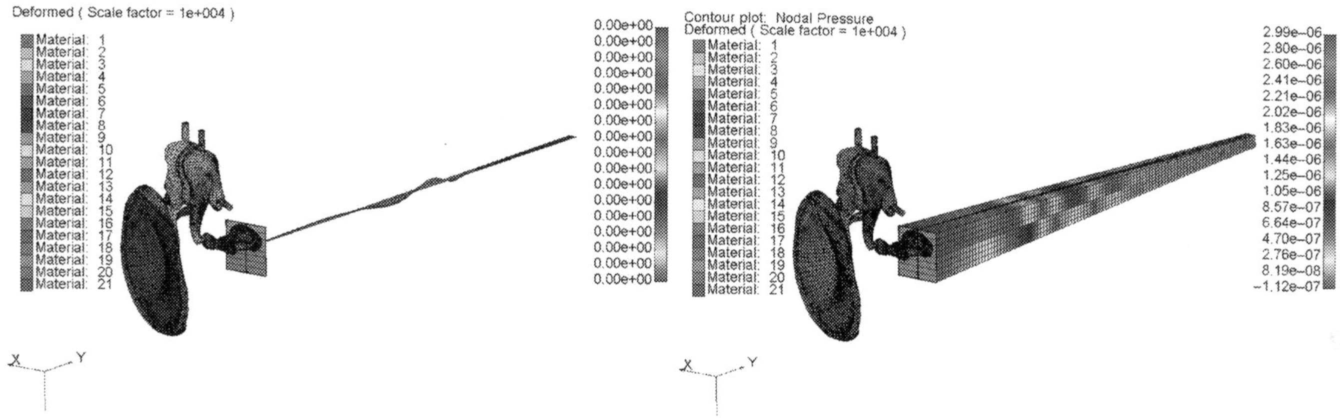

Fig. 9.20 Response of coupled model under outer excitation of 1 kHz: basilar membrane displacement (*left*) and fluid pressure in cochlear chambers (*right*).

$$\mathbf{F}_{OHC(x)} = \alpha\mathbf{K}(x)^*(\gamma^*\mathbf{U}(x-d) - \mathbf{U}(x+d)) \quad (9.21)$$

where x represents the distance from the stapes along the cochlear partition, with $x=0$ at the stapes and the round window and $x=L$ (uncoiled cochlear duct length) at the apex and the helicotrema.

The force $\mathbf{F}_{OHC(x)}$ combines feedforward and feedbackward OHC forces, expressed as a fraction α of the BM stiffness (i.e., OHC motility factor). γ is the ratio of the feedforward to the feedbackward coupling, representing relative strengths of the OHC forces exerted on the BM segment through the Deiters cells (DC), directly, and through the tilted phalangeal processes. d denotes the tilt distance, which is the horizontal displacement between the source and the recipient of the OHC force, assumed to be equal for feedforward and feedbackward cases.

We compared the results for BM velocity with respect to input stapes velocity for the passive cochlea with the measurement by Stenfelt et al. (2003) at 12 mm from the RW, the data from Gundersen et al. (1978), and the data from Wang et al. (2014) (Fig. 9.21).

Velocity magnitude for both passive and active tapered model with 5 kHz frequency is presented in Fig. 9.22. We can observe that the active model has higher intensity of velocity magnitude than the passive model. The velocity magnitude for different frequencies—1, 2, 4, 6, 8, and 10 kHz—and the passive and active rectangular box model with nonlinear elasticity modulus has been shown in Figs. 9.23 and 9.24.

The acoustic pressure of fluid in the scala vestibuli (SV) on the surface of the BM along the cochlear length induced by unit pressure load for the passive cochlea and frequencies 1, 2, 4, 6, 8, and 10 kHz has been shown in Fig. 9.25.

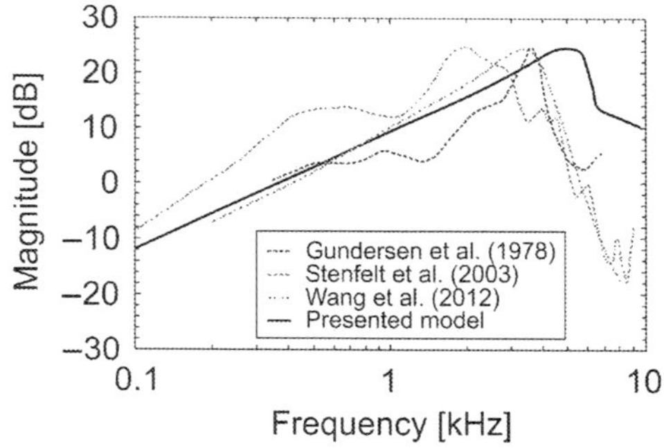

Fig. 9.21 BM velocity with respect to input stapes velocity for the passive cochlea. The measurement by Stenfelt et al., 2003 was at 12 mm from the RW, and the data from Gundersen et al. (1978) were measured at a similar position.

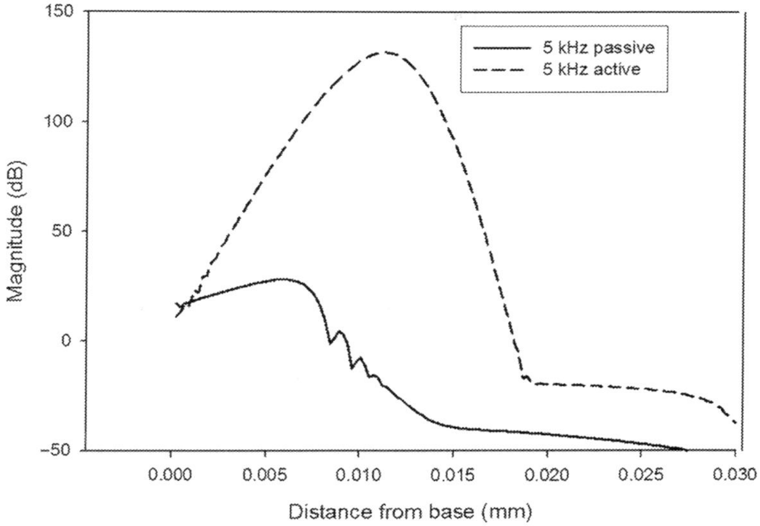

Fig. 9.22 Velocity magnitude for 5 kHz for passive and active (using feedforward and feedbackward OHC forces) tapered model for applied unit pressure on the oval window.

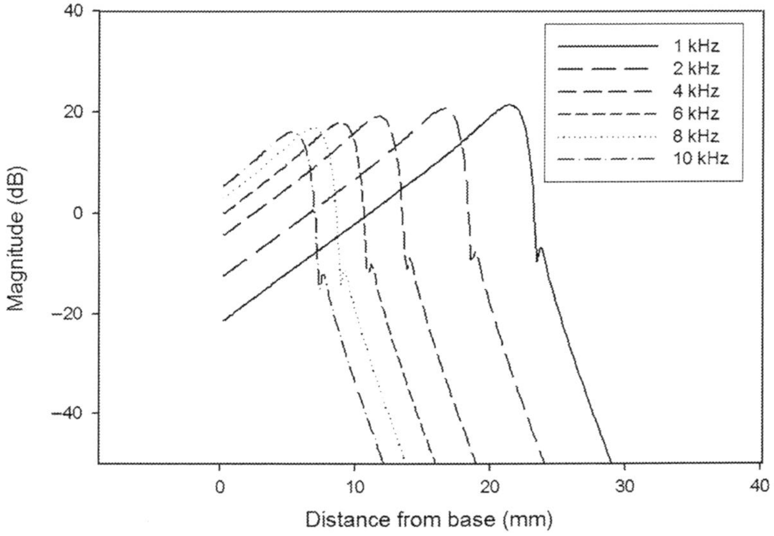

Fig. 9.23 Velocity magnitude for passive model in the box rectangular model. Unit pressure was applied on the oval window.

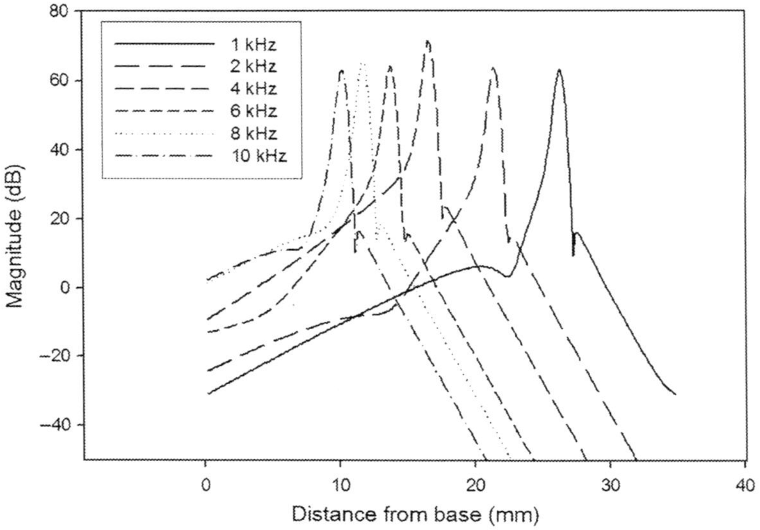

Fig. 9.24 Velocity magnitude for active box rectangular model (using feedforward and feedbackward OHC forces). Unit pressure was applied on the oval window.

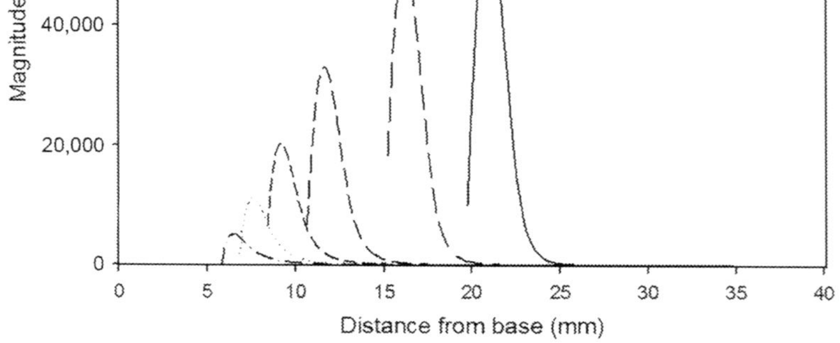

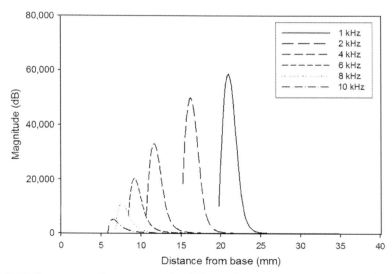

Fig. 9.25 Pressure passive response for different frequencies: 1, 2, 4, 6, 8, and 10 kHz where input unit pressure on the oval window was 1. Rectangular box model with nonlinear elasticity modulus was used.

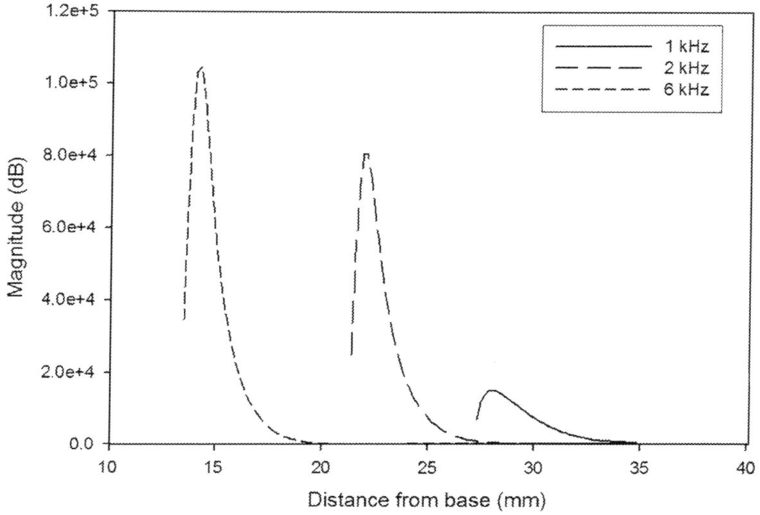

Fig. 9.26 Pressure distribution for different frequencies: 1, 2, and 6 kHz with active model (using feedforward and feedbackward OHC forces).

Pressure distribution for frequencies 1, 2, and 6 kHz with active coupling (using feedforward and feedbackward OHC forces) is presented in Fig. 9.26.

10 Conclusion

The human hearing system is very complex and interesting for investigation. In this chapter, we presented concepts of numerical modeling of mechanical processes occurring in the human hearing system using finite element method. Two different approaches are described and applied to simulate mechanical behavior of the human hearing system. Here, middle ear model and several different models of the inner ear were presented. Also presented is coupled model of the middle and inner ear. The developed models are fitted with experimental and semianalytic results found in literature.

The model presented here can be used for education purpose, to provide a better understanding of mechanical and acoustical processes taking place in the outer, middle, and inner ear. The traveling wave obtained by box model, coiled model, and coupled model (the middle ear and inner ear), described in this chapter, presents one of the most important properties of the cochlear macromechanics. Three-dimensional simplified box and spiral model are presented. Also, the software developed during this investigation can be very useful for people who are engaged in the development of hearing devices and cochlear implants.

Medical doctors may also be potential users of the software developed here. Using this model, medical doctors can have virtual view inside physics of hearing process. It can be very useful in investigations regarding specific hearing problems. Some disorders or physical damages can be included in the model to look at the problem from the inside.

We believe our results demonstrate the promise of large-scale computational modeling approach to the study of middle and inner ear mechanics.

Acknowledgment

This work was supported in part by grants from Serbian Ministry of Education and Science III41007, ON174028 and FP7 ICT SIFEM 600933 project.

References

Gan, R.Z., Feng, B., Sun, Q., 2004. Three-dimensional finite element modeling of human ear for sound transmission. Ann. Biomed. Eng. 32, 847–859.

Gundersen, T., Skarstein, O., Sikkeland, T., 1978. A study of the vibration of the basilar membrane in human temporal bone preparations by the use of the Mössbauer effect. Acta Otolaryngol. 86 (3–4), 225–232.

Kim, N., Homma, K., Puria, S., 2011. Inertial bone conduction: symmetric and anti-symmetric components. J. Assoc. Res. Otolaryngol. 12, 261–279.

Kojic, M., Filipovic, N., Stojanovic, B., Kojic, N., 2008. Computer Modeling in Bioengineering—Theoretical Background, Examples and Software. John Wiley and Sons, England, ISBN: 978-0-470-06035-3.

Kojic, N., Milosevic, M., Petrovic, D., Isailovic, V., Sarioglu, A.F., Haber, D., Kojic, M., Toner, M., 2015. A computational study of circulating large tumor cells traversing microvessels. Comput. Biol. Med. 0010-4825.63 (C), 187–195.

Kojic, M., Slavkovic, R., Zivkovic, M., Grujovic, N., Filipovic, N., 2009. PAK, Finite Element Software. BioIRC Kragujevac, University of Kragujevac, Kragujevac, Serbia.

Filipovic, N., Velibor, I., Tijana, D., Mauro, F., Milos, K., 2012. Multiscale modeling of circular and elliptical particles in laminar shear flow. IEEE Trans. Biomed. Eng. 59 (1), 50–53. ISSN: 0018-9294.

Guangjian Ni, 2012. Fluid Coupling and Waves in the Cochlea. PhD Thesis, University of Southampton, Faculty of Engineering and the Environment, Institute of Sound and Vibration Research.

Pickles, J.O., 1988. An Introduction to the Physiology of Hearing, second ed. Academic Press, London.

Steele, C.R., Taber, L.A., 1979. Comparison of WKB calculations and experimental results for three-dimensional cochlear models. J. Acoust. Soc. Am. 65 (4), 1007–1018.

Stenfelt, S., Puria, S., Hato, N., Goode, R.L., 2003. Basilar membrane and osseous spiral lamina motion in human cadavers with air and bone conduction stimuli. Hear. Res. 181, 131–143.

Tachos, N.S., Sakellarios, A.I., Rigas, G., Isailovic, V., Ni, G., Böhnke, F., Filipovic, N., Bibas, T., Fotiadis, D.I., 2016a. Middle and inner ear modelling: from microCT images to 3D reconstruction and coupling of models. EMBC, 5961–5964.

Tachos, N.S., Sakellarios, A.I., Rigas, G., Isailovic, V., Ni, G., Böhnke, F., Filipovic, N., Bibas, T., Fotiadis, D.I., 2016b. Middle and inner ear modelling: from microCT images to 3D reconstruction and coupling of models. In: IEEE 38th Annual International Conference of the IEEE Engineering in Medicine and Biology Society (EMBC), August 16–20. ISBN: 978-1-4577-0220-4, pp. 5961–5964. https://doi.org/10.1109/EMBC.2016.7592086.

Wang, X., Wang, L., Zhou, J., Hu, Y., 2014. Finite element modelling of human auditory periphery including a feed-forward amplification of the cochlea. Comput. Methods Biomech. Biomed. Engin. 17 (10), 1096–1107. https://doi.org/10.1080/10255842.2012.737458.

Zienkievicz, O.C., 1983. The finite element method, third ed. McGraw-Hill Book Co., New York. ISBN: 10:0070840725/13:9780070840720.

Further reading

Dallos, P., 1992. The active cochlea. J. Neurosci. 12 (12), 4575–4585.

Howard, D., Angus, J., 1996. Acoustics and Psychoacoustics. Focal Press/Elsevier, Linacre House, Jordan Hill, Oxford, UK, ISBN: 978-0-240-52175-6.

CHAPTER 10

Numerical modeling of cell separation in microfluidic chips

Contents

1 Introduction

Isolation of cells is an important preparatory step that has a wide range of applications in biology and medicine. One of the examples is the isolation of circulating tumor cells (CTCs) from a blood sample. These cells have been identified in the bloodstream of cancer patients; they are shed from a primary tumor and are the cause of metastases (Cristofanilli et al., 2004; Steeg, 2006). Cancer is clinically diagnosed using appropriate screening, that is, using computed tomography (CT) scans and biopsy procedures. This approach is expensive and uncomfortable for patients, but more importantly, it can sometimes cause a false-negative diagnosis, and it has limited potential when it is necessary to diagnose the disease in the early stage (Gress et al., 2001). Therefore, it could be appropriate to use newer approaches such as the detection and isolation of CTCs from the blood to identify and monitor nonhematologic cancer, using a noninvasive procedure (Cristofanilli et al., 2004; Mocellin et al., 2006; Rolle et al., 2005). Isolated CTCs can be subjected to a detailed biomolecular analysis, and this can not only improve the diagnostics of disease but also help to choose the most adequate

Computational Modeling in Bioengineering and Bioinformatics
https://doi.org/10.1016/B978-0-12-819583-3.00010-2

therapy that is adapted to the specific patient and tumor type. Also, this analysis can potentially provide information about the mechanisms that the cancer uses while spreading and forming metastasis. Out of 10^9 cells in a blood sample of patients with metastatic cancer, only one is the CTC, which makes these cells very rare and their isolation very challenging (Zieglschmid et al., 2005; Kahn et al., 2004).

During the past decades, many different approaches were proposed for the isolation of cells. Seal (Seal, 1964) first applied a simple sieve as a filter to separate the CTCs from other blood constituents, taking into consideration the observation that tumor cells are larger than the other cells in blood (erythrocytes, leukocytes, and platelets). Afterward, Fleischer et al. (Fleischer et al., 1964) presented irradiated and etched plastic filters for the separation of cells. These filters had holes that were positioned in a precisely controlled manner. The first microfluidic device was presented by Nagrath et al. (Nagrath et al., 2007). This device had microposts on silicon chips that were coated with a specific antibody, to optimize the capture of CTCs. Since then, many diverse microfluidic chips were proposed. Proposed microfluidic techniques for separating cells can be divided into two categories: active and passive methods. Active methods use some type of external force field (electric field (Choi and Park, 2005), magnetic field (Jung et al., 2010), acoustic waves (Li et al., 2015), optical interaction (Lin and Lee, 2008), or fluorescent labeling (Baret et al., 2009)), and the specific properties of cells are thus considered. Passive methods use the characteristics of particular cells, to ensure the separation. Some of these characteristics include size, shape, mass, mechanical stiffness, or deformability (Park and Jung, 2009; Wang et al., 2013; Lv et al., 2013; Godin et al., 2010; Mohamed et al., 2009).

The main characteristics that measure the performance of a microfluidic chip are cell recovery rate, output purity, cell viability, and throughput. Cell recovery rate is also called capture efficiency, and it defines the amount of target cells that are collected at the outlet of the microfluidic chip, when compared with all target cells that entered the inlet. Output purity defines accuracy of separation and is calculated as the amount of target cells, when compared with the overall number of cells that are collected at the outlet. Cell viability is mostly defined by the number of cells at the output that are not dead. Throughput (flow rate) determines the number of processed cells per unit time. In this sense, the advantages and drawbacks of mentioned approaches can be analyzed. Microfluidic chips based on active methods have higher separation efficiency and selectivity, but they have low throughput, because the interaction of cells with the chip requires smaller flow rates. Also, some of the presented methods require the cells to be labeled, which could lead to reduced viability of the cells. Microfluidic chips based on passive methods on the other hand have a much simpler design, and the blood sample can have bigger flow rates, ensuring higher throughput and the cells are viable and not affected by any previous labeling. The main drawback of these methods is that they have somewhat lower selectivity and efficiency.

Lv et al. (2013) also discussed the advantages and disadvantages of mentioned approaches for the separation of CTCs and compared the performance of several proposed devices. They concluded that the microfluidic chips that are isolating CTCs only based on the physical properties of cells (size and deformability) are generally more effective because the CTCs are larger and more rigid than the other blood constituents. This means that a microfluidic chip with specially positioned microfabricated pillars should be applied. This microfluidic filter would allow cells that have smaller diameter and/or are highly deformable to pass through the filter, while larger and more rigid cells that are of interest would remain captured. The size, geometry, and spatial distribution of the pillars have to be carefully designed, to ensure high efficiency of the microfluidic chip. Many researchers investigated the impact of different pillar shapes on the capture efficiency. Ranjan et al. (2014) studied various micropillar shapes, such as circular, square, I-shaped, T-shaped, and L-shaped. Karabacak et al. (2014) used circular pillars in their study, while Liu et al. (2013) concluded that triangular pillars are better suited for cells that undergo large deformations. Also, if the distance between micropillars is varied, more deformable and smaller cells (e.g., RBCs) can pass through the obstacles, while other more rigid cells (e.g., CTCs) are captured. White blood cells (WBCs) are of comparable dimension with the CTCs, but nevertheless, they are more deformable than CTCs, so they can pass through the microfluidic chip, while the CTCs will remain captured. To analyze different design strategies and achieve the highest possible capture efficiency, it would be appropriate to use numerical simulations.

Numerical simulations have been successfully applied in many areas, including drug delivery (Rakocević et al., 2013), motion of LDL (Filipović et al., 2014), motion of RBCs through the bloodstream (Đukić et al., 2016), and motion of otoconia particles in the semicircular canals of the inner ear (Đukić and Filipović, 2017). This type of simulations can also be used to study the processes within microfluidic chips (Đukić et al., 2015).

Several methods for studying the fluid flow in microfluidic chips have been proposed in literature. Some authors proposed different methods to experimentally measure the fluid flow (e.g., using particle image velocimetry (PIV) (Klank et al., 2002) and electroosmotic flow (Sun et al., 2007)). Researchers also used different methods to simulate the fluid flow numerically, including commercial multiple physics software package based on the finite volume method (Sun et al., 2007), the Nernst-Planck equation and the full Navier-Stokes equation (Lin et al., 2002), the Poisson equation (Qiao and Aluru, 2002), a dynamic model based on the transient momentum equation (Song et al., 2011). Also, some papers in literature analyzed different possible designs and determined the most appropriate design of the microfluidic chip (Sun et al., 2007; Taylor et al., 2008). The motion of deformable objects immersed in fluid domain has also been the topic of several research papers (Dupin et al., 2007; Feng and Michaelides, 2004; Krüger et al., 2011), and these methods can be applied to the motion of cells in a microfluidic chip. But these methods have not been validated using real experimental data.

Tracking the motion of individual cells in a microfluidic chip is enabled with the numerical model proposed in this chapter. It would be hard to perform this type of a detailed analysis of the behavior of both RBCs and CTCs in laboratory conditions and during experiments. The cited numerical methods mostly observe the domain as a continuum, while the method proposed in this chapter observes the fluid as a set of discrete particles and also models the interaction of individual cells with the boundaries of the microfluidic chip and with blood plasma. Due to these characteristics, the proposed model can be successfully applied for the simulation of separation of cells based on their physical properties. Another benefit of these simulations is that they enable the study of the effect of design changes on the behavior of CTCs and capture efficiency. The results of the developed software were compared with some numerical and experimental results published in literature, and good agreement demonstrated the accuracy of the proposed model.

The chapter is organized as follows: In Section 2, numerical methods are explained in detail. Section 3 explains the simulation setup and all the parameters that had to be defined. Results of the numerical simulations are presented in the same section. This includes examples of analysis of design changes and comparison with experimental data, to validate the proposed model. Section 4 concludes the chapter.

2 Numerical model

The numerical model presented in this chapter comprises three mayor parts—fluid flow simulation, modeling of the deformation of the immersed cell, and interaction between the immersed cell and surrounding fluid. Fluid flow is modeled at the microscale level, using the lattice Boltzmann method. The deformation of the immersed cell is divided in four different categories—change of volume, change of surface area, surface strain, and bending. Each of these deformations causes a reaction force, and the sum of these forces represents the actual force that is a result of the deformation. Interaction between immersed cell and surrounding fluid is modeled using a strong coupling approach, called immersed boundary method, where the forces between two domains are interpolated over the contact surface. Details of the numerical model will be described in the sequel of this section.

2.1 Modeling fluid flow

There are many numerical methods used for the simulation of fluid flow. Some of the standard methods that are used in computational fluid dynamics (CFD) are finite element method, finite difference method, and finite volume method. The essence of all mentioned methods is solving a system of equations (Navier-Stokes equation and continuity equation), to obtain the resulting field of macroscopic physical quantities—velocity, pressure, and shear stress. Besides standard methods that are based on equations derived for continuum domain, there are also discrete methods, such as lattice Boltzmann method (LB), dissipative particle dynamics (DPD), and smoothed-particle hydrodynamics (SPH). In this chapter, the LB method was used. This method is chosen because of its efficiency in simulations of fluid flow and motion of solid bodies through fluid domain (Dupin et al., 2007; Wu and Shu, 2010; Sun et al., 2003). Also, this method was already successfully applied to model the motion of low-density lipoprotein (LDL) particles through the bloodstream (Filipović et al., 2014) and to analyze and predict the motion of nanodrugs through the bloodstream (Filipović et al., 2012). The main advantage of this method is that it is not necessary to solve differential equations or a system of equations and thus the implementation is relatively simple and it is possible to parallelize the software based on LB method.

In LB method, the physical system is observed in an idealized manner, such that space and time are discretized and the whole domain is made of a large number of identical cells (Wolfram, 1986). Fluid is observed as a set of fictional particles. These particles can move along the predefined directions, and the dynamics of their motion is modeled through their mutual collisions and further propagation. However, in numerical simulations of real physical problems, that would mean tracking a large number of particles. If every particle was analyzed separately, that would mean that the fluid is completely analyzed at a microscale level. This approach is not suitable because the individual tracking is hard to achieve. On the other hand, for problems concerning fluid dynamics, following the changes in the system at the macroscale level is of the greatest interest. That is why the basic quantity in LB method is the distribution function, which describes the behavior and motion of particles in space. This function has an identical form for all the particles and considers the state of considered particle and particles that are its closest neighbors. The state of all particles is updated synchronously in a series of iterations, representing discrete time steps.

Equations of LB method are derived from a continuous Boltzmann equation. It can be demonstrated that it is possible to model the fluid flow at macroscale level by modeling the fluid at a microscale level and by tracking the dynamics of discrete particles. The derivation procedure will be briefly explained in the sequel of this section, while the details of this procedure can be found in literature (Malaspinas, 2009; Đukić, 2015).

Boltzmann equation represents the equation of balance of the distribution function. If external force is taken into consideration, the Boltzmann equation is given by

$$\frac{\partial f}{\partial t} + \boldsymbol{\xi} \cdot \frac{\partial f}{\partial \mathbf{x}} + \frac{\mathbf{g}}{m} \cdot \frac{\partial f}{\partial \boldsymbol{\xi}} = \Omega \tag{10.1}$$

where $\mathbf{g}$ denotes the external force vector and Ω is the collision operator that represents the change of distribution function due to particle collisions.

Collision operator Ω is defined with a very complex function, and therefore, a simplified model was proposed by Bhatnagar et al. (1954). It is assumed that the effect of particle collisions is to bring the fluid "closer" to the equilibrium state. This model is known as the single relaxation time approximation (SRT) or the Bhatnagar-Gross-Krook (BGK) model. Operator Ω is defined as

$$\Omega = -\frac{1}{\tau}\left(f - f^{(0)}\right) \tag{10.2}$$

where τ is the relaxation time, that is, the average time between two collisions, and $f^{(0)}$ is the equilibrium distribution function, the so-called Maxwell-Boltzmann distribution function.

LB method is based on principles that are valid for gases, but certain limitations are introduced to apply these principles in fluid dynamics. Maxwell-Boltzmann probability distribution function is taken from statistical mechanics, where it is used to determine the velocity of molecules. The details related to definition and derivation of Maxwell-Boltzmann distribution function are omitted here but can be found in literature (Carter, 2001; Kennard, 1938). The final expression for the equilibrium distribution function is given by

$$f^{(0)}(\mathbf{x}, \boldsymbol{\xi}, t) = \frac{\rho(\mathbf{x}, t)}{(2\pi\theta(\mathbf{x}, t))^{D/2}} \exp\left(-\frac{(\mathbf{u}(\mathbf{x}, t) - \boldsymbol{\xi})^2}{2\theta(\mathbf{x}, t)}\right) \tag{10.3}$$

In this expression,

D is the number of physical dimensions ($D=3$ for three-dimensional space),

$$\theta = k_B T / m,$$

$k_B = 1.38 \cdot 10^{-23} \frac{J}{K}$ is the Boltzmann constant,

T is absolute temperature in Kelvins (K),

m is particle mass (in the sequel it is considered that $m=1$).

Finally, BGK model of the Boltzmann equation, that is valid for continuum, is given by

$$\frac{\partial f}{\partial t} + \boldsymbol{\xi} \cdot \frac{\partial f}{\partial \mathbf{x}} + \mathbf{g} \cdot \frac{\partial f}{\partial \boldsymbol{\xi}} = -\frac{1}{\tau}\left(f - f^{(0)}\right) \tag{10.4}$$

The Boltzmann equation can be transformed to a system of equations that is used in fluid dynamics—continuity equation and Navier-Stokes equation. There are two approaches to obtain the required equations. One is to perform the Hermite polynomials basis projection on Eq. (10.4) and then to further transform obtained expressions. This procedure was first proposed and performed by Grad (Grad, 1949a; Grad, 1949b), and hence, it is also called Grad's 13-moment equations. The other approach was proposed by Chapman and Cowling (Chapman and Cowling, 1970), and it is called the Chapman-Enskog multiscale expansion. Within this procedure, the equations are separated on several scales such that all the derivatives and distribution function and all the quantities are expanded in terms of different order of the Knudsen number. The obtained expressions are further analyzed, and the required equations are obtained. There is also a simpler procedure that preserves the previously described basic idea and was first applied by Huang (Huang, 1987), and it is also explained in literature (Malaspinas, 2009; Đukić, 2015; Đukić, 2010). In this section, the derivation procedure is briefly discussed, to emphasize only some details that are relevant for the simulation setup.

Using the Hermite polynomial basis projection, the distribution function is expanded into an infinite series. However, in macroscopic equations (the mass and momentum conservation laws), only three quantities are of interest—the fluid density, momentum, and energy. Therefore, it can be concluded that members in the series after certain order can be neglected. This way the distribution function is truncated, that is, certain information that the original distribution function contained are lost. In literature, it is shown that this information is actually not required for fluid flow simulations (Malaspinas, 2009). With truncation of the distribution function up to order two, the most frequently used lattice Boltzmann model is obtained. This model is valid only for isothermal weakly compressible fluid, and hence, the energy conservation equation has no influence in the model. This is the most commonly used model because it represents the form of Navier-Stokes equations that is used in computational fluid dynamics.

In the derivation procedure, two important assumptions are introduced. The first assumption is related to the Knudsen number. Knudsen number is defined as the ratio between mean free path of a particle and the characteristic length scale of the domain:

$$Kn = \frac{l}{L} \tag{10.5}$$

In the expression (10.5), L denotes the characteristic length scale of the considered domain, and $l = c_s \tau$ is the mean free path of a particle (average distance that a particle travels between two collisions). Value c_s is defined as the characteristic velocity of particles, and it is also called the speed of sound in lattice units. Value of this parameter is defined within the discretization procedure, which will be discussed in the sequel. If Knudsen number would be close to 1 or greater than 1, the mean free path of a particle would be of the same order of magnitude as the characteristic length scale of the considered domain, and thus, the assumption that the fluid is simulated as a continuum would not be valid. Hence, the entire LB method and the BGK model are valid only in the limit of small Knudsen number ($Kn << 1$).

The second assumption is related to the Mach number. Mach number is defined as the ratio between characteristic fluid velocity and the "speed of sound" c_s:

$$\mathrm{Ma} = \frac{u}{c_s} \tag{10.6}$$

During the calculation of coefficients for the projection of equations on the Hermite polynomials basis, higher-order members are neglected, and thus, the entire LB method and BGK model are valid only in the limit of small Mach number.

These assumptions have to be taken into consideration when defining the parameters for the numerical simulation that will be discussed in the sequel.

BGK Boltzmann equation, which is defined in Eq. (10.1), is valid for continuum, and it is related to a continuous velocity field. This equation form is not suitable for numerical solving, and thus, it is necessary to perform discretization. The discretization procedure is performed in two steps. The first step is the discretization of the velocity space, and then, the space and time discretization is performed. The velocity space is discretized such that all integrals are calculated as weighted sums using the Gauss-Hermite quadrature (Krylov, 1962; Davis and Rabinowitz, 1984). Gauss-Hermite quadrature tends to calculate the value of integral as accurately as possible, by defining the optimal set of abscissae $\boldsymbol{\xi}_i$, $i = 1, 2, \ldots, q$ and weight coefficients ω_i, $i = 1, 2, \ldots, q$. These parameters define the lattice structure that is used in numerical simulations, and they actually define the directions along which the fictional particles are moving.

Lattice structure with q different velocity directions in d-dimensional space is denoted as $DdQq$ lattice. For simulations of isothermal incompressible fluid flow, several 2D and 3D lattice structures can be used, and for two most commonly used structures, an illustration of velocity directions is given in Fig. 10.1. Also the weight coefficients ω_i and vectors defining the abscissae $\boldsymbol{\xi}_i$ for both structures are listed in Table 10.1.

As it can be observed from Table 10.1, vectors that define abscissae don't have unit coordinates, and it is preferred to be so. Since individual values are irrational numbers $\xi_{ij} = \sqrt{3}$, a correction factor has to be introduced to obtain more appropriate values. For this purpose, the already mentioned

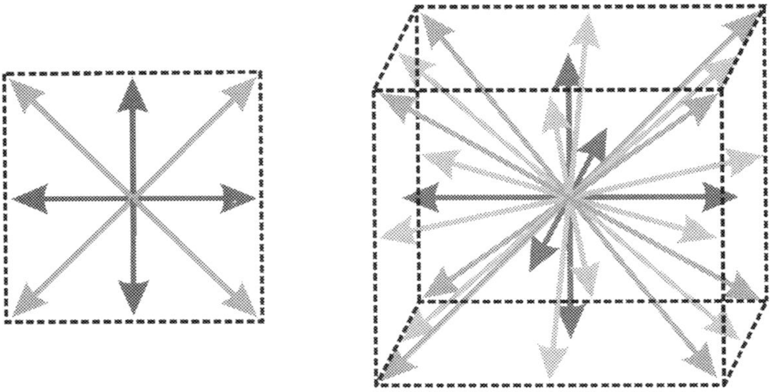

(A) (B)

Fig. 10.1 Illustration of velocity directions for lattice structures; (A) D2Q9; (B) D3Q27.

constant named speed of sound in lattice units c_s is used. This constant is actually a scale factor. Its value depends on the choice of discrete velocity set. For lattice structures that are considered in this chapter and that are listed in Table 10.1, the value of the constant is given by

$$c_s^2 = \frac{1}{3} \tag{10.7}$$

To obtain unit values of abscissae components, the following change of variables is introduced:

$$\bar{\boldsymbol{\xi}}_i = c_s \boldsymbol{\xi}_i \tag{10.8}$$

After the discretization of the velocity field, time and space discretizations are performed, based on the explanation that is given in literature (Malaspinas, 2009; He et al., 1998; Dellar, 2001). All equations are approximated using the trapezoidal rule and transformed to the explicit form, to make them more suitable for numerical solving.

When LB method is implemented, the basic equation is most commonly separated into two steps—collision step and propagation step. Two values of distribution function are defined—$\bar{f}_i^{in}$ and $\bar{f}_i^{out}$—and they represent values of the discretized distribution function before and after collision.

The collision step is given by

$$\bar{f}_i^{out}(\mathbf{x}, t) = \bar{f}_i^{in}(\mathbf{x}, t) - \frac{1}{\bar{\tau}}\left(\bar{f}_i^{in}(\mathbf{x}, t) - f_i^{(0)}(\rho, \mathbf{u})\right) + \left(1 - \frac{1}{2\bar{\tau}}\right)F_i \tag{10.9}$$

Table 10.1 Lattice structures—weight coefficients and vectors defining abscissae.

Structure	Weight coefficients	Vectors defining abscissae
D2Q9	$\omega_i = \begin{cases} 4/9 & i=0 \\ 1/9 & i=1,2,3,4 \\ 1/36 & i=5,6,7,8 \end{cases}$	$\xi_i = \begin{cases} (0,0) & i=0 \\ (\pm\sqrt{3},0);(0,\pm\sqrt{3}) & i=1,2,3,4 \\ (\pm\sqrt{3},\pm\sqrt{3}) & i=5,6,7,8 \end{cases}$
D3927	$\omega_i = \begin{cases} 8/27 & i=0 \\ 2/27 & i=1,\dots,6 \\ 1/216 & i=7,\dots,14 \\ 1/54 & i=15,\dots,26 \end{cases}$	$\xi_i = \begin{cases} (0,0,0) & i=0 \\ (\pm\sqrt{3},0,0);(0,\pm\sqrt{3},0);(0,0,\pm\sqrt{3}) & i=1,\dots,6 \\ (\pm\sqrt{3},\pm\sqrt{3},\pm\sqrt{3}) & i=7,\dots,14 \\ (\pm\sqrt{3},\pm\sqrt{3},0);(0,\pm\sqrt{3},\pm\sqrt{3});(\pm\sqrt{3},0,\pm\sqrt{3}) & i=15,\dots,26 \end{cases}$

The propagation step is given by

$$\bar{f}_i^{in}\left(\mathbf{x}+\bar{\boldsymbol{\xi}}_i, t+1\right)=\bar{f}_i^{out}(\mathbf{x}, t) \tag{10.10}$$

In Eq. (10.9), F_i denotes the external force term. To accurately implement the effect of external force in LB method, it is necessary to be careful and to preserve all the assumptions and not damage the derivation procedure. Detailed analysis of the inclusion of external force was conducted by Guo et al. (2002), and based on these conclusions, the external force is implemented in this chapter. The force term is given by

$$\mathbf{F}_i=\omega_i\rho\left(\frac{\bar{\boldsymbol{\xi}}_i-\mathbf{u}}{c_s^2}+\frac{\left(\bar{\boldsymbol{\xi}}_i\cdot\mathbf{u}\right)\bar{\boldsymbol{\xi}}_i}{c_s^4}\right)\cdot\mathbf{g}(\mathbf{x}, t) \tag{10.11}$$

In numerical simulation, the domain of interest is divided in a lattice mesh, containing a predefined number of nodes. Collision and propagation steps described in Eqs. (10.9) and (10.10) are repeated in a series of iterations, whereas each of these two steps must be applied to all nodes of the mesh, before the next step starts.

Macroscopic quantities relevant for the fluid flow, such as density, pressure, and velocity, are defined in terms of the calculated components of the distribution function.

Fluid density is given by

$$\rho=\sum_i\bar{f}_i \tag{10.12}$$

Fluid viscosity is defined as

$$\mu=c_s^2\rho\left(\bar{\tau}-\frac{1}{2}\right) \tag{10.13}$$

The expression for fluid velocity is given by

$$\bar{\mathbf{u}}=\frac{1}{\rho}\sum_i\bar{\boldsymbol{\xi}}_i\bar{f}_i=\mathbf{u}-\frac{\mathbf{g}}{2} \tag{10.14}$$

From Eq. (10.14), it is clear that velocity that is calculated in numerical simulations and that is denoted by $\bar{\mathbf{u}}$ does not represent "physical velocity" of the fluid flow. To determine the physical velocity (which is one of the most important characteristics of the flow at macroscale level), it is necessary to perform the following conversion:

$$\mathbf{u}=\bar{\mathbf{u}}+\frac{\mathbf{g}}{2} \tag{10.15}$$

Of course, in case that there is no external force field (when $\mathbf{g}=0$), velocity is the same regardless of the mentioned change of variables ($\mathbf{u}=\overline{\mathbf{u}}$).

Pressure is calculated in terms of fluid density in the considered mesh node, applying the following equation:

$$p=c_s^2\rho \tag{10.16}$$

Within LB method, a special principle is applied for the calculation of all physical quantities, such that all macroscopic quantities required for the simulation (and all quantities obtained as the result of the simulation) are defined in the so-called system of lattice units. Hence, it is necessary to determine the values of all parameters in dimensionless form before starting the simulation. Also, when the results are visualized and printed out, it is necessary to transform dimensionless quantities back to physical quantities.

One of characteristic quantities that is defined before starting LB simulations is the characteristic velocity (in dimensionless lattice units) that is denoted by u_{lb}. This parameter is very important because the appropriate choice of its value ensures that the assumptions made during the derivation procedure remain valid, that is, that the flow remains in the limit of small Mach and Knudsen number. This also practically means that the fluid remains weakly compressible. The value of characteristic velocity has to satisfy the following inequation:

$$u_{lb}<\frac{1}{c_s} \tag{10.17}$$

Another important parameter is the relaxation time. The relaxation time can be calculated as

$$\overline{\tau}=\frac{1}{c_s^2}\nu_p\frac{\Delta x^2}{\Delta t}+\frac{1}{2} \tag{10.18}$$

where ν_p is the kinematic viscosity in physical units, Δx is the scale factor for length, and Δt is the scale factor for time.

To ensure good stability of numerical solution and LB simulation, it is necessary to ensure that the obtained value of relaxation time satisfies the following inequation:

$$0.5\leq\overline{\tau}\leq 1 \tag{10.19}$$

and that this value is as close to 1 as possible. If this condition is not satisfied during the definition of parameters, then some other parameter has to be changed adequately (either the characteristic velocity u_{lb} or the number of nodes [resolution] N), to obtain a suitable relaxation time.

2.2 Modeling solid deformation

Deformable bodies that are considered in this chapter are frequently subjected to large deformations, and they consist of a highly deformable membrane and an inner structure. It is assumed that the inner structure behaves like a fluid. Also, it is considered that the membrane of the deformable body has negligible thickness and that it is made of a certain number of points that are interconnected. The discretization of the membrane is performed such that the entire membrane is divided on a defined number of triangles. Each node of this mesh is subjected to external deformation, due to the surrounding fluid. This deformation causes a reaction force with which the particular node opposes the deformation. The deformation can be divided in four categories—change of volume within membrane, change of surface area of the membrane, surface strain of the membrane, and bending of the membrane. This approach was used by Dupin et al. (2007). In the sequel of this section, it will be described how to model the reaction of deformable body to each of these four types of deformation, according to the expressions given in literature (Dupin et al., 2007).

2.2.1 Modeling reaction to the surface strain of the membrane

When modeling the surface strain of the membrane, it is considered that the membrane is a set of nodes that are interconnected with springs. Every movement of the nodes causes a spring force that represents the reaction to this type of deformation. If nodes j_1 and j_2 are analyzed, the unit vector $\mathbf{l}_{12}$ can be calculated as the vector that connects these nodes. If the current distance between these nodes is denoted by L_{12} and the initial distance between considered nodes is denoted by ${}^0L_{12}$, then the reaction force caused by the movement of nodes can be defined as

$$\mathbf{F}^S_{j_1} = -\mathbf{F}^S_{j_2} = K^S \frac{L_{12} - {}^0L12}{{}^0L12} \mathbf{l}_{12} \tag{10.20}$$

where K^S represents the parameter of resistance to surface strain that is defined during simulation setup for the particular type of deformable body.

All the mentioned variables are schematically shown in Fig. 10.2.

2.2.2 Modeling the reaction to the change of volume

Internal volume of the deformable body can be calculated using the expression

$$V = \sum_i A_i \mathbf{n}_i \cdot \mathbf{h}_i \tag{10.21}$$

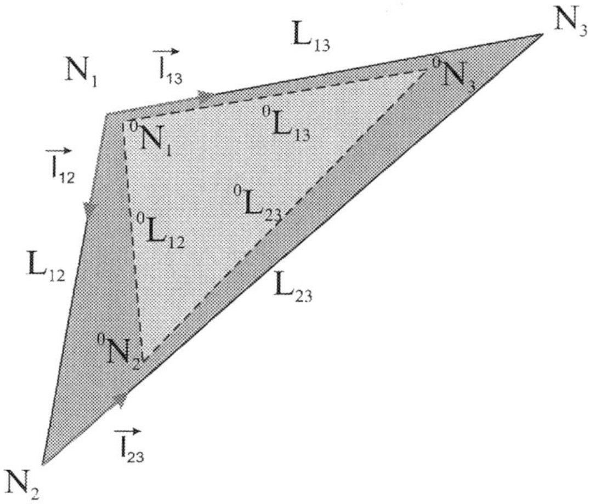

Fig. 10.2 Illustration of quantities necessary for defining the reaction force due to the surface strain.

Here, the summation is performed over all elements of the mesh $i=1, 2, \ldots, M$, where M is the number of elements in the mesh, $\mathbf{h}_i$ represents the vector that connects the center of gravity of the considered element with the closest point in the chosen coordinate plane (in this case, the x–y plane is chosen), $\mathbf{n}_i$ represents the normal unit vector of the considered element, and A_i is the area of the considered element. All mentioned variables are schematically shown in Fig. 10.3.

To define the reaction force due to the change of volume, it is assumed that the deformable body tends to keep the internal volume constant over time. If the current volume of a deformable body is denoted by V and the initial volume is denoted by ${}^0V$, then the reaction force in considered element can be defined as

$$\mathbf{F}_i^V = -K^V V - {}^0V^0 V A_i \mathbf{n}_i \tag{10.22}$$

where K^V represents the parameter of resistance to change of volume that is defined during simulation setup for the particular type of deformable body.

In this case, the entire membrane is divided in triangular elements, and hence, the calculated reaction force for each element is equally distributed on all three nodes of the considered element:

$$\mathbf{F}_{i1}^V = \mathbf{F}_{i2}^V = \mathbf{F}_{i3}^V = \frac{1}{3}\mathbf{F}_i^V \tag{10.23}$$

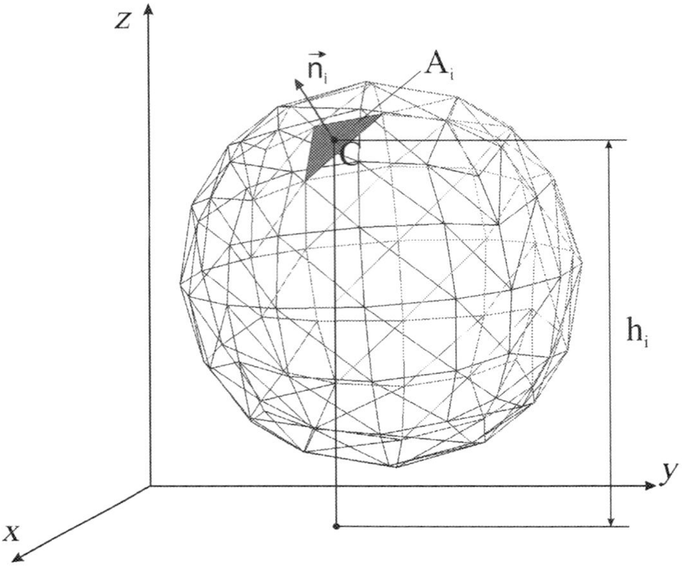

Fig. 10.3 Illustration of quantities necessary for defining the reaction forces due to the change of volume.

2.2.3 Modeling the reaction to the change of surface area of the membrane

Change of surface area is considered for two cases—the change of surface area of each element separately and the change of overall surface area of the membrane. It was analyzed in literature, for the case of erythrocyte, that the change of overall surface area has a greater influence on the reaction force than the change of surface area of individual elements (Dupin et al., 2007).

If the current surface area of *i*th element of the membrane is denoted by A_i and the initial surface area of *i*th element is denoted by ${}^0A_i$ and if the current overall surface area of the membrane is denoted by A and the initial overall surface area is denoted by ${}^0A$, then the reaction force can be defined as

$$\mathbf{F}_{ij}^{A} = -\left(K^{Al}\frac{A_i - {}^0Ai}{A_i} + K^{At}\frac{A - {}^0A}{A}\right)\mathbf{w}_j \tag{10.24}$$

where i denotes the index of the element and j is the local index of the node within the considered element ($j = 1, 2, 3$). Unit vector whose direction

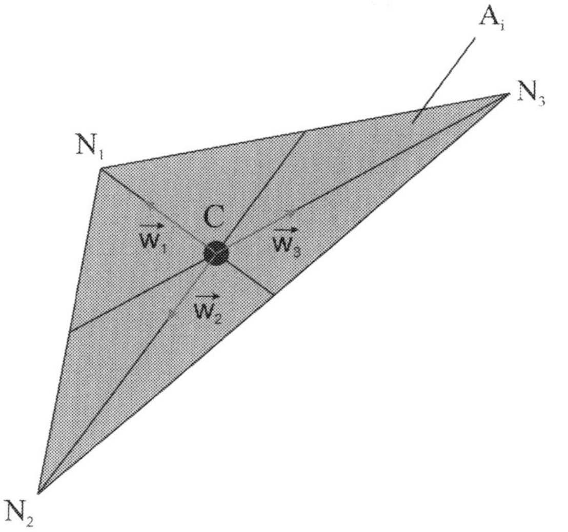

Fig. 10.4 Illustration of quantities necessary for defining the reaction forces due to the change of surface area.

coincides with the line that connects the center of gravity of the considered element with the particular node is denoted by $\mathbf{w}_j$. All mentioned variables are schematically shown in Fig. 10.4.

K^{Al} is the parameter of resistance to the local change of surface area, and K^{At} is the parameter of resistance to the overall (total) change of surface area. These parameters are defined during simulation setup for the particular type of deformable body. The previously mentioned condition (that the influence of overall surface area of the membrane on each element independently is greater than the influence of change of surface area of individual elements) is taken into consideration when defining the parameters K^{Al} and K^{At} such that $K^{At} >> K^{Al}$.

2.2.4 Modeling the reaction to the bending of the membrane

When the bending of the membrane is modeled, the change of angle between two elements is analyzed. All relevant variables are schematically shown in Fig. 10.4. The normal unit vectors of the considered elements i_1 and i_2 are denoted by $\mathbf{n}_{i1}$ and $\mathbf{n}_{i2}$. If the current angle between elements i_1 and i_2 is denoted by $\varphi_{i_1 i_2}$ and the initial angle between considered elements is denoted by ${}^0\varphi_{i_1 i_2}$, then the reaction force appears in all three nodes

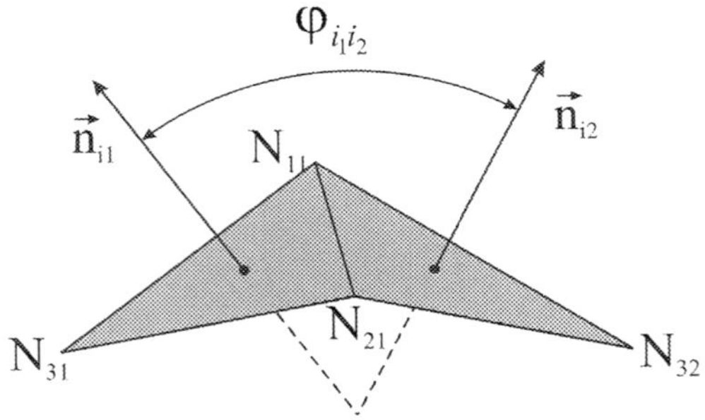

Fig. 10.5 Illustration of quantities necessary for defining the reaction forces due to bending.

of both elements (Dupin et al., 2007). The reaction force in the node that is not common to both elements (the nodes denoted by N_{31} and N_{32} in Fig. 10.5) can be defined as

$$\mathbf{F}_{j3}^{B} = K^{B}\frac{\varphi_{i_1 i_2} - {}^{0}\varphi_{i_1 i_2}}{{}^{0}\varphi_{i_1 i_2}}\mathbf{n}_{i_1} \tag{10.25}$$

and the force in common nodes can be defined as

$$\mathbf{F}_{j12}^{B} = \mathbf{F}_{j21}^{B} = -\frac{1}{2}\mathbf{F}_{j3}^{B} \tag{10.26}$$

where K^B represents the parameter of resistance to the bending that is defined during simulation setup for the particular type of deformable body.

The total reaction force of the deformable body caused by the defined deformation is calculated in each node as a sum of forces calculated for all four types of deformation.

$$\mathbf{F}_i(t) = \mathbf{F}_i^S + \mathbf{F}_i^V + \mathbf{F}_i^A + \mathbf{F}_i^B \tag{10.27}$$

2.3 Modeling solid-fluid interaction

In the phenomena that are considered in this chapter, the surrounding fluid and immersed cells form a coupled mechanical system. The fluid is acting on the immersed cell via external forces and causes the motion and deformation of this deformable immersed body and vice versa—the cell is opposing the deformation and influence of fluid, and this way alters the fluid flow. For modeling this type of solid-fluid interaction, two approaches can be used: loose and strong coupling. In the case of loose coupling, solid and fluid are simulated separately, and all the required parameters obtained in the solver for one domain are transferred to the solver for the other domain. A separate external program manages the solvers for both domains and ensures the appropriate exchange of information. In the case of strong coupling, both domains are simulated at the same time, as if it was a single mechanical system. All equations of the numerical model are solved in every time step, and this ensures that all required quantities (both for solid and fluid) are changing at the same time. Loose coupling is simpler to implement, but one of the main drawbacks of this approach is the fact that it is necessary to use same time steps for numerical solving of both domains, which is not always possible because the characteristics of solid and fluid are different. Strong coupling is applied when it is necessary to precisely model the motion of objects within the fluid domain. The main drawback of the second approach is that the solver is much more complex and the execution of the program requires more time. Strong coupling is used in the numerical model presented in this chapter, due to the higher accuracy and possibility to apply different time steps and different discretization meshes for solid and fluid domains.

There are several approaches to numerically model the motion of immersed body within the fluid domain. Some of them are using coordinate transformations (Mateescu et al., 1996), the moving mesh finite element approach (Li et al., 2002), or applying the arbitrary Lagrangian-Eulerian (ALE) formulation (Hu et al., 2001), or the immersed boundary method (IBM) (Feng and Michaelides, 2004). Each of these approaches has its advantages and drawbacks. In the numerical model presented in this chapter, the immersed boundary method is used. The reason for this choice is that this method does not require any changes in the system of equations for the fluid domain, since an external force field is included to model the effect of immersed body on the fluid.

Immersed boundary method was presented by Peskin (1977). A fixed Cartesian mesh that consists of Eulerian points is used to represent the fluid domain. This representation corresponds with the representation used in LB method and presented in Section 2.1. A set of Lagrangian points is used to represent the boundary surface between the fluid and immersed body. This boundary is assumed to be easily deformable, with high stiffness (Wu and Shu, 2010). This method was successfully applied for modeling the motion of rigid (Feng and Michaelides, 2004; Wu and Shu, 2010; Đukić, 2012) and deformable bodies immersed in fluid (Krüger et al., 2011; Murayama et al., 2011).

The overall number of Lagrangian points on the boundary is denoted by G, and the coordinates of the lth point ($l = 1, 2, \ldots, G$) are denoted by $\mathbf{X}^l_B(t)$.

Since the discretization of fluid and solid domain is different, all quantities that are required for the modeling of solid-fluid interaction are interpolated over the boundary points. This practically means that for each boundary point of the immersed object, the influence of several close points from the fluid domain is considered and vice versa. The interpolation in immersed boundary method is performed using the Dirac delta function that is given by

$$\delta(\mathbf{x} - \mathbf{X}_B(t)) = D_{ij}\left(\mathbf{x}_{ij} - \mathbf{X}^l_B\right) = \delta\left(x_{ij} - X^l_B\right)\delta\left(y_{ij} - Y^l_B\right)\delta\left(z_{ij} - Z^l_B\right) \tag{10.28}$$

The value of function $\delta(r)$ is defined in literature (Peskin, 1977)

$$\delta(r) = \begin{cases} \frac{1}{4h}\left(1 + \cos\left(\frac{\pi r}{2}\right)\right), & |r| \leq 2 \\ 0, & |r| > 2 \end{cases} \tag{10.29}$$

where the distance between two points of the fixed fluid mesh is denoted by h. In this case, since LB method is considered, this distance is defined to be equal to 1.

Fluid affects the solid, and this influence is modeled considering the no-slip boundary condition on the boundary between these two domains. To satisfy the no-slip boundary condition, the velocity of the fluid in a boundary point has to be equal to the velocity of solid in the same boundary point. Velocity in Lagrangian points is interpolated over the surrounding Eulerian points of the fluid domain. In this way, the velocity of boundary points of the immersed object can be calculated as follows:

$$\mathbf{u}^l_B\left(\mathbf{X}^l_B, t\right) = \sum_{i,j} \mathbf{u}\left(\mathbf{x}_{ij}, t\right) D_{ij}\left(\mathbf{x}_{ij} - \mathbf{X}^l_B\right) \tag{10.30}$$

The Euler Forward method can then be applied on Eq. (10.30), to obtain the new position of boundary points:

$$^{t+\Delta}t\mathbf{X}B^l = {}^t\mathbf{X}B^l + \mathbf{u}^l_B \Delta t \tag{10.31}$$

Using the new coordinates of the immersed object, the deformation caused by the surrounding fluid can be calculated, as well as the reaction forces that this new deformation has caused, like it was described in Section 2.2.

The reaction force in time point t, in the lth point of the boundary, can be denoted by $\mathbf{F}_l(t)$.

The external force field that acts on the fluid is interpolated over the boundary surface. The external force field in a particular point of the fixed Cartesian mesh is given by

$$\mathbf{g}\left(\mathbf{x}_{ij}, t\right) = \sum_{l=1}^{m} \mathbf{F}_l(t) D_{ij}\left(\mathbf{x}_{ij} - \mathbf{X}_B^l(t)\right) \tag{10.32}$$

This external force is included in the equations for the fluid domain (i.e., it is used to calculate the force term using Eq. (10.11) and afterwards incorporated in the Eq. (10.9) of the LB method), and this force causes changes in the fluid flow. The surrounding fluid again causes new deformation of the immersed body, and the entire process is repeated in iterations.

3 Numerical simulations

In this section, the results obtained in numerical simulations will be presented. First, the details of the simulation setup are discussed. Besides the parameters related to the fluid domain, to simulate the behavior of deformable objects correctly, during the simulation setup, it is necessary to define the values of parameters of resistance to the four types of deformation that were defined in Section 2.2, for each type of deformable body. In this chapter, the motion of two deformable bodies is considered: the motion of red blood cell (RBC) and CTC.

RBC has a biconcave shape. It can be considered that this shape is formed such that one big sphere is deformed by two smaller spheres on both sides along the same axis and this makes the surface of the RBC membrane concave. Due to this shape, RBC has around 50% greater surface area than a spherical particle of an equal diameter. The equation that describes the shape of the RBC was empirically defined by Evans and Fung (Evans and Fung, 1972), and this equation is used to create the triangular mesh of the membrane of the RBC for numerical simulations. Values of parameters of resistance to the four types of deformation are taken to be equal to the ones proposed in literature (Bagchi et al., 2005; Dao et al., 2003) and with values that were used by Dupin et al. (Dupin et al., 2007). These values are given in Table 10.2.

Since CTCs are shed from a primary tumor, their characteristics depend on the type of primary tumor. In this chapter, the behavior of a breast cancer

Table 10.2 Values of parameters of resistance to deformation for the RBC.

Parameter	Value
K^V	$50 \mathrm{pN\mu m}^{-1}$
K^{At}	$1.67 \mathrm{pN\mu m}^{-1}$
K^{Al}	$1.67 \cdot 10^{-1} \mathrm{pN\mu m}^{-1}$
K^B	$10^{-1} \mathrm{pN\mu m}^{-1}$
K^S	$5 \cdot 10^{-1} \mathrm{pN\mu m}^{-1}$

Table 10.3 Values of parameters of resistance to deformation for the CTC.

Parameter	Value
K^V	$14.5 \mathrm{pN\mu m}^{-1}$
K^{At}	$0.48 \mathrm{pN\mu m}^{-1}$
K^{Al}	$0.48 \cdot 10^{-1} \mathrm{pN\mu m}^{-1}$
K^B	$0.29 \cdot 10^{-1} \mathrm{pN\mu m}^{-1}$
K^S	$1.4375 \cdot 10^{-1} \mathrm{pN\mu m}^{-1}$

cell is modeled. This particular type was chosen because relevant experimental data were published in literature (Guck et al., 2005; Gusenbauer et al., 2011) and the characteristics of CTC were taken to be equal to the ones proposed in cited literature. Values of parameters of resistance to the four types of deformation are listed in Table 10.3. The CTC has a spherical shape, and the diameter of CTC is two times greater than the diameter of the RBC.

First, only the fluid flow through several diverse microfluidic chips was simulated. The flow velocity is one of the important parameters that influence the capture efficiency of the chip, because it affects the duration of the contact between the cell and pillars. Fig. 10.6 shows the simulated fluid flow for four different chosen shapes of pillars: circular (Fig. 10.6A), squared (Fig. 10.6B), triangular (Fig. 10.6C), and a special shape proposed in literature (Lv et al., 2013) (Fig. 10.6D). As it can be seen from the plotted velocity streamlines, the shape of pillars affects the flow and thus also affects the deformation of immersed cells, like it was also concluded in literature (Ranjan et al., 2014; Karabacak et al., 2014; Liu et al., 2013).

The motion and deformation of the RBC in bloodstream were already successfully modeled within the caudal vein plexus of the animal species called zebrafish. These numerical results are validated using experimental data, and the good agreement of results was reported in literature (Đukić et al., 2016). This approach was used in this chapter as well, for the modeling of motion of deformable cells within microfluidic chip. The microfluidic

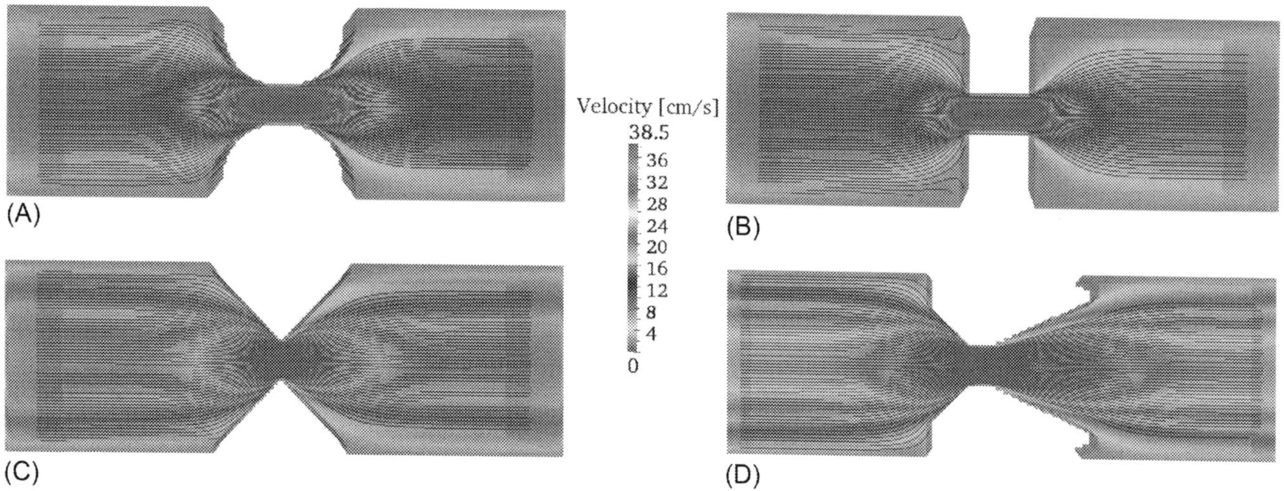

Fig. 10.6 Fluid flow through microfluidic chip with different pillar shapes: (A) circular pillar; (B) squared pillar; (C) triangular pillar; and (D) shape of pillar proposed in literature (Lv et al., 2013).

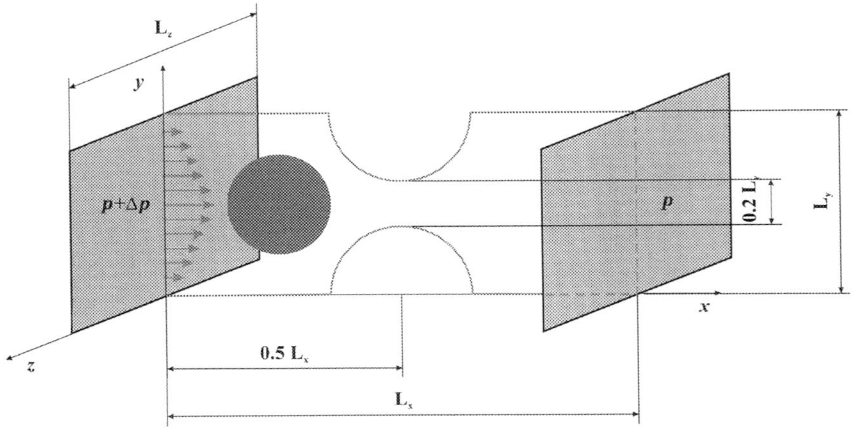

Fig. 10.7 Geometry of the microfluidic chip with circular pillars.

chip with circular pillars was chosen for these simulations. The geometry of the particular design of the chip is schematically shown in Fig. 10.7. Dimensions of the domain in all three directions are defined such that the following relations are valid: $L_y = L_z$ and $L_x = 3L_y$. Parameters related to the fluid flow simulation using LB method are given in Table 10.4. Boundary condition is defined such that the velocity on the walls is equal to zero and the initial condition is defined as a prescribed pressure difference between inlet and outlet (along the x-axis).

Both RBC and CTC are initially placed such that the coordinates of its center of gravity are given by $x_C = \frac{2L_y}{5}$, $y_C = \frac{L_y}{2}$, and $z_C = \frac{L_y}{2}$. Distance between the pillars is set to be five times smaller than the overall length of the domain along the y-axis. The resolution of the domain is set such that the diameter of the RBC is 1.35 times greater than the mentioned distance between the fixed pillars.

Fig. 10.8A shows the simulated distribution of the velocity through the microfluidic chip, with several positions of the CTC highlighted with red lines. These red lines represent the cross sections of the CTC along the x-y plane in several moments in time. More detailed shapes of the CTC

Table 10.4 Parameters for the simulation of fluid flow using LB method.

N	50
u_{lb}	0.01
Re	1

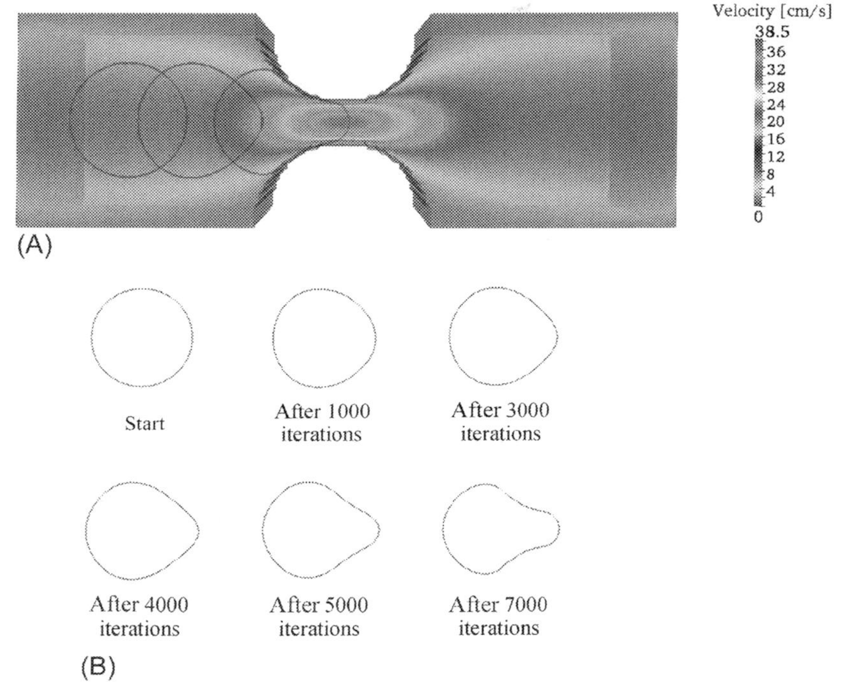

Fig. 10.8 Numerical simulation of motion of CTC through a microfluidic chip with circular pillars; (A) distribution of velocity and several positions of CTC represented with red lines; (B) shapes of the CTC during simulation.

are shown in Fig. 10.8B, where it is also noted the exact number of iterations necessary to reach this shape. Similar results for the RBC are shown in Figs. 10.9A and B. Fig. 10.9A shows the motion of RBC through the microfluidic chip, while Fig. 10.9B shows the detailed shapes of the RBC after certain number of iterations during the simulation. Figs. 10.8 and 10.9 show that numerical results are in accordance with the main idea of the microfluidic chips for separation of cells based on size and deformability. Due to the higher deformability of the RBC, it can pass through the space between pillars, while the less deformable CTC remains captured.

As it was mentioned before, numerical simulations can be used for the analysis of potential design of microfluidic chip and can help to determine the most efficient geometric arrangement of pillars within the chip. One of the examples of such analysis is the examination of the most adequate pillar radius. The previously described simulation setup was used for a sensitivity analysis, more precisely, for the analysis of influence of the pillar radius on

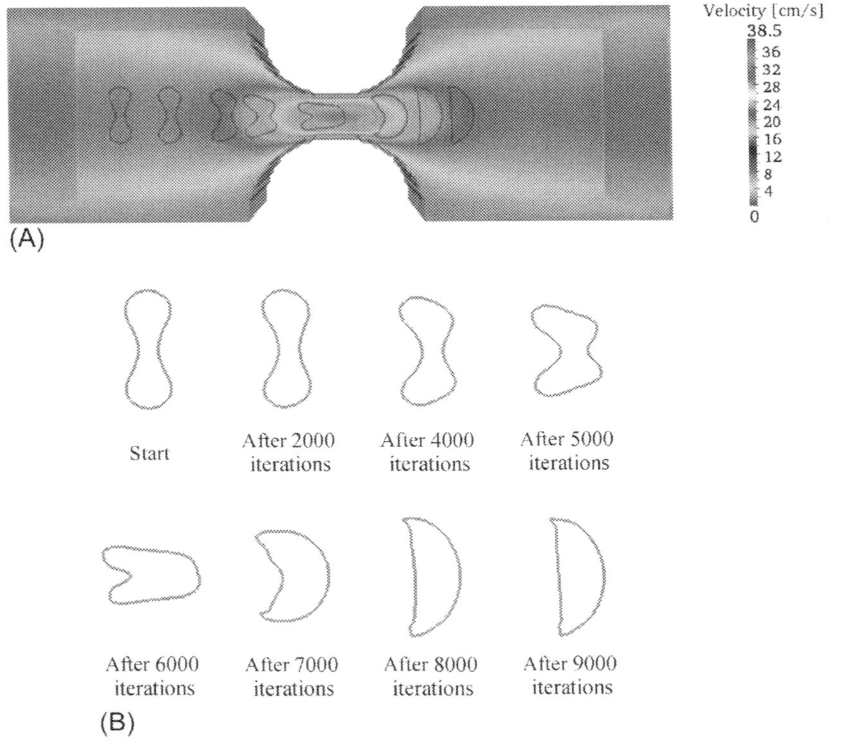

Fig. 10.9 Numerical simulation of motion of RBC through a microfluidic chip with circular pillars; (A) distribution of velocity and several positions of RBC represented with red lines; (B) shapes of the RBC during simulation.

the isolation of the CTC. The position of the pillars was always the same, the one defined in Fig. 10.7. As the radius of the pillar grows, the distance between pillars decreases, and the passage of the CTC is not possible and vice versa. This can be observed from the graph in Fig. 10.10.

The proposed numerical model was also validated using the experimental data. In already cited paper, Lv et al. (2013) presented a microfluidic chip system to isolate CTCs from the blood of cancer patients. This system was used to verify the results of the numerical model, and a more detailed validation and analysis of the simulation setup was discussed in literature (Đukić et al., 2015). The simulation setup is very similar to the previously described setup. Instead of only two pillars, here, the pillars are organized in a specific manner, with predefined spaces between them. Part of the simulated domain is shown in Fig. 10.11. The distribution of velocity, with drawn velocity

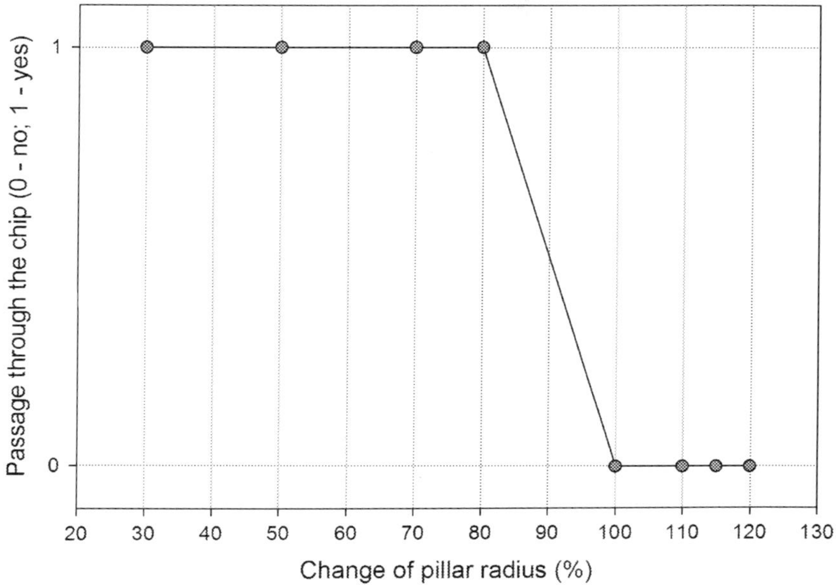

Fig. 10.10 Influence of pillar radius on the capture of the CTC within microfluidic chip with circular pillars.

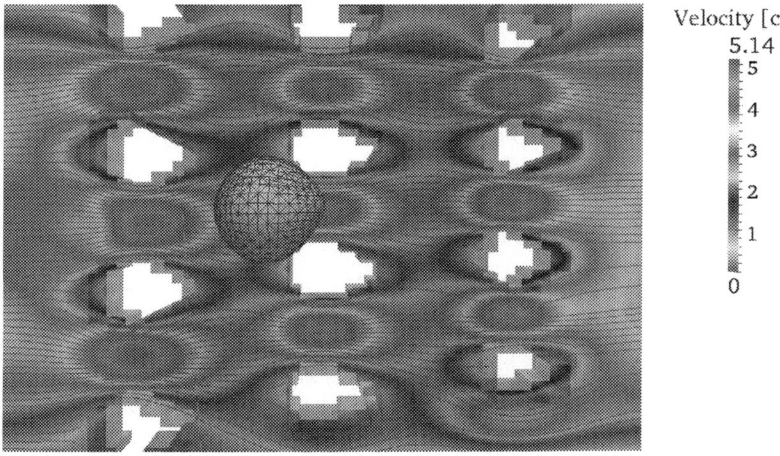

Fig. 10.11 Result of numerical simulation of motion of CTC through a microfluidic chip with pillars as proposed in literature (Lv et al., 2013).

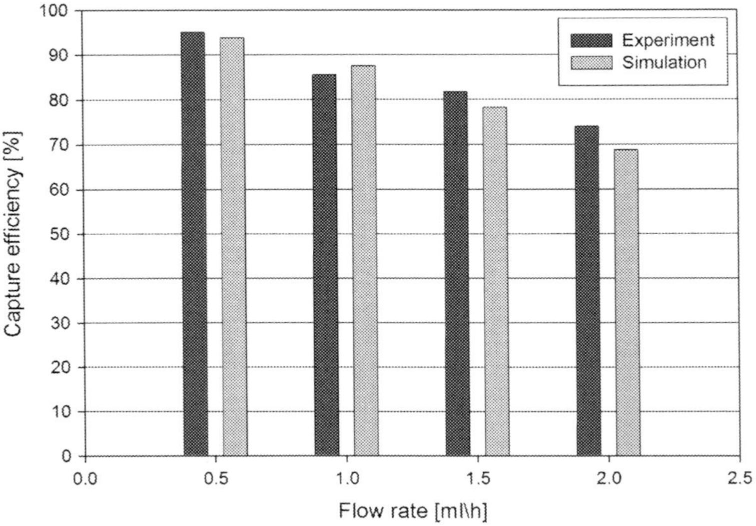

Fig. 10.12 Comparison of capture efficiency obtained in experiment (Lv et al., 2013) and numerical simulation, for different flow rates.

streamlines and a captured CTC, are also shown in Fig. 10.11. The comparison of results obtained in experiment and numerical simulation is shown in Fig. 10.12. The capture efficiency is compared with different flow rates. As it can be observed, for higher flow rates, the number of captured CTCs is smaller, because due to faster fluid flow, greater forces act on the deformable cells and cause higher deformations, thus ensuring the passage of some CTCs that under lower flow rates stay captured. The good agreement of obtained numerical results with the experimental values can be observed from Fig. 10.12, but it is also demonstrated through the value of standard deviation that is equal to 6.78%.

4 Conclusion

Separation of CTCs in a microfluidic chip based on physical properties such as size, shape, and deformability is a promising technique for separation of CTCs from a blood sample, because they don't require any previous treatment of the sample, that is, antibody labeling or any type of chemical reagents. This is a great advantage, because these chips offer a less expensive clinical tool that requires shorter time for the preparation of the blood sample and also ensures that all isolated cells remain viable and ready for further analysis and diagnostics.

The process of separation of CTCs in a microfluidic chip can be hardly analyzed in detail in experimental conditions. Due to the small dimensions of the system, detailed measurements and tracking of individual cells are difficult. Numerical simulations offer a good alternative for studying and better understanding of the behavior of individual cells. The information about the deformation of these cells and their exact path through the chip could be very useful for the improvements of the design of microfluidic chips to obtain higher capture efficiency and selectivity.

In this chapter, a numerical model is presented that is capable of simulating the motion of deformable cells through a microfluidic chip. The drawback of the presented results is that simulations included only a single deformable cell and no interaction between one CTC and other CTCs and RBCs was considered. This numerical model has the capability to simulate multiple CTCs and RBCs, as well as their interaction, but this type of simulation would require higher amount of computer resources, and the execution time would significantly increase. In the future improvements of the model, more attention should be dedicated to a better optimization and parallelization of the source code, to enable these additional capabilities. The good agreement with experimental results demonstrated the capability of the proposed model to simulate the complex process of isolation of CTCs. This enables clinicians and researchers to observe what will occur with the suspended cells in complex geometries of the microfluidic chip. The sensitivity analysis showed another advantage of the proposed model—using these simulations, it is possible to examine the effect of changes in the design of the microfluidic chip. The proposed numerical model can be used as an effective tool to improve microfluidic chip design and to precisely control and define the most optimal parameters that ensure high purity, throughput, and capture efficiency.

Acknowledgments

This study is supported by the Ministry of Education, Science and Technological Development of the Republic of Serbia (projects number III41007 and ON174028). This study is also supported by the SCOPES project (IP:Z74Z0_137357). We acknowledge that the results of this research have been achieved using the PRACE-3IP project (FP7 RI-312763) resource FIONN based in Ireland at The Irish Centre for High-End Computing (ICHEC).

References

Bagchi, P., Johnson, P.C., Popel, A.S., 2005. Computational fluid dynamic simulation of aggregation of deformable cells in a shear flow. J. Biomech. Eng. 127 (7), 1070–1080.
Baret, J.C., Miller, O.J., Taly, V., Ryckelynck, M., El-Harrak, A., Frenz, L., et al., 2009. Fluorescence-activated droplet sorting (FADS): efficient microfluidic cell sorting based on enzymatic activity. Lab Chip 9, 1850–1858.
Bhatnagar, P.L., Gross, E.P., Krook, M., 1954. A model for collision processes in gases. I. small amplitude processes in charged and neutral one-component systems. Phys. Rev. E 77 (5), 511–525.
Carter, A.H., 2001. Classical and Statistical Thermodynamics. Prentice-Hall, Inc, New Jersey.
Chapman, S., Cowling, T.G., 1970. The Mathematical Theory of Nonuniform Gases. Cambridge University Press, Cambridge.
Choi, S., Park, J.K., 2005. Microfluidic system for dielectrophoretic separation based on a trapezoidal electrode array. Lab Chip 5, 1161–1167.
Cristofanilli, M., Budd, G.T., Ellis, M.J., Stopeck, A., Matera, J., Miller, M.C., et al., 2004. Circulating tumor cells, disease progression, and survival in metastatic breast cancer. N. Engl. J. Med. 351, 781–791.
Dao, M., Lim, C.T., Suresh, S., 2003. Mechanics of human red blood cell deformed by optical tweezers. J. Mech. Phys. Solids 51, 2259–2280.
Davis, P.J., Rabinowitz, P., 1984. Methods of Numerical Integration. New York.
Dellar, P.J., 2001. Bulk and shear viscosities in lattice Boltzmann equations. Phys. Rev. E 64 (3):031203.
T. Đukić, 2010. Lattice Boltzmann Method, BSc Thesis, University of Kragujevac, Serbia.
T. Đukić, 2012. Modeling Solid-Fluid Interaction Using LB Method, MSc Thesis, University of Kragujevac, Serbia.
T. Đukić, 2015. Modeling Motion of Deformable Body Inside Fluid and Its Application in Biomedical Engineering, PhD dissertation, University of Kragujevac, Serbia.
Đukić, T., Filipović, N., 2017. Numerical modeling of the cupular displacement and motion of otoconia particles in a semicircular canal. Biomech. Model. Mechanobiol. 16 (5), 1669–1680.
Đukić, T., Karthik, S., Saveljić, I., Djonov, V., Filipović, N., 2016. Modeling the behavior of red blood cells within the caudal vein plexus of zebrafish. Front. Physiol. https://doi.org/10.3389/fphys.2016.00455.
Đukić, T., Topalović, M., Filipović, N., 2015. Numerical simulation of isolation of cancer cells in a microfluidic chip. J. Micromech. Microeng. 25 (8), 084012.
Dupin, M., Halliday, I., Care, C., Alboul, L., Munn, L., 2007. Modeling the flow of dence suspensions of deformable particles in three dimensions. Phys. Rev. E Stat. Nonlinear Soft Matter Phys. 75 (6), 066707.
Evans, E.A., Fung, Y.C., 1972. Improved measurements of the erythrocyte geometry. Microvasc. Res. 4 (4), 335–347.
Feng, Z., Michaelides, E., 2004. The immersed boundary-lattice Boltzmann method for solving fluid-particles interaction problem. J. Comput. Phys. 195 (2), 602–628.
Filipović, N., Isailović, V., Đukić, T., Ferrari, M., Kojić, M., 2012. Multi-scale modeling of circular and elliptical particles in laminar shear flow. IEEE Trans. Biomed. Eng. 59 (1), 50–53.
Filipović, N., Živić, M., Obradović, M., Đukić, T., Marković, Z., Rosić, M., 2014. Numerical and experimental LDL transport through arterial wall. Microfluid. Nanofluid. 16 (3), 455–464.
Fleischer, R.L., Price, P.B., Symes, E.M., 1964. Novel filter for biological materials. Science 143, 249–250.

Godin, M., Delgado, F.F., Son, S.M., Grover, W.H., Bryan, A.K., Tzur, A., et al., 2010. Using buoyant mass to measure the growth of single cells. Nat. Methods 7, 387–390.
Grad, H., 1949a. Note on the N-dimensional Hermite polynomials. Commun. Pure Appl. Math. 2 (4), 325–330.
Grad, H., 1949b. On the kinetic theory of rarefied gases. Commun. Pure Appl. Math. 2 (4), 331–407.
Gress, F., Gottlieb, K., Sherman, S., Lehman, G., 2001. Endoscopic ultrasonography-guided fine-needle aspiration biopsy of suspected pancreatic cancer. Ann. Intern. Med. 134, 459–464.
Guck, J., Schinkinger, S., Lincoln, B., Wottawah, F., Ebert, S., Romeyke, M., et al., 2005. Optical deformability as an inherent cell marker for testing malignant transformation and metastatic competence. Biophys. J. 88 (5), 3689–3698.
Guo, Z., Zheng, C., Shi, B., 2002. Discrete lattice effects on the forcing term in the lattice Boltzmann method. Phys. Rev. E 65 (4), 046308.
Gusenbauer, M., Cimrak, I., Bance, S., Exl, L., Reichel, F., Oezelt, H., et al., 2011. A tunable cancer cell filter using magnetic beads: cellular and fluid dynamic simulations. arXiv. 1110.0995v1 [physics.flu-dyn].
He, X., Shan, X., Doolean, G.D., 1998. Discrete Boltzmann equation model for non-ideal gases. Phys. Rev. E 57 (1), R13–R16.
Hu, H.H., Patankar, N.A., Zhu, M.Y., 2001. Direct numerical simulations of fluid–solid systems using arbitrary Lagrangian–Eulerian Technique. J. Comput. Phys. 169, 427–462.
Huang, K., 1987. Statistical Mechanics. John Wiley, New York.
Jung, Y., Choi, Y., Han, K.H., Frazier, A.B., 2010. Six-stage cascade paramagnetic mode magnetophoretic separation system for human blood samples. Biomed. Microdevices 12, 637–645.
Kahn, H.J., Presta, A., Yang, L.Y., Blondal, J., Trudeau, M., Lickley, L., et al., 2004. Enumeration of circulating tumor cells in the blood of breast cancer patients after filtration enrichment: correlation with disease stage. Breast Cancer Res. Treat. 86, 237–247.
Karabacak, N.M., Spuhler, P.S., Fachin, F., Lim, E.J., Pai, V., Ozkumur, E., et al., 2014. Microfluidic, marker-free isolation of circulating tumor cells from blood samples. Nat. Protoc. 9, 694–710.
Kennard, E.H., 1938. Kinetic Theory of Gases. McGraw-Hill, New York.
Klank, H., Goranović, G., Kutter, J.P., Gjelstrup, H., Michelsen, J., Westergaard, C.H., 2002. PIV measurements in a microfluidic 3D-sheathing structure with three-dimensional flow behavior. J. Micromech. Microeng. 12 (6), 862.
Krüger, T., Varnik, F., Raabe, D., 2011. Efficient and accurate simulations of deformable particles immersed in a fluid using a combined immersed boundary lattice Boltzmann finite element method. Comput. Math. Appl. 61, 3485–3505.
Krylov, V.I., 1962. Approximate Calculation of Integrals. Macmillan, New York.
Li, P., Mao, Z., Peng, Z., Zhou, L., Chen, Y., Huang, P.-H., et al., 2015. Acoustic separation of circulating tumor cells. Proc. Natl. Acad. Sci. U. S. A. 112, 4970–4975.
Li, R., Tang, T., Zhang, P.W., 2002. A moving mesh finite element algorithm for singular problems in two and three space. J. Comput. Phys. 177, 365–393.
Lin, J.-Y., Fu, L.-M., Yang, R.-J., 2002. Numerical simulation of electrokinetic focusing in microfluidic chips. J. Micromech. Microeng. 12 (6), 955.
Lin, Y.-H., Lee, G.-B., 2008. Optically induced flow cytometry for continuous microparticle counting and sorting. Biosens. Bioelectron. 24, 572–578.
Liu, Z., Huang, F., Du, J., Shu, W., Feng, H., 2013. Rapid isolation of cancer cells using microfluidic deterministic lateral displacement structure. Biomicrofluidics 7, 011801.
Lv, P., Tang, Z., Liang, X., Guo, M., Han, R.P.S., 2013. Spatially gradated segregation and recovery of circulating tumor cells from peripheral blood of cancer patients. Biomicrofluidics 7 (3), 034109.

O. P. Malaspinas, 2009. Lattice Boltzmann Method for the Simulation of Viscoelastic Fluid Flows, Ph.D. dissertation, Ecole Polytechnique Fédérale de Lausanne, Switzerland.
Mateescu, D., Mekanik, A., Paidoussis, M.P., 1996. Analysis of 2-D and 3-D unsteady annular flows with oscillating boundaries, based on a time-dependent coordinate transformation. J. Fluids Struct. 10, 57–77.
Mocellin, S., Hoon, D., Ambrosi, A., Nitti, D., Rossi, C.R., 2006. The prognostic value of circulating tumor cells in patients with melanoma: a systematic review and meta-analysis. Clin. Cancer Res. 12, 4605–4613.
Mohamed, H., Murray, M., Turner, J.N., Caggana, M., 2009. Isolation of tumor cells using size and deformation. J. Chromatogr. A 1216, 8289–8295.
Murayama, T., Yoshino, M., Hirata, T., 2011. Three-dimensional lattice Boltzmann simulation of two-phase flow containing a deformable body with a viscoelastic membrane. Commun. Comput. Phys. 9 (5), 1397–1413.
Nagrath, S., Sequist, L.V., Maheswaran, S., Bell, D.W., Irimia, D., Ulkus, L., et al., 2007. Isolation of rare circulating tumour cells in cancer patients by microchip technology. Nature 450, 1235–1239.
Park, J.S., Jung, H.I., 2009. Multiorifice flow fractionation: continuous size-based separation of microspheres using a series of contraction/expansion microchannels. Anal. Chem. 81, 8280–8288.
Peskin, C.S., 1977. Numerical analysis of blood flow in the heart. J. Comput. Phys. 25 (3), 220–252.
Qiao, R., Aluru, N.R., 2002. A compact model for electroosmotic flows in microfluidic devices. J. Micromech. Microeng. 12 (5), 625.
Rakocević, G., Đukić, T., Filipović, N., Milutinović, V., 2013. Computational Medicine in Data Mining and Modeling. Springer, New York, USA.
Ranjan, S., Zeming, K.K., Jureen, R., Fisher, D., Zhang, Y., 2014. DLD pillar shape design for efficient separation of spherical and non-spherical bioparticles. Lab Chip 14, 4250–4262.
Rolle, A., Günzel, R., Pachmann, U., Willen, B., Höffken, K., Pachmann, K., 2005. Increase in number of circulating disseminated epithelial cells after surgery for nonsmall cell lung cancer monitored by MAINTRAC(R) is a predictor for relapse: a preliminary report. World J. Surg. Oncol. 3, 18.
Seal, S.H., 1964. A sieve for the isolation of cancer cells and other large cells from the blood. Cancer 17, 637–642.
Song, H., Wang, Y., Pant, K., 2011. System-level simulation of liquid filling in microfluidic chips. Biomicrofluidics 5 (2), 24107.
Steeg, P.S., 2006. Tumor metastasis: mechanistic insights and clinical challenges. Nat. Med. 12, 895–904.
Sun, Y., Lim, C.S., Liu, A.Q., Ayi, T.C., Yap, P.H., 2007. Design, simulation and experiment of electroosmotic microfluidic chip for cell sorting. Sensors Actuators A Phys. 133, 340–348.
Sun, C., Migliorini, C., Munn, L.L., 2003. Red blood cells initiate leukocyte rolling in post-capillary expansions: a lattice Boltzmann analysis. Biophys. J. 85 (1), 208–222.
Taylor, J.K., Ren, C.L., Stubley, G.D., 2008. Numerical and experimental evaluation of microfluidic sorting devices. Biotechnol. Prog. 24, 981–991.
Wang, G., Mao, W., Byler, R., Patel, K., Henegar, C., Alexeev, A., et al., 2013. Stiffness dependent separation of cells in a microfluidic device. PLoS One 8, e75901.
Wolfram, S., 1986. Cellular automaton fluids 1: basic theory. J. Stat. Phys. 3 (4), 471–526.
Wu, J., Shu, C., 2010. Particulate flow simulation via a boundary condition-enforced immersed boundary-lattice Boltzmann scheme. Commun. Comput. Phys. 7 (4), 793–812.
Zieglschmid, V., Hollmann, C., Bocher, O., 2005. Detection of disseminated tumor cells in peripheral blood. Crit. Rev. Clin. Lab. Sci. 42, 155–196.

CHAPTER 11

Computational analysis of abdominal aortic aneurysm before and after endovascular aneurysm repair

Contents

1 Introduction

Abdominal aortic aneurysm (AAA) is a permanently dilatation 50% greater than normal aortic diameter (McGloughlin and Doyle, 2010) that is frequently observed in the aging population (LeFevre, 2014), and it affects

Computational Modeling in Bioengineering and Bioinformatics
https://doi.org/10.1016/B978-0-12-819583-3.00011-4

6%–9% of the people in the industrialized world. It is a major health problem that typically affects men after the age of 50 (O'Gara, 2003), and it is the 13th leading cause of death in Western societies (Minino et al., 2004; Schermerhorn, 2009). AAAs cause 15,000–30,000 deaths per year in the United States only (Kent et al., 2004) and 1.3% of all deaths among men aged 65–85 in developed countries (Thompson, 2003; Sakalihasan et al., 2005). In the United States alone, 1.5 million have undiagnosed AAAs (Schermerhorn, 2009). They can remain asymptomatic for most of their development, and if left untreated, they can enlarge and eventually rupture with catastrophic mortality rate of 80% (Bown et al., 2002) to 90% (LeFevre, 2014).

AAA is formed preferentially in infrarenal area, before aortic bifurcation. In most cases, AAA requires treatment that includes standard open repair (OR) or endovascular aneurysm repair (EVAR). The traditional OR involves an incision in the abdomen and the exclusion of the diseased aneurysm with a synthetic graft. The EVAR is an alternative and minimally invasive treatment to standard OR (Prinssen et al., 2004) that implies exclusion of AAA by endovascular stent-graft implantation through femoral arteries under fluoroscopic control. After stent-graft implantation, a weakened aortic wall is protected from pulsatile blood flow by forming a new blood flow through the stent graft. Potentially fatal complications in case of high-risk patients are avoided using the EVAR procedure. Improved critical care, preoperative optimization, and the advent of EVAR have improved 30-day mortality for elective AAA repair, but on the other hand, 5-year survival remains poor despite advances in medical care and treatment strategies (Bahia et al., 2015). This clearly demonstrates the need of AAAs diagnosing and monitoring in order to make progress in both medical and economic domain.

According to the biomechanical perspective, it is generally accepted that the expansion of aneurysms is driven by flow-induced progressive degradation of the vessel wall (Vorp, 2007; Fillinger et al., 2002; Scotti and Finol, 2007). Since the internal mechanical forces in the aorta are maintained by dynamic action of blood flow, the quantification of the AAA hemodynamics and mechanical stress is essential for characterization of its biomechanical behavior. Furthermore, it is important to include both the dynamics of blood flow and the wall motion response associated with the pulsatile nature of the flow to accurately model the AAA. Thus, fluid-solid interaction (FSI) should be considered in studying the hemodynamics and biomechanics of AAA. Beside AAA modeling, it is of great importance to model and examine mechanical parameters of AAA after EVAR. Most research work focused on

AAA does not consider FSI between lumen blood, stent-graft wall, ILT, and AAA wall. There are studies that consider an individual patient's geometry coupled with FSI, such as iliac bifurcation (Chandra et al., 2013), and ILT (Di Martino et al., 2001). Also, effects of drag forces on stent graft, which may influence stent-graft migration and potentially failure, have been investigated (Li and Kleinstreuer, 2006). However, there is a lack of information about stent-graft influence on blood vessel wall. In clinical conditions and in vivo experiments, it is difficult to examine the interactions between different structures and determine the mechanical parameters. Last years, computational modeling and analysis of AAA and stent grafts have become useful methods for determining mechanical parameters, which cannot be measured in vivo. With that aim, this study analyzed the patient-specific AAA before and after EVAR procedure. The computational simulations, based on the finite element method (FEM), were performed considering the blood flow through the AAA and interaction with the unstented aortic wall, as well as the blood flow through the implanted stent graft and interaction with the stent graft, ILT, and aortic wall.

The rest of the paper is organized as follows: Risk factors and surgical treatments for AAA are shortly described in the Sections 2. Section 3 covers main computational methods applied for AAA. Thereafter, creation of the 3D patient-specific models is presented in the Section 4. The CT scans of patient, aged 65, who had the AAA extended to the left common iliac artery, were used for geometrical reconstruction that required usage of several software packages. The material properties and boundary conditions that were applied for numerical models' creation and FSI analysis are given in the Section 5. General finite element procedure and fluid-structure interaction are shortly explained in the Section 6. Section 7 covers discussion of results (velocity field of blood, blood pressure and shear stress distribution, von Mises stress distribution, and displacements of stent graft and aortic wall) for employed unstented and stented AAA model. Finally, discussion and conclusion of presented work are given in the Sections 8 and 9, presenting the main findings and limitations of this study, and followed by plans for further improvements.

2 Risk factors and surgical treatments for AAA

Recently, Golledge summarized the research findings and current theories on AAA pathogenesis and highlighted potential medical therapies for AAA, including clinical trials also (Golledge, 2019). Nowadays, population is exposed to risk factors, which have influence on aneurysmal creation and growth. Numerous prior studies had identified many factors that potentially contribute to the development, enlargement, and rupture of AAA (Domagala et al., 2016). Namely, a substantial amount of research on AAA expansion and rupture focused on different biological and biomechanical factors, and lately, special attention was put to genes and chemical influences. Among biological factors and risk factors, authors discussed the influence of diameter, sex, blood pressure, and smoking (Choke et al., 2005). It was found that history of smoking was the most important risk factor for developing an AAA and had been suggested as a possible criterion for selective AAA screening (Greco et al., 2010). The recent study confirmed the correlation between the AAA and its major risk factors (male sex, smoking cigarettes, coronary artery disease, and history of myocardial infarct) as well as a weak association between AAA and hypertension and a negative correlation with diabetes (Tkaczyk et al., 2019).

From a biomechanical point of view, major factors contributing to AAA expansion and rupture are the wall stress (Fillinger et al., 2003; Fillinger et al., 2002) and strength (Upchurch Jr. and Schaub, 2006), wall stiffness (Kolipaka et al., 2016), vessel asymmetry (Doyle et al., 2009; Doyle et al., 2010), intraluminal thrombus (ILT) (Biasetti et al., 2010), entire geometry (Georgakarakos et al., 2010), etc. However, the links between these factors and the underlying mechanisms responsible for the formation, growth, and stabilization or rupture of AAAs are still insufficiently investigated. AAA leads to various complications such as pressure on the surrounding structures and organs, thrombosis, embolism, and rupture, which are followed by significant morbidity, disability, and mortality. Therefore, it is important not only to understand the basic pathological mechanism and current treatments but also to recognize the different variants and their complications and to know the indications for aortic repair (Bhave et al., 2019).

The therapeutic indication for a nonruptured AAA is usually set at a diameter of 5.0–5.5 cm (for men) and 4.5–5.0 cm (for women) (Kühnl et al., 2017), while treatment consists of open surgical repair, the replaced of affected vessel segment with a prosthesis (OR, open repair), or endovascular repair, by implanting a stent graft (EVAR). Development of minimally invasive EVAR technique is supported by efforts to reduce surgical insult, mortality, and morbidity associated with OR. A number of studies had reported that EVAR was associated with improved postoperative mortality rates compared with OR. Thus, EVAR had become more common for nonruptured AAA as well as ruptured AAA (Kühnl et al., 2017). Although, recent large randomized trials showed that EVAR was associated with a significantly lower operative mortality compared with OR, it was also associated with higher rates of graft-related complications (stent-graft migration, stenosis, occlusion, perforation, stent fracture, endoleak, secondary rupture, etc.) and reinterventions (De Bruin et al., 2010; Bahia et al., 2015).

3 Computational methods applied for AAA

Numerous studies have been performed in the analysis of relationship between rupture risk estimation and biomechanics of AAA, as well as stent-graft behavior used for EVAR treatment. For many years, diameter was taken to be the primary parameter associated with the rupture risk estimation. Namely, the threshold of $\geq$5.5 cm was applied as indicative for AAA repair (Venkatasubramaniam et al., 2004). However, continuous studies in the past showed that 10%–24% of small aneurysms (<5.5 cm) may rupture (McGloughlin and Doyle, 2010), while aneurisms, which diameter exceeds the threshold, remain stable. It led to the development of new tools for risk stratification and detection of additional biomechanical parameters that affect AAA rupture risk. For example, Stenbaek et al. described that the development of ILT may be a better predictor than the maximum diameter of the AAA as a rupture risk parameter (Stenbaek et al., 2000). Numerical analyses that coupled computational fluid dynamics (CFD) and finite element analysis (FEA) were used for studying the hemodynamic effects on patient-specific aortic wall and AAA rupture risk assessment, concerning the influence of aortic diameter, geometry, wall stress, ILT, etc. (Vorp, 2007; Fillinger et al., 2002). Recently, Erhart at al. examined biomechanical rupture risk parameters of asymptomatic, symptomatic, and ruptured AAA using FEA (Erhart et al., 2015). Shum et al. showed that in addition to maximum diameter, sac length, sac height, volume, surface area, bulge height, and ILT volume were all highly correlated with rupture status (Shum et al., 2011). In this study machine learning method was used with aim to develop a model that was capable to discriminate between ruptured and unruptured aneurysms. Beside this study, FEA coupled with machine learning approach gave promising results in cardiovascular modeling (Filipovic et al., 2011b; Robnik-Šikonja et al., 2018).

Numerical modeling of AAA and stent grafts that has been used to assess biomechanical parameters (strain, stress, forces, etc.) occurring in vivo can play an important role in research on stent graft and blood vessel mechanics. Various stent-graft designs used for EVAR in cardiovascular treatments were also examined using FSI analysis. Based on this approach, the displacement forces on the device and stress reduction on the aneurysm wall by stent-graft placement were assessed. Previous studies examined the stresses on the aneurysm sac before and after EVAR and the stresses placed on the stent graft and its attachment system, giving the report that the majority of stress was absorbed by the stent graft after repair reducing AAA wall stresses by 90% (Scotti et al., 2004; Di Martino et al., 2004). Kleinstreuer et al. investigated a stent graft under cyclic loading and its deployment in a circular AAA neck, which was a step forward in stent-graft modeling (Kleinstreuer et al., 2008).

Jayendiran et al. performed the computational study, which served as an additional tool to vascular surgeons in choosing the material and design for stent-graft recipients (Jayendiran et al., 2018). They changed the material and structural properties of stent graft. Capability of FEM computer simulations to predict the deformed configuration of stent graft after EVAR was demonstrated by virtual evaluation of stent-graft deployment within the AAA (De Bock et al., 2012).

All previously stated studies demonstrated the importance of biomechanical modeling of AAA using the FEM. Nevertheless, there is a lack of information about stent-graft influence on AAA wall. In clinical conditions and in vivo experiments, it is difficult to examine the interactions between different structures and determine the mechanical parameters. Therefore, FEM approach provides an additional understanding of potential indicators for EVAR procedure, as well as AAA biomechanical behavior after EVAR procedure.

4 Geometrical model of AAA

This section describes the reconstruction of the geometry of the abdominal aortic area and common iliac arteries, ILT, as well as the stent graft. Reconstruction and modeling were carried out in several stages with aim to create a real anatomical 3D geometry of different entities. To accomplish this, several different software packages were used, whose various tools and functions contributed to the optimization of surface elements, 3D elements, and preparation of the model for blood flow simulation and fluid-solid interaction.

Two cases were considered, and the corresponding domains were modeled accordingly:

1. Computational analysis of blood flow through the AAA and interaction with the unstented aortic wall
2. Computational analysis of blood flow through the implanted stent graft and interaction with the stent graft, ILT, and aortic wall.

With aim to create the geometries of different domains and their assembly as a whole, the following software packages were used:

1. *Materilise Mimics* (Mimics 10.01, Materialise, Leuven, Belgium)
2. *Geomagic Studio* (Geomagic Studio 2013, Raindrop Geomagic Inc., Research Triangle Park, NC, the United States)
3. *Femap* (Femap v.11, Siemens PLM Software, the United States)

Modeling of the aortic lumen, aortic wall, ILT, and stent graft and testing their mechanical properties was based on the finite element method (FEM). The basic idea of the FEM-based modeling is assuming that the spatial domain of the problem is divided, that is, discretized into the subdomains, which are subjected to the continuum mechanics and numerical mathematics. The subdomains are terminologically designated as finite elements. The finite elements are connected with common surfaces and nodes. The analysis of the finite elements, obtained by discretization of the continuum, allows the numerical simulation of the system's response to the prescribed load. In the following sections, the main steps and functions in creation of the geometrical models will be described.

4.1 Creation of 3D geometry based on 2D images data

In order to perform the computational analysis of AAA, the patient-specific models were created based on CT scan images. *Mimics* software (Materialise Interactive Medical Image Control System) was used to reconstruct the 3D geometry of AAA based on 2D CT scan images. *Mimics* is an image processing software for 3D design and modeling from a multitude of 2D images. The 3D models can be used for various engineering applications. *Mimics* creates 3D model through image segmentation of CT, micro-CT, magnetic resonance imaging (MRI), ultrasound (US), etc. The region of interest, selected in the segmentation process, is converted to the 3D surface model using an adapted algorithm. The most used input files are DICOM, but other formats (TIFF, JPEG, BMP, and RAV) are supported. Output formats differ, depending on the field of application. The most common output formats are STL, VRML, PLI, and DKSF. The obtained models can be optimized for the further finite element analysis or computer fluid dynamics. *Mimics* is accepted by biomedical engineers and manufacturers of cardiovascular, orthopedic, maxillofacial, and pulmonary medical devices.

In this study, we analyzed an individual male patient, aged 65. The patient-specific cases before and after EVAR procedure were modeled. The set of DICOM images, with resolution of 512 × 512 mm and the slice increment of 0.625 mm, were imported (*File≫Import images*) in the segmentation software *Mimics*. In case of the model creation before and after EVAR, geometries of two domains (blood flow and aortic wall) and four domains (blood flow, stent graft, ILT, and aortic wall) were generated, respectively. After importing and converting the DICOM files, the system planes were adjusted. Coronal, sagittal, and transversal cross sections were on disposal, as well as the visualization of calculated 3D model. A user-friendly interface enabled extraction of the region of interest such as part from the infrarenal aorta to the end of the common iliac arteries. Considering this, the different solid and fluid domains were created. Manipulation in three different planes, which are presented in Fig. 11.1, enabled the creation of more accurate 3D model of the patient's anatomy. This specific patient had the AAA, which was extended toward the left common iliac artery. The AAA can be extended toward one or both iliac arteries, and approximately 20%–40%

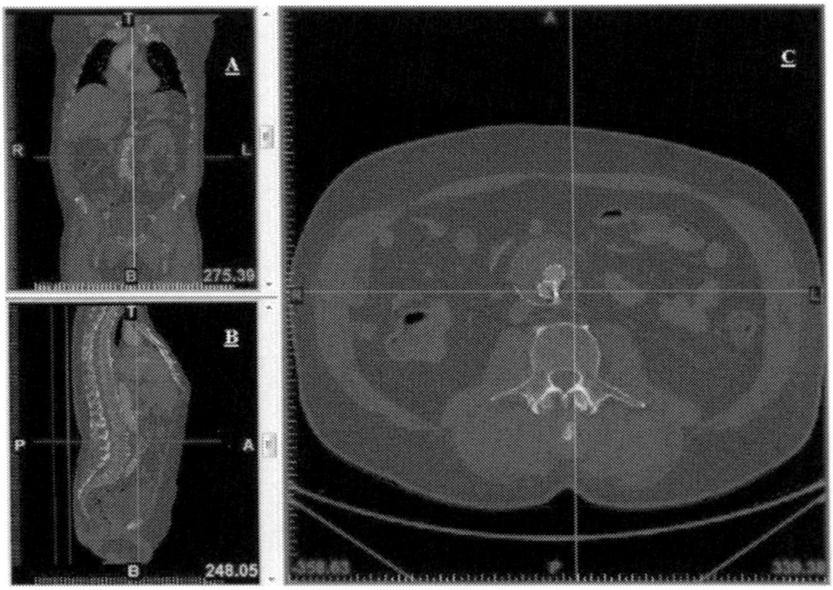

Fig. 11.1 Coronal (A), sagittal (B), and transversal (C) cross section of modeled AAA.

of patients with AAA have unilateral or bilateral iliac artery aneurysms (Hobo et al., 2008).

Using the *thresholding* option from the *segmentation* menu, visibility, that is, contrast, should be adjusted in order to make the region of interest visible to the user. The visibility of certain tissues depends on the quality of the image and the amount of injected contrast during CT procedure. Dense tissues, such as bones, are displayed with white color, less dense tissues, such as brain and muscles, are shown in gray color, while black color presents the regions with the smallest density.

Reconstruction of different domains required several phases of segmentation, depending on the region of interest. Images were segmented from the infrarenal aorta to the end of the common iliac arteries. As previously mentioned, the reconstructed AAA was extended distal to the left common iliac artery. Removing the blood vessels from the surrounding area was accomplished with the *Edit Masks* and *Multiple Slice Edit* commands. After extracting the region of interest, the 3D model was calculated with *Segmentation≫Calculate 3D* option. The initial 3D model was exported in the STL format using the *Export≫ASCII STL* command. The stereolithography (STL) representation set the boundary surface for all domains.

The same procedure was repeated during reconstruction of implanted stent-graft geometry, but in this case, the area of aneurysm was removed, and the stent-graft geometry was kept and calculated. The stent graft is easily visible within the CT, due to enhanced contrast caused by metal wires. In this case, the metal wires were not reconstructed and modeled due to their complicated geometry. Reconstruction of metal wires causes significantly increasing of finite element mesh density, which affects the duration of computer simulation and its performance. The 3D geometry of the branched stent-graft tube was created and exported in the STL format. Thus, after performed work in *Mimics*, the geometry of AAA and the stent-graft geometry were separately saved. The surface of these initial 3D models of different domains was quite rough, due to overlapping and distorted elements. Therefore, the STLs had to be processed to further optimization of the surface meshes that was performed in the *Geomagic Studio* software.

4.2 Creation of the 3D models with smooth surfaces

In order to create suitable meshes and models for computational simulations, *Geomagic Studio* software was employed. This software focuses on the reconstruction of 3D surface model and its real representation, which is widely used for product design, engineering, art, and different fields of bioengineering. Obtained model can be prepared for further analysis, simulation, and animation and used in other engineering programs.

In case of creation of smooth surfaces of AAA and stent-graft model, the same functions were used, considering that those models were separately manipulated. The last saved STL file from *Mimics* was imported in the *Geomagic Studio*. Firstly, the residues of the surrounding tissues were removed with option *Polygons≫Manifold*. Each model consisted of triangular elements making the finite element mesh. By activating the *Display≫ Geometry Display≫Edges*, the edges of the finite elements become visible, making easy to manipulate the mesh. In order to create the model with smooth surface mesh (but keeping the patient-specific form) and with the optimal number of finite elements due to later calculations, the commands from *Polygons* menu were used (*Decimate, Remove Spikes, Remesh, Refine, Defeature, Fill hole,* etc.). As representation, the finite element mesh before and after using *Remove Spikes* command can be seen in Fig. 11.2.

1. In case of the first simulation—analysis of blood flow through the AAA and interaction with the unstented aortic wall—two domains were taken into consideration: blood flow domain (lumen) and unstented AAA wall. The finite element mesh of the aortic wall was created on the basis of lumen surface, as previously described. Therefore, the boundary surface between solid and fluid domain had coincident nodes and elements. Representation of the entire aortic lumen model with AAA (left) and the separated AAA model (right) used for the simulation can be seen in Fig. 11.3. A smooth finite element mesh of the AAA wall can be seen in Fig. 11.4.

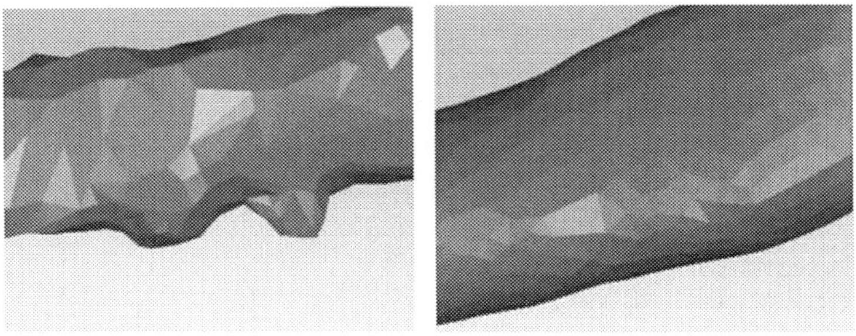

(A) (B)

Fig. 11.2 Representation of the finite element mesh (A) before and (B) after using *Remove Spikes* command.

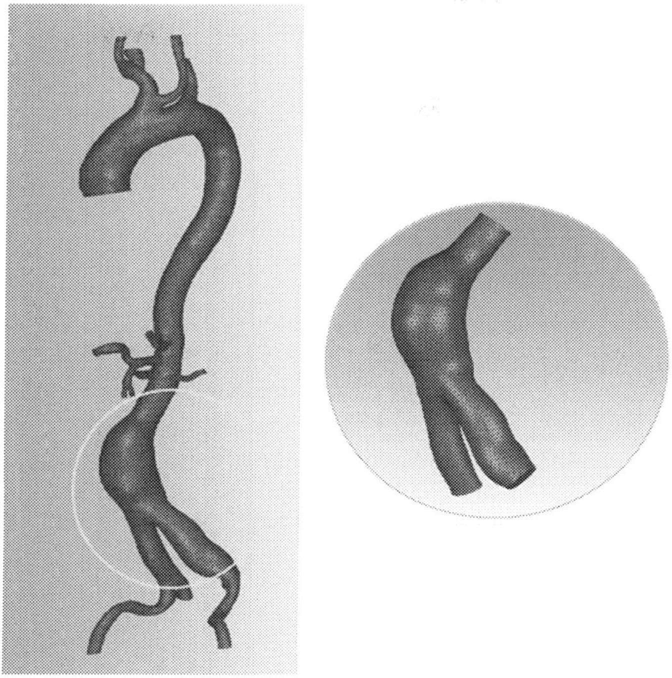

Fig. 11.3 Representation of the entire aortic lumen model with AAA (*left*) and the separated AAA model (*right*) used for the simulation (anterior view).

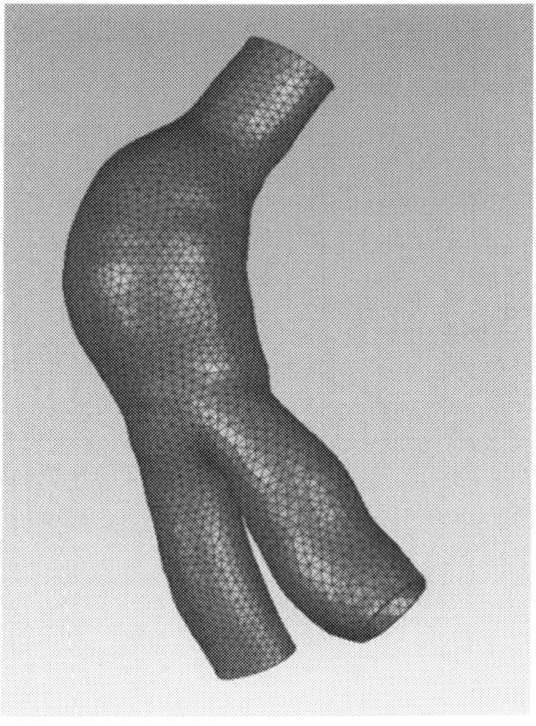

Fig. 11.4 Geometrical model of the AAA wall (anterior view).

It was difficult to accurately capture the aortic wall thickness in the patient-specific CT images due to unclear boundary between inner and outer wall surfaces. Therefore, a uniform and average aortic wall thickness was prescribed in the modeling of AAA. Based on created smooth surface mesh of AAA, which was considered as boundary surface between lumen (fluid domain) and aortic wall, the AAA wall was modeled using option *Polygons≫Shell*. The aortic wall had prescribed thicknesses of 2 mm (Di Martino et al., 2001; Frauenfelder et al., 2006). Also, top surface of infrarenal aorta and bottom surfaces of common iliac arteries had the orientation approximately normal to the blood flow. Those flat top and flat bottom surfaces of the AAA wall and lumen were coincident (i.e., planar).

2. In case of the second simulation—analysis of blood flow through the implanted stent graft and interaction with the stent graft, ILT, and aortic wall—four domains were taken into consideration: blood flow domain (lumen), stent graft, ILT, and aortic wall. As previously described, lumen was created based on CT scans in *Mimics*, whereby the initial geometry was improved in *Geomagic* software. Based on the blood flow domain within the stent graft, the geometry of stent graft was generated with prescribed uniform and average thickness of 0.2 mm. It should be underlined that the stent-graft geometry was simplified, due to difficulty to reconstruct a real geometry of graft tube and nitinol wires from the CT data.

   In order to create continuum, the geometry of ILT was created considering the outer boundary surface of the stent graft and inner boundary surface of the aortic wall. Due to computational requirements, the boundary surfaces between the blood flow domain, stent graft, ILT, and aortic wall had coincident nodes and elements. Representation of the entire aortic lumen model (left) after stent-graft implantation and the separated region of interest (right) used for the simulation can be seen in Fig. 11.5.

Geometrical model of the (A) stent-graft wall and (B) aortic wall can be seen in Fig. 11.6. The estimated diameters of the stent graft proximally and distally (flared branch end) to the AAA were approximately 24 (D_1) and 25 mm (D_3), respectively. The diameter of the stent-graft branch within the healthy right common iliac artery was approximately 20 mm (D_2), while the both stent-graft branches had average diameters of 14 mm (D_4). The estimated lengths of the stent-graft fixation proximally and distally to the AAA were 15 (L_1) and 25 mm (L_2), respectively.

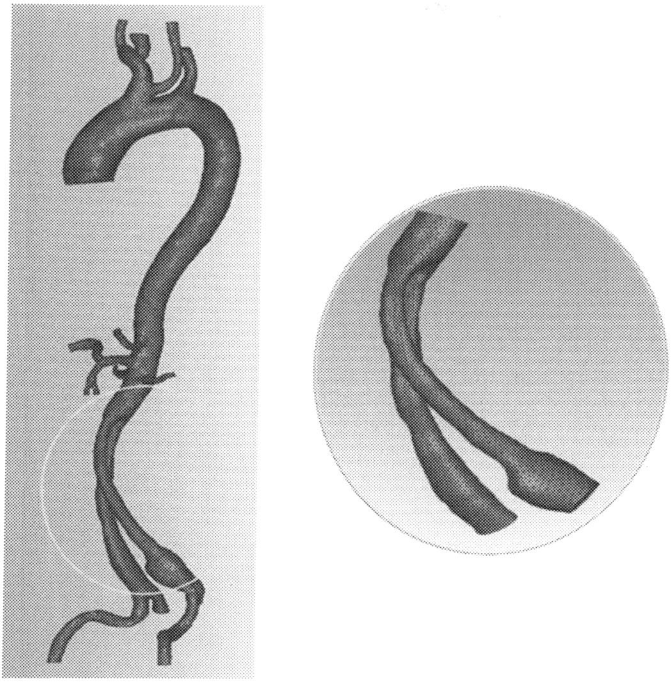

Fig. 11.5 Representation of the entire aortic lumen model (*left*) after the stent-graft implantation and the separated region of interest (*right*) used for the simulation (anterior view).

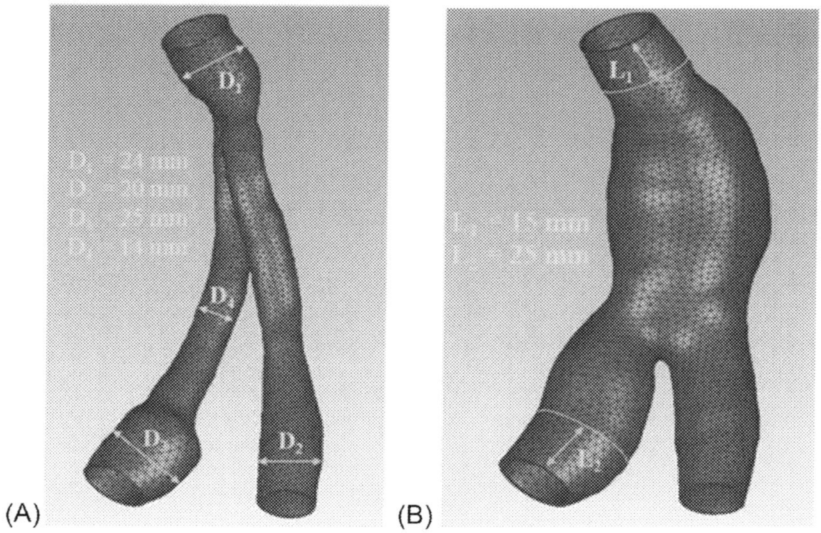

Fig. 11.6 Geometrical model of the (A) stent-graft wall and (B) aortic wall (posterior view).

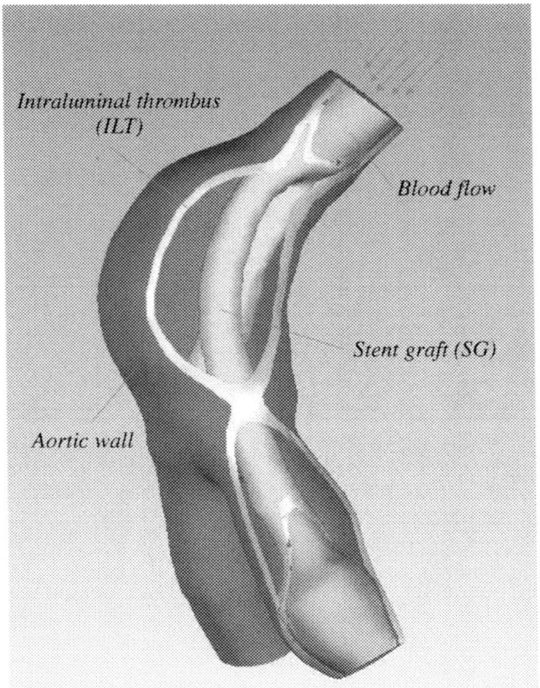

Fig. 11.7 Model of the AAA with domains of the blood flow, stent graft, intraluminal thrombus, and aortic wall.

The difference between the blood flow, stent graft, ILT, and aortic wall domain can be seen in Fig. 11.7. Also, direction of the prescribed blood flow at inlet (top) surface is marked with an arrow. It should be noted that top surface of infrarenal aorta and bottom surfaces of common iliac arteries, as well as top and bottom surfaces of the stent graft, have the orientation approximately normal to the blood flow direction. Those flat top and flat bottom surfaces of the blood flow domain, aortic wall, and stent graft are coincident (i.e., planar).

4.3 Creation of the volumetric 3D models

After surface model creation, the obtained models were exported in the STL format from *Geomagic* to *Femap* software. *Femap* is engineering analysis program used for modeling of complex engineering problems, based on the FEM, including pre- and postprocessing. *Femap* is suitable for structural optimization and comprehends different types of analysis, as well as assessment of system's characteristics. Beside plenty functionalities, several tools were used to prepare the models for further computational analyses. All modeled domains for the simulation of unstented and stented AAA were separately imported (STL format), and therefore, quality of the meshes was checked (*Meshing Toolbox≫Quality*). The nodes and elements of unstented and stented model of AAA were numbered in increasing order, taking into account the merging of coincident nodes and elements on the boundary surfaces between neighboring domains. Further, the 3D volumes were created by filling in the surface models with the tetrahedral elements. In order to obtain the model consisted of eight-node brick elements, suitable for numerical simulations, *Femap* was used in combination with our *in-house* software for the conversion of tetrahedral elements to eight-node brick elements. The corresponding data sets were imported into the PAK (solver for analysis of constructions) software, for further computational FSI simulations of unstented and stented AAA (Kojic et al., 2008).

5 Numerical model of AAA

In order to perform computational analysis of unstented and stented AAA, which included different domains, we employed the PAK solver, for performing CFD (Filipovic and Schima, 2011) and FSI simulations on the computational meshes. It solves the mass and momentum conservation (Navier-Stokes) equations for the fluid domain and the equilibrium equations of elasticity for the solid domain using a finite element discretization method. Also, it implements an arbitrary Lagrangian-Eulerian (ALE) approach. The fluid and solid domain meshes were independently converted and then imported into PAK for further analysis. The following sections describe applied material characteristics and boundary conditions employed within the computational simulations.

5.1 Material properties

It is very important to use arterial wall with proper material model. The mechanical properties of soft biological tissues depend greatly on their microstructure integration attained in the constitutive model (Zhang et al., 2016). Although the aneurysmal wall has been found to be mechanically anisotropic (Rodriguez et al., 2009), assumption of isotropic properties is a reasonable in most cases. Wall thickness is also very important. Since aneurysmal rupture occurs at a specific site of the aortic lumen, the properties of the wall affect the computed solution. The aortic wall thickness in computational methods is usually assumed to be in the range of 1.5–2 mm. Since the patient-specific aortic wall thickness was unknown, we prescribed the constant aortic wall thickness of 2 mm.

Solid domains (aortic wall, stent graft, and ILT) were modeled as linearly elastic isotropic material, under the assumption that they are affected by average blood pressure (80–120 mmHg) (Di Martino et al., 2001; Thubrikar et al., 2001). Although it is known that Young's modulus of 1.2 MPa is present in healthy arteries, experimental results shows that Young's modulus has significantly higher values in case of AAA (between 3 and 5 MPa) (Thubrikar et al., 2001). In this work, Young's modulus of 5 MPa and Poisson coefficient of 0.45 were assumed for the aortic wall.

Considering the difficulties in reconstruction of stent graft, which has metal wires and graft tube, the stent-graft model was simplified, and material characteristics of equivalent composite material were considered (Suzuki et al., 2001). In that order, equivalent Young's modulus of 10 MPa and Poisson coefficient of 0.27 were adopted (Li and Kleinstreuer, 2006). The stent-graft thickness varies between 0.1 and 0.2 mm (Suzuki et al., 2001). Thus, for the simulation, stent-graft thickness of 0.2 mm was adopted. Furthermore, Young's modulus of 0.1 MPa and Poisson coefficient of 0.49 were adopted for ILT. The density of $1.12\cdot10^{-3}$, $6\cdot10^{-3}$, and $1.121\cdot10^{-3}\,g/mm^3$ was adopted for the aortic wall, stent graft, and ILT, respectively (Di Martino et al., 2001; Li and Kleinstreuer, 2006).

In order to perform analysis, which reflects the realistic human blood flow, we used the average blood properties: a dynamic viscosity of 0.0035 Pa·s and a density of $1.05\cdot10^{-3}\,g/mm^3$ (Gao et al., 2006). Flow is considered to be laminar, homogeneous (Newtonian), and viscous incompressible. Material characteristics of blood domain, ST, ILT, and aortic wall are given in Table 11.1.

Table 11.1 Mechanical properties of different domains

Parameters	Blood domain	Aortic wall	SG	ILT
Wall thickness [mm]	/	2	0.2	/
Dynamic viscosity [Pa·s]	$3.5\cdot10^{-3}$	/	/	/
Young's modulus [MPa]	/	5	10	0.1
Poisson coefficient	/	0.45	0.27	0.49
Density [g/mm^3]	$1.05\cdot10^{-3}$	$1.12\cdot10^{-3}$	$6\cdot10^{-3}$	$1.121\cdot10^{-3}$

5.2 Boundary conditions

The boundary conditions for the simulation of unstented and stented AAA are depicted in Table 11.2. The boundary conditions vary depending on analysis and domain included within the model. In case of simulation of unstented AAA, the employed boundary conditions for fluid domain consisted of prescribed velocity at the inlet of infrarenal aorta (top surface), zero velocity at the aortic walls, and physiological resistance pressure at the aortic outlets. In case of simulation of stented AAA, the employed boundary conditions for fluid domain consisted of prescribed velocity at the inlet of stent graft (top surface), zero velocity at the stent-graft walls and physiological resistance pressure at the stent-graft outlets. In both simulations at the inlet of infrarenal aorta/stent graft, a parabolic velocity profile was used together with a pulsatile waveform based on the measured data (Fig. 11.8). Flow rate as the inlet boundary condition was in correspondence with previous publications (Frauenfelder et al., 2006; Olufsen et al., 2000). The simulated cardiac cycle was discretized into 40 time steps.

Table 11.2 Boundary conditions for simulation of AAA before and after EVAR

Computational analysis of blood flow through the AAA and interaction with the unstented aortic wall	**Computational analysis of blood flow through the implanted stent graft and interaction with the stent graft, ILT, and aortic wall**
Boundary conditions for fluid domain	
1. Prescribed parabolic velocity profile at the inlet of infrarenal aorta (top surface), as time-dependent function $v(t)$ **2.** Zero velocity at the aortic walls (no slipping boundary condition) **3.** Prescribed physiological resistance pressure at the aortic outlets (bottom surfaces)	**1.** Prescribed parabolic velocity profile at the inlet of stent graft (top surface), as time-dependent function $v(t)$ **2.** Zero velocity at the stent-graft wall (no slipping boundary condition) **3.** Prescribed physiological resistance pressure at the stent-graft outlets (bottom surfaces)
Boundary conditions for solid domains	
Fixed top surface of infrarenal aorta and bottom surfaces of iliac arteries	
/	Fixed top surface of stent graft and bottom surfaces of stent-graft branches

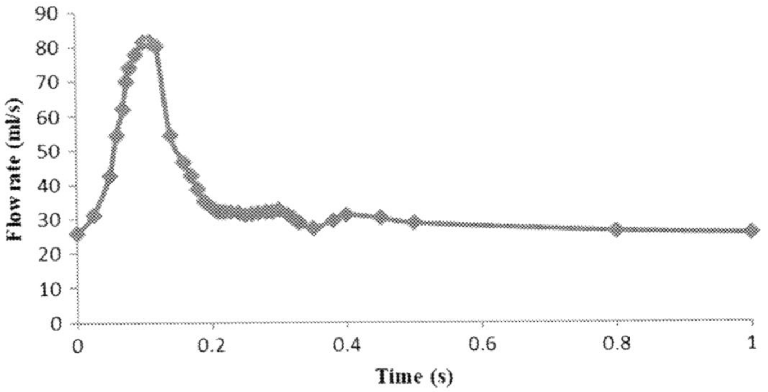

Fig. 11.8 Prescribed blood flow rate.

The employed boundary conditions for solid domains in the simulation of unstented AAA consisted of fixed top surface of infrarenal aorta and bottom surfaces of iliac arteries, while in the simulation of stented AAA beside mentioned boundary conditions for the aorta, top surface of the stent graft and bottom surfaces of stent-graft branches were also fixed. The stent graft and aortic wall proximally and distally to the AAA were assumed to be tied together, simulating attachment of the stent graft to the aortic wall, thus ignoring the possibility of endoleaks and local dislodgement.

After applied appropriate boundary conditions and material characteristics for the both models, the PAK software for FSI analysis was used for the computational simulations, which calculated the velocity field of blood, pressure distribution, shear stress, wall displacements, and von Mises stress distribution.

6 Finite element procedure and fluid-structure interaction

Blood flow in aorta has a time-dependent 3D flow, so the 3D flow was governed by the Navier-Stokes equations and the continuity equation, which can be written as follows:

$$\rho\left(\frac{\partial v_i}{\partial t}+v_j\frac{\partial v_i}{\partial x_j}\right)=-\frac{\partial p}{\partial x_i}+\mu\left(\frac{\partial^2 v_i}{\partial x_j \partial x_j}+\frac{\partial^2 v_j}{\partial x_j \partial x_i}\right) \tag{11.1}$$

$$\frac{\partial v_i}{\partial x_i}=0 \tag{11.2}$$

where v_i denotes the blood velocity in direction x_i, p is the pressure, ρ is the fluid density, and μ is the dynamic viscosity. Summation is assumed on the repeated indices $i,j=1,2,3$. Eq. (11.1) represents the balance of linear momentum, while Eq. (11.2) expresses the incompressibility condition. The laminar flow condition appropriate for this type of analysis (Filipovic et al., 2011a) was used. The finite element code was validated using the analytical solution for shear stress and velocities through the curved tube (Filipovic et al., 2006a). A penalty formulation was used (Filipovic et al., 2006b). The incremental-iterative form of the equations for time step and equilibrium iteration "i" is

$$\begin{bmatrix} \frac{1}{\Delta t}\mathbf{M_v} + {}^{t+\Delta t}\mathbf{K}_{\mathbf{vv}}^{(i-1)} + {}^{t+\Delta t}\mathbf{K}_{\mu\mathbf{v}}^{(i-1)} + {}^{t+\Delta t}\mathbf{J}_{\mathbf{vv}}^{(i-1)} & \mathbf{K_{vp}} \\ \mathbf{K}_{\mathbf{vp}}^{\mathbf{T}} & \mathbf{0} \end{bmatrix} \begin{Bmatrix} \Delta\mathbf{v}^{(i)} \\ \Delta\mathbf{p}^{(i)} \end{Bmatrix} = \begin{Bmatrix} {}^{t+\Delta t}\mathbf{F}_{\mathbf{v}}^{(i-1)} \\ {}^{t+\Delta t}\mathbf{F}_{\mathbf{p}}^{(i-1)} \end{Bmatrix} \tag{11.3}$$

The left upper index "$t+\Delta t$" denotes that the quantities are evaluated at the end of time step. The matrix $\mathbf{M_v}$ is mass matrix, $\mathbf{K_{vv}}$ and $\mathbf{J_{vv}}$ are convective matrices, $\mathbf{K_{\mu v}}$ is the viscous matrix, $\mathbf{K_{vp}}$ is the pressure matrix, and $\mathbf{F_v}$ and $\mathbf{F_p}$ are forcing vectors. The pressure is eliminated at the element level through the static condensation. For the penalty formulation, the incompressibility constraint is defined in the following manner:

$$\mathrm{div}\,\mathbf{v} + \frac{p}{\lambda} = 0 \tag{11.4}$$

where λ is a relatively large positive scalar so that p/λ is a small number (practically zero).

6.1 Shear stress calculation

At the regions of stasis and flow reversals, there is an important role of motion and deformation in the shear stress calculation (Perktold and Rappitsch, 1995; Figueroa et al., 2009). The distribution of stresses within the blood was estimated. The stresses ${}^t\sigma_{ij}$ at time "t" are equal:

$${}^t\sigma_{ij} = -{}^t p\delta_{ij} + {}^t\sigma_{ij}^{\mu} \tag{11.5}$$

where

$${}^t\sigma_{ij}^{\mu} = {}^t\mu^t\left(v_{i,j} + v_{j,i}\right) \tag{11.6}$$

is the viscous stress and ${}^t\mu$ is viscosity corresponding to the velocity vector ${}^t\mathbf{v}$ at a spatial point within the blood domain. The viscous stresses are represented by Eq. (11.6).

The shear stress is calculated as

$$ {}^t\tau = {}^t\mu \frac{\partial^t v_t}{\partial n} \tag{11.7} $$

where ${}^t v_t$ denotes the tangential velocity and n is the normal direction at the vessel wall. At the integration points near the wall surface, the tangential velocity was estimated first, and then the velocity gradient $\partial^t v_t / \partial n$ was calculated. Finally, the viscosity coefficient ${}^t\mu$ using the average velocity was evaluated. Blood was taken as an incompressible Newtonian fluid, appropriate for larger arteries (Perktold and Rappitsch, 1995). The average blood properties were used: dynamic viscosity (μ) of 0.0035 Pa·s and density (ρ) of $1.05 \cdot 10^{-3}$ g/mm^3.

6.2 Modeling the deformation of blood vessels

In modeling blood flow in large blood vessels, we have recognized two distinct cases: (a) rigid walls and (b) deformable walls. It is very important to determine the stress-strain state in tissue, when blood and blood vessel system is analyzed, as well as the effects of the wall deformation on the blood flow characteristics.

There are complex mechanical characteristics of blood vessel tissue. The tissue can be modeled from linear elastic to nonlinear viscoelastic model. If the principle of virtual work is applied, the differential equations of motion of a finite element are

$$ \mathbf{M}\ddot{\mathbf{U}} + \mathbf{B}^w\dot{\mathbf{U}} + \mathbf{K}\mathbf{U} = \mathbf{F}^{ext} \tag{11.8} $$

where the element matrices are as follows: $\mathbf{M}$ is mass matrix; $\mathbf{B}^w$ is the damping matrix, in case when the material has a viscous resistance; $\mathbf{K}$ is the stiffness matrix; and $\mathbf{F}^{ext}$ is the external nodal force vector, which includes body and surface forces acting on the element. The dynamic differential equations of motion by the standard assembling procedure are obtained. These differential equations are integrated with a selected time step size Δt. The displacements ${}^{n+1}\mathbf{U}$ at end of time step are finally obtained according to the equation

$$ \hat{\mathbf{K}}_{tissue}\, {}^{n+1}\mathbf{U} = {}^{n+1}\hat{\mathbf{F}} \tag{11.9} $$

where the tissue stiffness matrix $\hat{\mathbf{K}}_{tissue}$ and vector ${}^{n+1}\hat{\mathbf{F}}$ are expressed in terms of the matrices and vector in Eq. (11.8). The previous equation is obtained under the assumption that the problem is linear, the viscous resistance is constant, displacements are small, and the material is linear elastic.

In many examples, the wall motions and displacements can be large; hence, the problem becomes nonlinear. Therefore, the linear formulation of the equation may not be appropriate. For a geometrically nonlinear problem, there is incremental-iterative equation:

$$ {}^{n+1}\hat{\mathbf{K}}_{tissue}^{(i-1)}\Delta\mathbf{U}^{(i)} = {}^{n+1}\hat{\mathbf{F}}(i-1) - {}^{n+1}\mathbf{F}^{\mathrm{int}(i-1)} \tag{11.10} $$

Here, $\Delta\mathbf{U}^{(i)}$ are the nodal displacement increments for the iteration "i," and the system matrix ${}^{n+1}\hat{\mathbf{K}}_{tissue}^{(i-1)}$, the force vector ${}^{n+1}\hat{\mathbf{F}}^{(i-1)}$ and the vector of internal forces ${}^{n+1}\mathbf{F}^{\mathrm{int}(i-1)}$ correspond to the previous iteration.

The geometrically linear part of the stiffness matrix, $({}^{n+1}\mathbf{K}_L)_{tissue}^{(i-1)}$, and nodal force vector, ${}^{n+1}\mathbf{F}^{\mathrm{int}(i-1)}$, are defined:

$$ \left({}^{n+1}\mathbf{K}_L\right)_{tissue}^{(i-1)} = \int_V \mathbf{B}_L^{T\,n+1}\mathbf{C}_{tissue}^{(i-1)}\mathbf{B}_L dV,\ \left({}^{n+1}\mathbf{F}^{\mathrm{int}}\right)^{(i-1)} = \int_V \mathbf{B}_L^{T\,n+1}\boldsymbol{\sigma}^{(i-1)} dV \tag{11.11} $$

Here, the consistent tangent constitutive matrix ${}^{n+1}\mathbf{C}_{tissue}^{(i-1)}$ of tissue and the stresses at the end of time step ${}^{n+1}\boldsymbol{\sigma}^{(i-1)}$ depend on the material model used.

6.3 FSI interaction

In many models of cardiovascular examples where deformation of blood vessel walls was taken into account, we can implement the loose coupling approach for the fluid-structure interaction (Veljkovic et al., 2012; Filipovic and Schima, 2011; Krsmanovic et al., 2012; Filipovic et al., 2012). The overall algorithm consists of the following steps:

a) For the current geometry of the blood vessel, determine blood flow (with arbitrary Lagrangian-Eulerian (ALE) formulation). The boundary condition for the fluid is wall velocities at the common blood-blood vessel surface.

b) Calculate the loads, which act on the walls from fluid domain (blood).

c) Determine deformation of the walls taking the current loads from the fluid domain (blood).

d) Check for the overall convergence, which includes fluid and solid domain. If convergence is reached, go to the next time step. Otherwise go to step (a).
e) Update blood domain geometry and velocities at the common solid-fluid boundary for the new calculation of the fluid domain. In case of large wall displacements, update the finite element mesh for the fluid domain. Go to step (a).

7 Results

After computational simulations, it is necessary to visualize the obtained results in order to better understand the distribution and the cause of the maximum values of mechanical parameters at the characteristic sites. In addition, the results of unstated and stented AAA will be discussed, which include the influence of fluid domain (distribution of velocity fields, pressures fields, and shear stress) and behavior of solid domains (von Mises stress distribution and displacements). In all cases, the results are presented at the peak systolic moment, as the worst-case scenario, which was reached after about 0.3 s of the simulation, since it is known that aortic substructures are under maximal loads at the peak systolic moment.

7.1 Unstented AAA

As result of the computational FSI simulation of unstented AAA, the velocity field (units: mm/s) of blood is presented in Fig. 11.9. The flow pattern showed areas of high velocity at the aortic inlet and infrarenal branch proximally to the AAA, while low velocity was present in the region of the greatest AAA diameter, which corresponds to the normal physical findings (law of Bernoulli). Comparing the left and right iliac artery, the velocity was decreased within the left one due to aneurysmal dilatation (extended AAA). Also, the cross section of blood velocity is separately depicted, which confirms its parabolic profile.

Blood circulation affects pressure and shear stress distribution that are presented in Fig. 11.10A and B, respectively. Blood pressure (units: MPa, 1 Pa = 0.0075 mmHg) reached the maximal value at the peak systolic moment, which corresponds to the maximal human blood pressure of around 124 mmHg. The highest pressure was present at the inlet, proximally to the AAA, thereafter distally decreased toward the aortic outlets. The shear

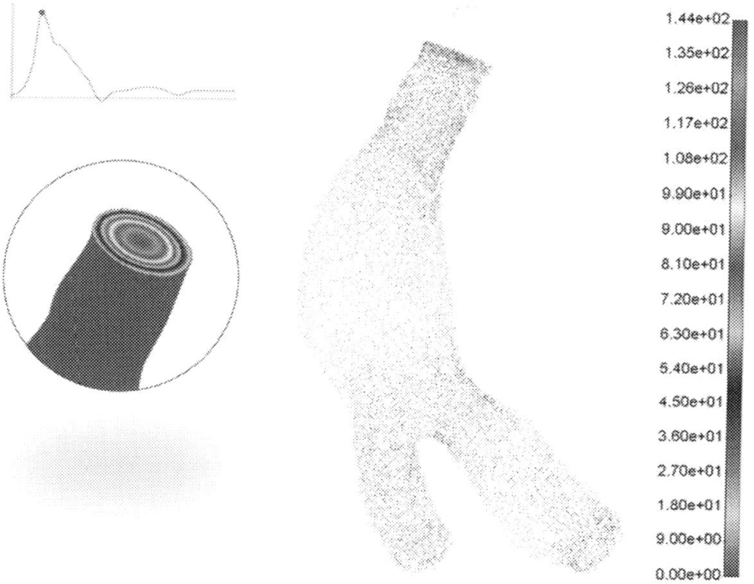

Fig. 11.9 Unstented AAA: velocity field of blood (units: mm/s) at the peak systolic moment (anterior view).

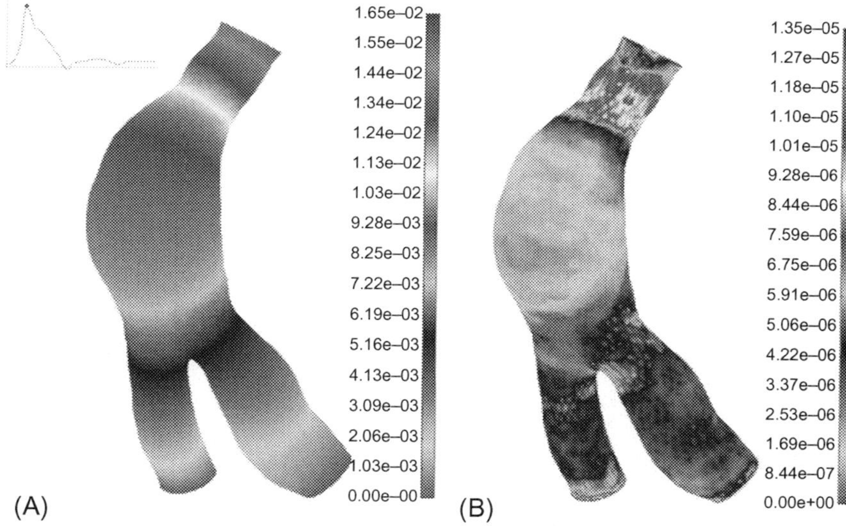

Fig. 11.10 Unstented AAA: (A) pressure distribution (units: MPa) and (B) shear stress distribution (units: MPa) at the peak systolic moment (anterior view).

stress distribution (units: MPa), which corresponds to the blood velocities, was increased proximally to the AAA, thereafter decreased toward the AAA and iliac arteries. The areas of high shear stress were mainly situated in regions of enhanced recirculation.

The von Mises stress distribution (units: MPa) within the unstented aortic wall is presented in Fig. 11.11. Blood circulation, which was considered in the simulation, and distribution of mechanical stress have significant influence on aneurysmal growth and aortic wall behavior. The von Mises stress, as the main parameter for evaluation of aortic wall degradation and rupture risk, was analyzed with accent on the areas of high stress in the aortic wall. The obtained stress distribution was affected by the shape of the aneurysm formation and the material characteristics of the aortic wall and by interaction between the fluid and solid domains. High stress at the aortic bifurcation indicated the need for operative treatment. Furthermore, the obtained blood flow characteristics showed the presence of stagnant blood flow at the maximal aneurysmal diameter, which affects the aneurysmal growth (Fig. 11.9).

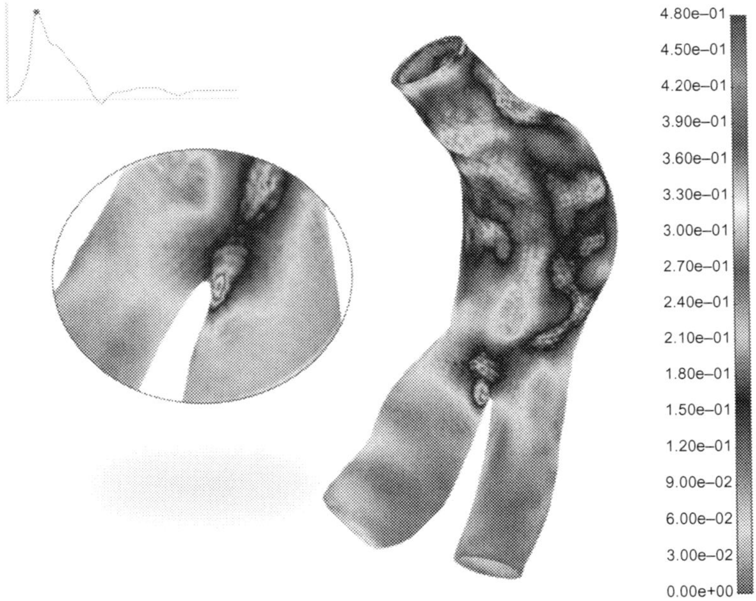

Fig. 11.11 Unstented AAA: von Mises stress distribution at the peak systolic moment (units: MPa) within the aortic wall (posterior view).

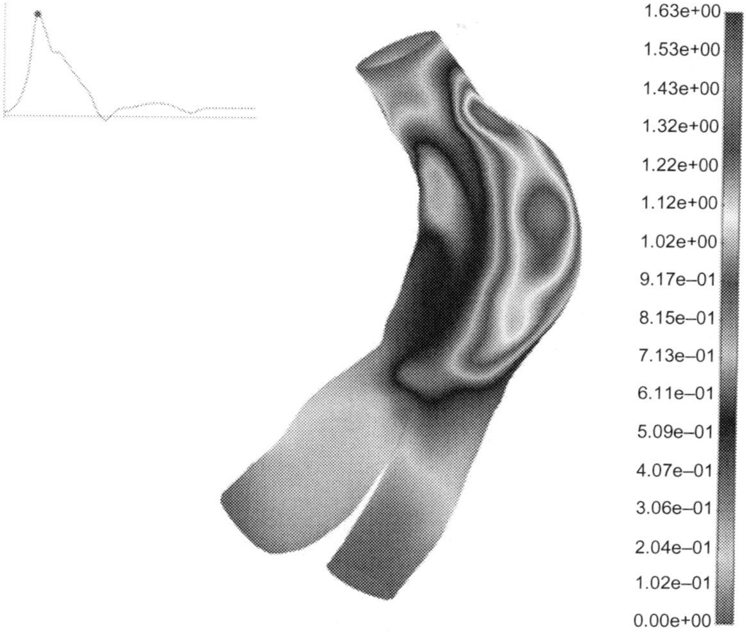

Fig. 11.12 Unstented AAA: aortic wall displacements (units: mm) at the peak systolic moment (posterior view).

Displacements of the unstented aortic wall are depicted in the Fig. 11.12 (units: mm). As result of the performed simulation, displacements were maximal at the aneurysmal region and up to the aneurysmal neck, which could affect on further growth of AAA. The presented displacements at aortic inlet and outlets depended on prescribed boundary conditions.

7.2 Stented AAA

As a result of the computational FSI simulation of stented AAA, the velocity field of blood (units: mm/s) is presented in Fig. 11.13. It can be seen that velocity field of blood in peak systolic moment had the highest values at the stent-graft bifurcation and gradually decreased going toward lower stent-graft parts, that is, the outlets of iliac arteries. Velocity field depends on stent-graft geometry and blood flow conditions and affects pressure and shear stress distribution that are presented in Fig. 11.14A and B, respectively. Blood pressure (units: MPa) reached the maximal value at the peak systolic moment, which corresponds to the maximal human blood pressure of around 128 mmHg. The highest pressure was present at the stent-graft

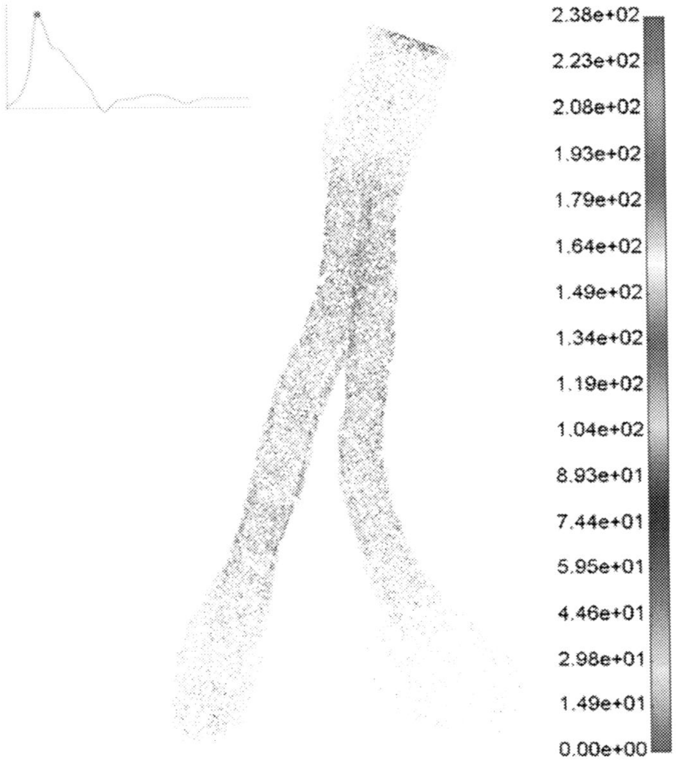

Fig. 11.13 Stented AAA: velocity field of blood (units: mm/s) at the peak systolic moment (anterior view).

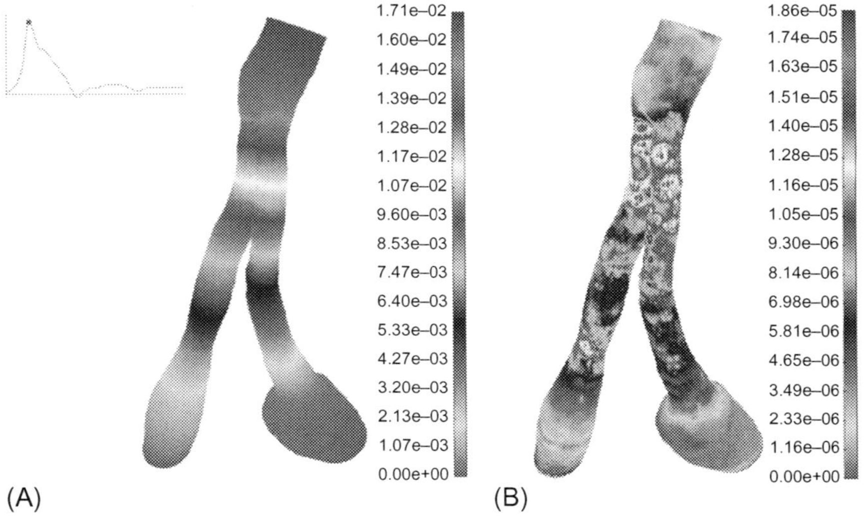

Fig. 11.14 Stented AAA: (A) pressure distribution (units: MPa) and (B) shear stress distribution (units: MPa) at the peak systolic moment (anterior view).

inlet, thereafter distally decreased toward the stent-graft outlets, which depended on the prescribed boundary conditions. The shear stress distribution (units: MPa), which corresponded to the blood velocities, was increased after the stent-graft bifurcation, thereafter decreased toward the stent-graft outlets. A jet of fluid, created during systole, provided high velocity and high shear stress at the stent-graft bifurcation.

The von Mises stress distribution within the stent graft and aortic wall is presented in Fig. 11.15. After stent-graft implantation, von Mises stress (units: MPa) within the aortic wall was reduced (Fig. 11.15C), which prevents aneurysm from further growing and rupture. On the other hand, stent graft was under the high stress (Fig. 11.15B), especially at the bifurcation (0.54 MPa, Fig. 11.15A) where jet of fluid, created during systole, provided high velocity. Also, in the EVAR procedure, higher Young's modulus of stent graft causes decreasing of von Mises stress in the aortic wall (Gawenda et al., 2004). Due to that fact, it is of great importance to choose appropriate dimensions and material characteristics of stent graft for treated patient, with aim to avoid postoperative complications.

Displacements (units: mm) of the stent graft and aortic wall are presented in Fig. 11.16. As well as the von Mises stress, aortic wall displacements were

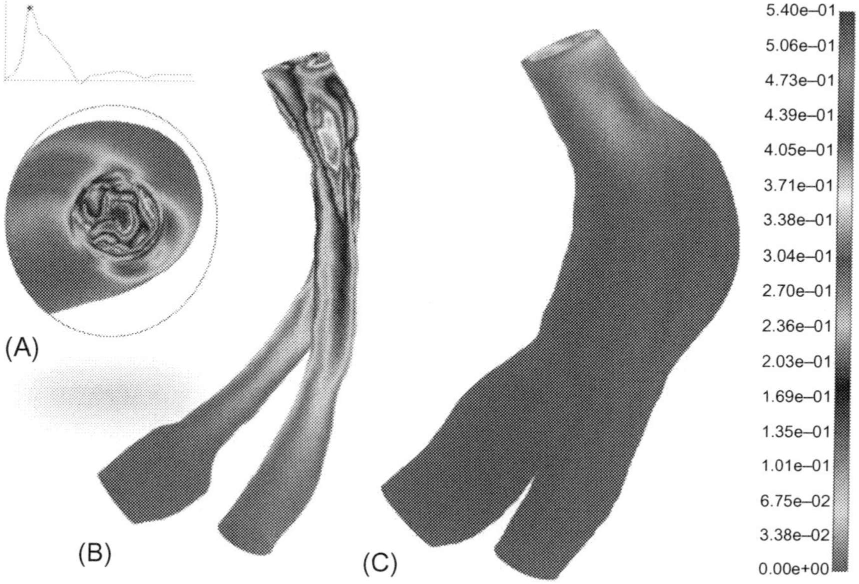

Fig. 11.15 Stented AAA: von Mises stress distribution at the peak systolic moment (units: MPa): (A) stent graft, top view; (B) stent graft; and (C) aortic wall, posterior view.

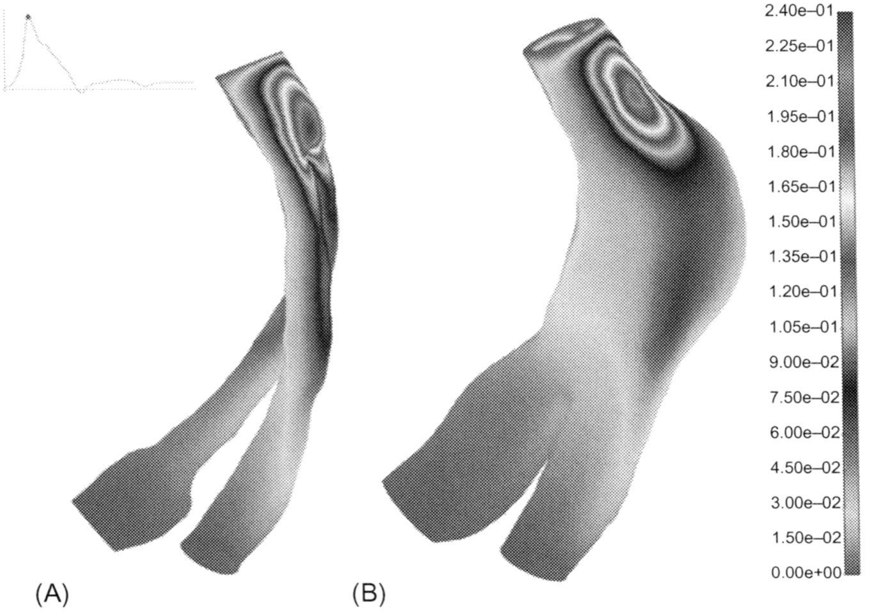

Fig. 11.16 Stented AAA: (A) stent graft and (B) aortic wall displacements (units: mm) at the peak systolic moment (posterior view).

also reduced after the stent-graft implantation, which is one more benefit of EVAR procedure. As result of performed simulation, displacements were maximal at the stent-graft neck proximally to the AAA (Fig. 11.16A), as well as at the proximal aneurysmal neck (Fig. 11.16B). The presented displacements proximal to AAA depended on prescribed boundary conditions. Displacements of stent graft distal branches and distal iliac arteries were low, which does not correspond to real physiological behavior.

8 Discussion

The development of an AAA is a complex interaction of biomechanical factors resulting in changes in composition of the aortic wall. Although it was believed that aneurysm diameter is a predominant predisposing factor in aneurysm rupture risk, studies have demonstrated that wall stress is more complicated to predict, as it is influenced by aneurysm shape and geometry, architecture, the presence of ILT, and other mechanical characteristics. In order to reduce the risk of AAA rupture, stent-graft implantation is an attractive minimally invasive procedure with well-known postoperative complications and benefits. This study investigated the mechanical behavior of the AAA before stent-graft implantation and thereafter the effects of stent-graft implantation. The analysis was performed using numerical methods and algorithms for FSI on computational finite element meshes. Instead of simplified parametric model (Shengmao et al., 2017), this study was based on the 3D finite element models using the patient-specific CT scan images (Djorovic et al., 2019). The computational simulations were performed considering the blood flow through the AAA and interaction with the unstented aortic wall, as well as the blood flow through the implanted stent graft and interaction with the stent graft, ILT, and aortic wall. In order to perform computational analysis of unstented and stented AAA, which included different domains, the PAK solver was employed prescribing the appropriate boundary conditions and equivalent material characteristics for included domains.

The employed FSI analysis provided a comprehensive insight into the examination of AAA's mechanical characteristics and, in chi-specific case, enabled computation of fluid domain influence (distribution of velocity fields, pressures fields, and shear stress) and behavior of solid domains (von Mises stress distribution and displacements). In case of analyzed mechanical parameters, the models were presented at the peak systolic moment, as the worst-case scenario. Before stent-graft implantation AAA wall was under higher stress and displacements, with stagnant blood flow at the maximal AAA diameter, which indicated possibility for further AAA growth and justification of EVAR procedure. After stent-graft implantation, stress within the aortic wall and displacements were reduced, which prevents aneurysm from further growing and rupture, but on the other hand, stent graft was under the high stress, especially at the bifurcation, which indicates that later complications may be possible. This analysis was built on experiences of other groups having performed FSI simulations of AAA. The obtained and analyzed results correspond to the previous works (Li and Kleinstreuer, 2006; Molony et al., 2009; Chandra et al., 2013; Di Martino et al., 2001).

It should be noted that the analyzed models used for this study involved some strong limitations. First, we simplified the stent-graft geometry, due to difficulty to reconstruct a real geometry of graft tube and nitinol wires from the CT data. Second, the linear material of AAA model did not fully describe biological aortic tissue, considering that aortic wall has nonlinear and anisotropic properties. The assumption of linear elasticity for the stent graft was also a limitation of the work. Third, average material properties and wall thickness were adopted from the literature, which do not replicate the patient-specific characteristics. Also, we assumed that the top surface of infrarenal aorta and stent graft and the bottom surfaces of stent graft and iliac arteries were fixed from motion. Due to that fact, the presented model did not replicate the realistic movement. Furthermore, we prescribed average blood flow rate at the inlet of infrarenal aorta. Hence, the presented results can be applied only for this study.

9 Conclusion

The presented FSI simulation was performed with the aim to examine the biomechanical characteristics of the patient-specific unstented AAA and benefits after EVAR procedure and stent-graft implantation. After presented results, the stent-graft implantation was reasonable in order to prevent further aneurysmal growth and rupture. Also, blood circulation had significant influence on the stent graft and aortic wall behavior. The employed computational simulation enabled better insight into the biomechanical processes due to the difficulties in obtaining such parameters experimentally.

Beside all benefits, which aortic endovascular approach takes with, various problems are still present, such as stent-graft migration, different types of endoleaks, and stent mechanism breakage. These phenomena depend on appropriate selection of stent-graft type and dimensions, which correlate with stent-graft durability. In that order, this will be an objective of our future studies. Our knowledge about local anatomical conditions for abdominal aortic stent-graft implantation can be additionally improved by FSI and image-based patient-specific modeling. These techniques are highly promising and able to realistically simulate different interventional options that might facilitate clinical decision-making and provide useful insights into the treatment of each individual patient.

Acknowledgment

This paper is supported by the Ministry of Education, Science and Technological Development of the Republic of Serbia with projects III41007 and OI174028.

References

Bahia, S., Holt, P., Jackson, D., Patterson, B., Hinchliffe, R., Thompson, M., Karthikesalingama, A., 2015. Systematic review and meta-analysis of long-term survival after elective infrarenal abdominal aortic aneurysm repair 1969–2011: 5 year survival remains poor despite advances in medical care and treatment strategies. Eur. J. Vasc. Endovasc. Surg. 50, 320–330.

Bhave, N., Isselbacher, E., Eagle, K., 2019. Aortic aneurysms. In: Toth, P., Cannon, C. (Eds.), Comprehensive Cardiovascular Medicine in the Primary Care Setting. In: Contemporary Cardiology, Humana Press, Cham, pp. 365–377.

Biasetti, J., Gasser, T., Auer, M., Hedin, U., Labruto, F., 2010. Hemodynamics of the normal aorta compared to fusiform and saccular abdominal aortic aneurysms with emphasis on a potential thrombus formation mechanism. Ann. Biomed. Eng. 38, 380–390.

Bown, M., Sutton, A., Bell, P., Sayers, R., 2002. A meta-analysis of 50 years of ruptured abdominal aortic aneurysm repair. Br. J. Surg. 89, 714–730.

Chandra, S., Raut, S., Jana, A., Biederman, R., Doyle, M., Muluk, S., Finol, E., 2013. Fluid-structure interaction modeling of abdominal aortic aneurysms: the impact of patient-specific inflow conditions and fluid/solid coupling. J. Biomech. Eng. 135, 0810011–08100114.

Choke, E., Cockerill, G., Wilson, W., Sayed, S., Dawson, J., Loftus, I., Thompson, M., 2005. A review of biological factors implicated in abdominal aortic aneurysm rupture. Eur. J. Vasc. Endovasc. Surg. 30, 227–244.

De Bock, S., Iannaccone, F., De Santis, G., De Beule, M., Van Loo, D., Devos, D., … Verhegghe, B., 2012. Virtual evaluation of stent graft deployment: a validated modeling and simulation study. J. Mech. Behav. Biomed. Mater. 13, 129–139.

De Bruin, J., Baas, A., Buth, J., Prinssen, M., Verhoeven, E., al., e., 2010. Long-term outcome of open or endovascular repair of abdominal aortic aneurysm. N. Engl. J. Med. 362, 1881–1889.

Di Martino, E., Bohra, A., Scotti, C., Finol, E., Vorp, D., 2004. Wall stresses before and after endovascular repair of abdominal aortic aneurysms. AAA Pre- and Post-EVAR, In: International Mechanical Engineering Congress and Exposition (IMECE).

Di Martino, E.S., Guadagni, G., Fumero, A., Ballerini, G., Spirito, R., Biglioli, P., Redaelli, A., 2001. Fluid-structure interaction within realistic three-dimensional models of the aneurysmatic aorta as a guidance to assess the risk of rupture of the aneurysm. Med. Eng. Phys. 23, 647–655.

Djorovic, S., Saveljic, I., Filipovic, N., 2019. Computational simulation of carotid artery: from patient-specific images to finite element analysis. J. Serbian Soc. Comput. Mech. 13 (1). In press.

Domagala, Z., Stepak, H., Drapikowski, P., al., e., 2016. Biomechanical factors in Finite Element Analysis of abdominal aortic aneurysms. Acta Angiol. 164–171.

Doyle, B.J., Callanan, A., Walsh, M., Grace, P., McGloughlin, T., 2009. A finite element analysis rupture index (FEARI) as an additional tool for abdominal aortic aneurysm rupture prediction. Vasc. Dis. Prev. 6, 114–121.

Doyle, B., Cloonan, A., Walsh, M., Vorp, D., McGloughlin, T., 2010. Identification of rupture locations in patient-specific abdominal aortic aneurysms using experimental and computational techniques. J. Biomech. 43, 1408–1416.

Erhart, P.H.-D., Geisbüsch, P., Kotelis, D., Müller-Eschner, M., Gasser, T., von Tengg-Kobligk, H., Böckler, D., 2015. Finite element analysis in asymptomatic, symptomatic, and ruptured abdominal aortic aneurysms: in search of new rupture risk predictors. Eur. J. Vasc. Endovasc. Surg. 49, 239–245.

Figueroa, C., Taylor, C., Yeh, V., al., e., 2009. Effect of curvature on displacement forces acting on aortic endografts: a 3-dimensional computational analysis. J. Endovasc. Ther. 284–294.

Filipovic, N., Ivanovic, M., Krstajic, D., Kojic, M., 2011b. Hemodynamic flow modeling through an abdominal aorta aneurysm using data mining tools. IEEE Trans. Inf. Technol. Biomed. 15, 189–194.

Filipovic, N., Kojic, M., Ivanovic, M., Stojanovic, B., Otasevic, L., Rankovic, V., 2006b. MedCFD, Specialized CFD Software for Simulation of Blood Flow Through Arteries. University of Kragujevac, Serbia.

Filipovic, N., Mijailovic, S., Tsuda, A., Kojic, M., 2006a. An implicit algorithm within the arbitrary Lagrangian-Eulerian formulation for solving incompressible fluid flow with large boundary motions. Comp. Methods Appl. Mech. Eng. 195, 6347–6361.

Filipovic, N., Milasinovic, D., Zdravkovic, N., Böckler, D., von Tengg-Kobligk, H., 2011a. Impact of aortic repair based on flow field computer simulation within the thoracic aorta. Comput. Methods Prog. Biomed. 101, 243–252.

Filipovic, N., Rosic, M., Tanaskovic, I., Milosevic, Z., Nikolic, D., Zdravkovic, N., … Parodi, O., 2012. ARTreat project: three-dimensional numerical simulation of plaque formation and development in the arteries. IEEE Trans. Inf. Technol. Biomed. 16, 272–278.

Filipovic, N., Schima, H., 2011. Numerical simulation of the flow field within the aortic arch during cardiac assist. Artif. Organs 35, 73–83.

Fillinger, M., Marra, S., Raghavan, M., Kennedy, F., 2003. Prediction of rupture risk in abdominal aortic aneurysm during observation: wall stress versus diameter. J. Vasc. Surg. 37, 724–732.

Fillinger, M., Raghavan, M., Marra, S., Cronenwett, J., Kennedy, F., 2002. In vivo analysis of mechanical wall stress and abdominal aortic aneurysm rupture risk. J. Vasc. Surg. 36, 589–597.

Frauenfelder, T., Lotfey, M., Boehm, T., Wildermuth, S., 2006. Computational fluid dynamics: hemodynamic changes in abdominal aortic aneurysm after stent graft implantation. Cardiovasc. Intervent. Radiol. 29, 613–623.

Gao, F., Watanabe, M., Matsuzawa, T., 2006. Stress analysis in a layered aortic arch model under pulsatile blood flow. Biomed. Eng. Online 5, 25.

Gawenda, M., Knez, P., Winter, S., Jaschke, G., Wassmer, G., Schmitz-Rixen, T., Brunkwall, J., 2004. Endotension is influenced by wall compliance in a latex aneurysm model. Eur. J. Vasc. Endovasc. Surg. 27, 45–50.

Georgakarakos, E., Ioannou, C., Kamarianakis, Y., Papaharilaou, Y., Kostas, T., Manousaki, E., Katsamouris, A., 2010. The role of geometric parameters in the prediction of abdominal aortic aneurysm wall stress. Eur. J. Vasc. Endovasc. Surg. 39, 42–48.

Golledge, J., 2019. Abdominal aortic aneurysm: update on pathogenesis and medical treatments. Nat. Rev. Cardiol. 16, 225–242.

Greco, G., Egorova, N., Gelijns, A., Moskowitz, A., Manganaro, A., Zwolak, R., al., e., 2010. Development of a novel scoring tool for the identification of large >/=5 cm abdominal aortic aneurysms. Ann. Surg. 252, 675–682.

Hobo, R., Sybrandy, J., Harris, P., Buth, J., 2008. Endovascular repair of abdominal aortic aneurysms with concomitant common iliac artery aneurysm: outcome analysis of the EUROSTAR Experience. J. Endovasc. Ther. 15, 12–22.

Jayendiran, R., Nour, B., Ruimi, A., 2018. Fluid-structure interaction (FSI) analysis of stent-graft for aortic endovascular aneurysm repair (EVAR): Material and structural considerations. J. Mech. Behav. Biomed. Mater. 87, 95–100.

Kent, K., Zwolak, R., Jaff, M., Hollenbeck, S., Thompson, R., Schermerhorn, M., … Cronenwett, J., 2004. Screening for abdominal aortic aneurysm: a consensus statement. J. Vasc. Surg. 39, 267–269.

Kleinstreuer, C., Li, Z., Basciano, C., Seelecke, S., Farber, M., 2008. Computational mechanics of Nitinol stent grafts. J. Biomech. 2370–2378.

Kojic, M., Filipovic, N., Stojanovic, B., Kojic, N., 2008. Computer Modeling in Bioengineering—Theoretical Background, Examples and Software. John Wiley and Sons, Chichester, England.

Kolipaka, A., Illapani, V., Kenyhercz, W., Dowell, J., Go, M., Starr, J., … White, R., 2016. Quantification of abdominal aortic aneurysm stiffness using magnetic resonance elastography and its comparison to aneurysm diameter. J. Vasc. Surg. 64, 966–974.

Krsmanovic, D., Koncar, I., Petrovic, D., Milasinovic, D., Davidovic, L., Filipovic, N., 2012. Computer modelling of maximal displacement forces in endoluminal thoracic aortic stent graft. Comput. Methods Biomech. Biomed. Eng. 17, 1012–1020.
Kühnl, A., Erk, A., Trenner, M., Salvermoser, M., Schmid, V., Eckstein, H., 2017. Incidence, treatment and mortality in patients with abdominal aortic aneurysms. Dtsch. Arztebl. Int. 114, 391–398.
LeFevre, M., 2014. Screening for abdominal aortic aneurysm: U.S. Preventive Services Task Force recommendation statement. Ann. Intern. Med. 161, 281–290.
Li, Z., Kleinstreuer, C., 2006. Analysis of biomechanical factors affecting stent-graft migration in an abdominal aortic aneurysm. J. Biomech. 39, 2264–2273.
McGloughlin, T., Doyle, B., 2010. New approaches to abdominal aortic aneurysm rupture risk assessment: engineering insights with clinical gain. Arterioscler. Thromb. Vasc. Biol. 30, 1687–1694.
Minino, A., Heron, M., Murphy, S., Kochanek, K., 2004. Deaths: final data for 2004. Natl. Vital Stat. Rep. 55, 1–119.
Molony, D., Callanan, A., Kavanagh, E., Walsh, M., McGloughlin, T., 2009. Fluid-structure interaction of a patient-specific abdominal aortic aneurysm treated with an endovascular stent-graft. Biomed. Eng. Online 8, 24.
O'Gara, T., 2003. Aortic aneurysm. Circulation, e43–e45.
Olufsen, M., Peskin, C., Kim, W., Pedersen, E., A, N., & J, L., 2000. Numerical simulation and experimental validation of blood flow in arteries with structured-tree outflow conditions. Ann. Biomed. Eng. 28, 1281–1299.
Perktold, K., Rappitsch, G., 1995. Computer simulation of local blood flow and vessel mechanics in a compliant carotid artery bifurcation model. J. Biomech. 28, 845–856.
Prinssen, M., Verhoeven, E., Buth, J., et al., 2004. A randomized trial comparing conventional and endovascular repair of abdominal aortic aneurysms. N. Engl. J. Med. 351, 1607–1618.
Robnik-Šikonja, M., Radović, M., Đorović, S., Anđelković-Ćirković, B., Filipović, N., 2018. Modeling ischemia with finite elements and automated machine learning. J. Comput. Sci. 99–106.
Rodriguez, J., Martufi, G., Doblare, M., Finol, E., 2009. The effect of material model formulation in the stress analysis of abdominal aortic aneurysms. Ann. Biomed. Eng. 37, 2218–2221.
Sakalihasan, N., Limet, R., Defawe, O., 2005. Abdominal aortic aneurysm. Lancet 365, 1577–1589.
Schermerhorn, M., 2009. A 66-year-old man with an abdominal aortic aneurysm: review of screening and treatment. JAMA 302, 2015–2022.
Scotti, C., Finol, E., 2007. Compliant biomechanics of abdominal aortic aneurysms: a fluid-structure interaction study. Comput. Struct. 85, 1097–1113.
Scotti, C., Finol, E., Viswanathan, S., Shkolnik, A., Amon, C., 2004. Computational fluid dynamics and solid mechanics analyses of a patient-specific AAA pre- and post-EVAR. In: ASME IMECE: Bioengineering.
Shengmao, L., Han, X., Bi, Y., Ju, S., Gu, L., 2017. Fluid-structure interaction in abdominal aortic aneurysm: effect of modeling techniques. Biomed. Res. Int. 2017.
Shum, J., Martufi, G., di Martino, E., al., e., 2011. Quantitative assessment of abdominal aortic aneurysm geometry. Ann. Biomed. Eng. 39, 277–286.
Stenbaek, J., Kalin, B., Swedenborg, J., 2000. Growth of thrombus may be a better predictor of rupture than diameter in patients with abdominal aortic aneurysms. Eur. J. Vasc. Endovasc. Surg. 20, 466–469.
Suzuki, K., Ishiguchi, T., Kawatsu, S., Iwai, H., Maruyama, K., Ishigaki, T., 2001. Dilatation of EVGs by luminal pressures: experimental evaluation of polytetrafluorothylene (PTFE) and woven polyester grafts. Cardiovasc. Intervent. Radiol. 24, 94–98.

Thompson, M., 2003. Controlling the expansion of abdominal aortic aneurysms. Br. J. Surg. 90, 897–898.

Thubrikar, M., Al-Soudi, J., Robicsek, F., 2001. Wall stress studies of abdominal aortic aneurysm in a clinical model. Ann. Vasc. Surg. 15, 355–366.

Tkaczyk, J., Przywara, S., Terpiłowski, M., Brożyna, K., Iłżecki, M., Terlecki, P., Zubilewicz, T., 2019. Prevalence and risk factors of abdominal aortic aneurysm among over 65 years old population in Lublin, Poland. Acta Angiol. 1–6.

Upchurch Jr., G.R., Schaub, T., 2006. Abdominal aortic aneurysm. Am. Fam. Physician, 1198–1204.

Veljkovic, D., Filipovic, N., Kojic, M., 2012. The effect of asymmetry and axial prestraining on the amplitude of mechanical stresses in abdominal aortic aneurysm. J. Mech. Med. Biol. 12.1250089.

Venkatasubramaniam, A., Fagan, M., Mehta, T., Mylankal, K., Ray, B., Kuhan, G., … McCollum, P., 2004. A comparative study of aortic wall stress using finite element analysis for ruptured and non-ruptured abdominal aortic aneurysms. Eur. J. Vasc. Endovasc. Surg. 28, 168–176.

Vorp, D., 2007. Biomechanics of abdominal aortic aneurysm. J. Biomech. 40, 1887–1902.

Zhang, Y., Barocas, V.H., Kassab, G., Lochner, D., McCulloch, A., Tran-Son-Tay, R., Trayanova, N., 2016. Multi-scale modeling of the cardiovascular system: disease development, progression, and clinical intervention. Ann. Biomed. Eng. 44, 2642–2660.

CHAPTER 12

Sport biomechanics: Experimental and computer simulation of knee joint during jumping and walking

Contents

1 Introduction

Biomechanic analysis, computer simulation, and modeling of jumping are very important fields of research and have been rapidly growing, especially because of their application in sports training process and prevention of sports injury/sport science, sport bioengineering, and sports medicine.

Computational Modeling in Bioengineering and Bioinformatics
https://doi.org/10.1016/B978-0-12-819583-3.00012-6

Jumping is one of the most common activities during sports. In the same time the maximal value of force and torque appeared. The consequences are very often injuries especially for ankle, knee joint, ligaments, meniscus, etc. One of the crucial goals in sport medicine and rehabilitation is to understand biomechanic process in the extremities and joints during jumping.

There are a lot of kinds of jump, which are performed during sport game like drop jump, squat jump, and counter movement jump. For this chapter, we considered other kinds of jumps: flywheel jump, jump without flywheel, left leg landing jump, and right leg landing jump. The standard method in experimental analysis of jumping parameters is using force plate (Cross, 1999; Filipovic et al., 2009). The force platform can measure force and torque during jump in different directions using force sensor. Improvement of electronic devices and gadgets enables us to measure jumping parameters using wearable devices such as sensor in footwear (Setuain et al., 2015).

The contact stress inside knee joint is directly related to its functional activities. It is very important to analyze the biomechanical characteristics of the motion and stress during high-flexion activities (Thambyah et al., 2005). Prediction of these two important characteristics is relatively difficult because of the limitation of ethic and measuring devices. Finite element analysis can be a good approach to predict the internal stress and distribution under dynamic loading conditions (Godest et al., 2002; Halloran et al., 2005; Fitzpatrick and Rullkoetter, 2012; Halloran et al., 2010). In the literature, there are some finite element models, which can simultaneously proceed with dynamic synchronous prediction for the patellofemoral joint and femorotibial joint (Halloran et al., 2005), and less have represented the whole joint with physiological soft tissue constraint (Baldwin et al., 2012).

The knee is considered one of the most complex and important joints in the human body (Fig. 12.1). The main role of the knee joint is to provide stability for human body during different types of activities. Knee stability is achieved by the anterior and posterior cruciate ligaments, the medial and lateral collateral ligaments, the menisci, the joint capsule, and the musculature that surrounds the joint. Knee injuries are often caused by sports. One of the most common injuries is meniscus tear.

The meniscus is a C-shape fibrocartilaginous tissue. The menisci reduce articular cartilage contact pressures, transmit tibiofemoral loads, and provide stability and lubrication. Lateral meniscus transmits more load then medial meniscus. Percent of load that is transmitted depends on the angle of flexion. When the knee is in extension, the menisci transmits 50% of the load, but when the knee is in 90 degree flexion, 90% of the joint load is transmitted. The medial and lateral menisci have distinguishing features. The body of the

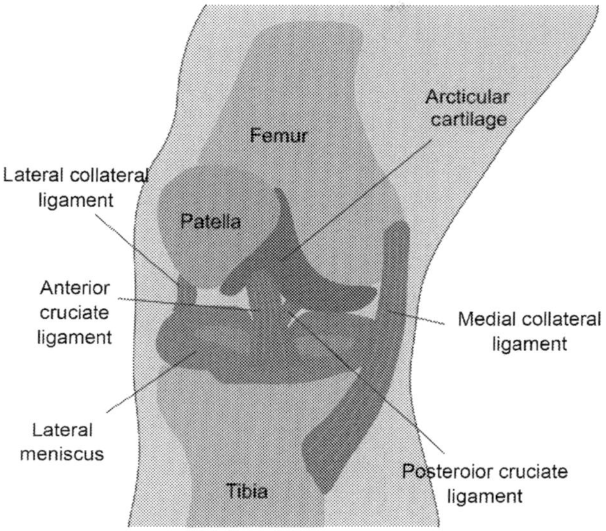

Fig. 12.1 Anatomy of the knee joint.

medial meniscus is both wider and thicker posteriorly than anteriorly, while the lateral meniscus is more symmetrical and wider on average.

Because injures of human knee joint are frequent, many researchers have been analyzing knee joint model using the finite element method. FE model allows us to change the model in a short period of time and helps us save time and money. The precisely developed FE model can accurately predict effects for a large number of different situations. This type of analysis can provide information that is not always possible to obtain from experiments. The accuracy of FE analysis depends on the accurate input of information. This is a problem because although a knee joint has been examined a lot, mechanical behavior of the knee joint is not completely known. This is especially true for parts of the knee joint such as menisci.

New video capture boards, automatic digitizing, real-time transformation, and filtering software have recently been developed. This allows for kinematic and kinetic data to be displayed simultaneously (Altmeyer et al., 1994; Ariel et al., 1997; Borelli, 1989). These data can be time synchronized so that a researcher, coach, and athlete could effectively evaluate athletic performance under various competitive conditions. They may also be of assistance for making adjustments to training (Braune and Fischer, 1987; Finch et al., 1998; Plagenhoef, 1968). In the current investigation, video analysis of images, together with the force plate measurement, was utilized to determine the deformation of the cartilage in the knee.

Video acquisition and parsing sequences of images significantly influences the performance of tracking algorithms. Careful planning of image acquisition is crucial for the success of the proposed system. Motion tracking of small objects can be quite difficult and depends on many parameters with special attention on environmental conditions and the type of video acquisition system. Although various tracking algorithms have been developed in the past, their robustness has not been clearly described in the literature. The existing methods can be roughly classified as high-tech hardware based and algorithm based. In the first group, high-tech hardware is used, that is, high-quality video, multiple high-tech cameras and lenses, powerful high-end computers, and best lighting. In such a case, applied algorithms are not very sophisticated. Often a blob tracker is sufficient (Pingali et al., 1998, 2000). The primary issue is to set up and coordinate multiple cameras or to visualize results of motion tracking, for example, in a 3D virtual environment (Pingali et al., 1998, 2000). However, special and very sophisticated algorithms that are able to resolve the task of compensation of demerits must be employed. In this chapter, a method is proposed that compromises between two previously described methods. One utilized only a high-speed camera, while our motion tracking software is based on a moving seed point algorithm. Using a high-speed camera for complex kinematic calculations in the motion tracking was avoided. A few similar advanced algorithms were able to address the issue when various experimental conditions are implemented. Many biomechanists argue that true stiffness of the human body is the combination of all the individual stiffness values contributed by the muscles, tendon, ligaments, cartilage, and bone. Sports and clinical biomechanics are typically interested in the role of stiffness because it is related to both performance and injury. Stiffness in human body can be described from the level of a single muscle fiber to modeling the entire body as a mass and spring. The standard method in jump analyzing is based on the use of a force sensor platform for ground reaction force measuring (Cross, 1999; Filipovic et al., 2009; Setuain et al., 2015). The force platform can measure force and torque during jump in different directions using a force sensor. The knee implants (Kutzner et al., 2010) were used to directly measure loads of participants during daily activities. In vivo knee measurements are very invasive and practically impossible for the case described in our study. There are many techniques for measuring an external variable, which can be further used to create biomechanical and mathematical models. Some researches (Li et al., 2008) were based on the registration of fluoroscopic images and computer model of the knee. The most recent method utilized force plate data, CT or MRI skeletal structure data, and motion capture obtained from the infrared position sensor (Reinschmidt et al., 1997; Yanga et al., 2010). In other research paper (Djurić-Jovičić et al., 2011), the accelerometer was used to estimate the angle of lower extremities. The measure performance can be improved using a combination of accelerometer and gyroscope sensor such as goniometer (Tao et al., 2012). This is a pretty cheap technique but requires additional processing of collected data, and the error of estimated angle is up to 6%. The measure performance can be improved using a combination of accelerometer and gyroscope sensor such as goniometer (Dejnabadi et al., 2005; Tao et al., 2012). Tendon injuries occur often in sports (Jósza and Kannus, 1997), especially in explosive sports where a high demand of force, speed, and power is utilized for optimal performance. Athletes are repeatedly subjected to continuously high stress during exercise, which makes the bone tendon junction, being the weakest point, susceptible to lesion. Cumulative microtrauma may occur, which weakens collagen cross-linking and finally results in mucoid degeneration of the tendon (Khan et al., 1998). As a result, noninvasive determination of cartilage deformation in the knee may be a very useful tool for prediction the of knee injury.

In this study, a combined image processing-tracking algorithm was used in order to measure kinematics and dynamics of athletes jumping on a force plate and foot pressure during walking, together with numerical calculation of cartilage deformation in the knee. In the following section, the procedure for force measurement, basic algorithm for image tracking the object, foot pressure measurement, and numerical finite element methods for cartilage deformation is explained. The results for force plate and foot pressure during walking, together with numerical calculation of cartilage deformation in the knee, are presented. Finally, some discussion and conclusions for the possibility and future application of biomechanical analysis for sportsman are considered.

2 Methods

2.1 Geometry of the model

Knee geometry is very complex, with many different elements. That is the reason why it is significant to have good scans as a basis, because the model and the results of simulation depend on that quality. Creating 3D models is done manually or automatically, and the result is the creation of structures that will become the finite element model in the end. MRI scans from right leg of a healthy person (40-year-old female) were used to create 3D model of the human joint. Scans were obtained, while patient's leg was at full extension. Resolution of scans was 320×320 pixels, and slice increment was 0.589 mm.

Three-dimensional knee joint model is consisted of these elements: femur, tibia, fibula, articular cartilage, menisci, and ligaments. All knee joint elements were segmented from scans. Surface mesh of these elements was automatically generated and was refined in order to improve mesh quality. Improved surface mesh was used for volumetric mesh creation. The knee joint model was modeled by 300531 nodes and 246632 eight-node brick elements (Fig. 12.2). Further information about used knee joint are given in paper.

The finite element analysis was done using the finite element solver (PAK-C) developed at the Faculty of Engineering, University of Kragujevac, Serbia (Filipovic et al., 2008).

2.2 Material properties

Material properties used for simulation were gathered from the literature. Table 12.1 lists all the materials used for calculation, as well as their properties (Young's modulus, Poisson's ration). All the materials used for calculation are considered to be linear elastic, homogenous, and isotropic.

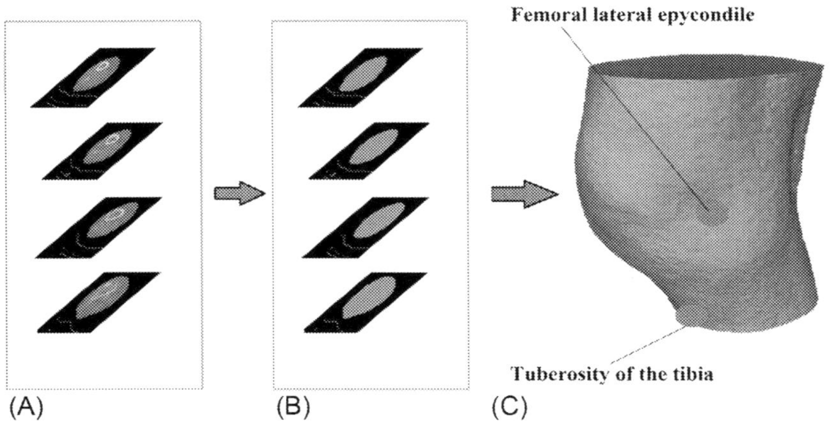

Fig. 12.2 Three-dimensional model of knee. (A) Raw CT slices, (B) segmented CT images, and (C) 3D model with marker position.

Table 12.1 Material properties of knee joint elements (Lee et al., 2004; Bei and Fregly, 2004)

Material	Young's modulus [MPa]	Poisson's ratio
Femur—cortical	14317	0.315
Femur—spongy	295	0.315
Tibia—cortical	20033	0.315
Tibia—spongy	295	0.315
Fibula	17000	0.3
Articular cartilage	12	0.45
Menisci	80	0.3
Ligaments	345	0.22

2.3 Boundary conditions

Boundary conditions were important part of the finite element analysis. Appropriate constraints and loading have significant effect on the results. The nodes on the lower surface of tibia and fibula were fixed, while all other nodes had one degree of freedom. These nodes were allowed to move along the axis that is normal to upper surface of the femur (in this case translation along y-axis).

Vertical compression force was applied to the upper surface of the femur. The contacts between various parts of the model were modeled as bonded contacts. They were detected automatically by the software, assuming that the contact should be defined on the place where faces of the various materials overlap. The applied force corresponds to the force obtained during previously explained experiments (Kutzner et al., 2010).

2.4 Mechanical model of knee joint and assessment of cartilage stress distribution

Knee joint motion represents a complex combination of rotations and translations. The major parts involved in the knee kinematical behavior include femur, tibia, and patella. Forces that act at a knee joint are given in Fig. 12.3.

The dominant forces that act at a knee joint are body weight and ground reaction force, which are opposite to each other. Muscle forces and contact forces between femur and tibia are also included in the model. Equilibrium equations of the knee joint are given in the succeeding text (Wismans et al., 1980):

$$F_r + F_b + F_{p1} \cdot N_1 + F_{p2} \cdot N_2 + \sum_{i=1}^{7} F_i = 0 \tag{12.1}$$

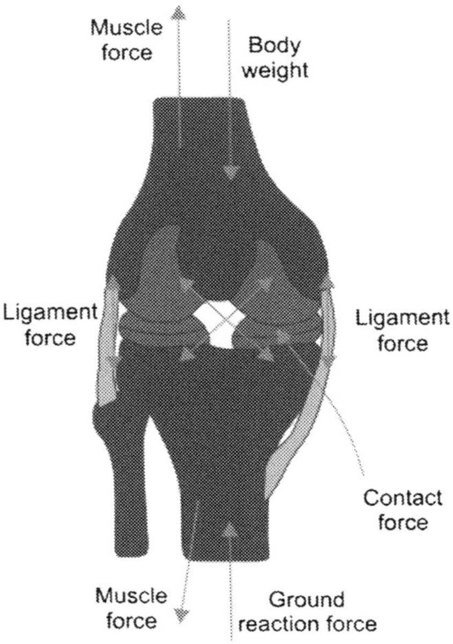

Fig. 12.3 Forces on a knee joint.

$$M_r + F_r \times P_r + F_b \times P_b + F_{p1} \cdot N_1 \times P_1 + F_{p2} \cdot N_2 \times P_2 + \sum_{i=1}^{7} F_i \times V_i = 0 \quad (12.2)$$

where F_r is the ground reaction force, F_bs the body weight, F_{p1} and F_{p2} are the two contact point forces, N_1 and N_2 are the two contact points normal, $F(i=1\ldots7)$ are the ligament and capsule forces, M_r is the knee joint driver moment along local z axis, and P_r, P_b, P_1, P_2, and $V\,(i=1\ldots7)$ are the position vectors where the corresponding forces are being applied (Wismans et al., 1980). We assumed that femur and tibia could be represented as rigid bodies and that their deformations were very small in contrast to relatively large deformations of cartilage and ligaments. Note that synovial fluid significantly reduces the friction between cartilage surfaces and menisci.

2.5 Spring-damper-mass model

In order to simulate jumping of the subject with the whole body, a simplified spring-damper-mass model was used (Fig. 12.4). It consisted of four masses. The upper body was modeled using two masses, one representing its rigid mass, m_3, and the other representing its wobbling mass, m_4. The thigh, leg, and foot of the supporting leg were modeled using two masses,

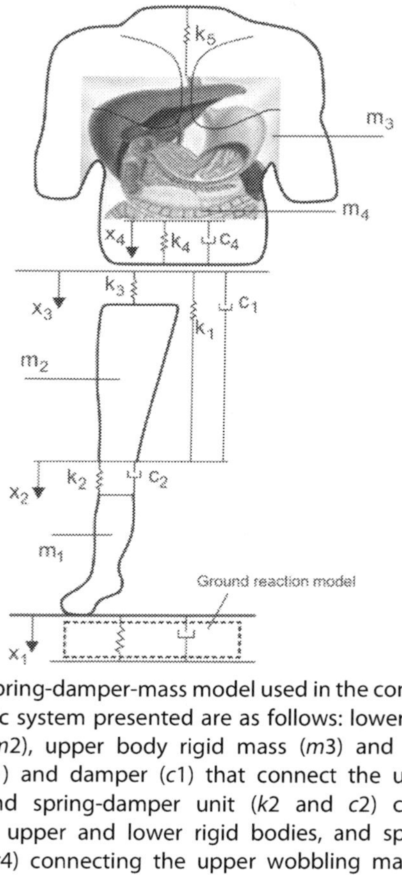

Fig. 12.4 A simplified spring-damper-mass model used in the computer simulation. The components of dynamic system presented are as follows: lower body rigid mass (*m*1) and wobbling mass (*m*2), upper body rigid mass (*m*3) and wobbling mass (*m*4), compressive spring (*k*1) and damper (*c*1) that connect the upper and lower rigid bodies, spring (*k*3) and spring-damper unit (*k*2 and *c*2) connecting the lower wobbling mass to the upper and lower rigid bodies, and spring (*k*5) and spring-damper unit (*k*4 and *c*4) connecting the upper wobbling mass to the upper rigid mass (Liu et al., 1998).

one representing its rigid mass, m_1, and the other representing its wobbling mass, m_2.

The total body mass was taken from real sportsman weight. In the following system of equations (Eq. 12.3), dynamic system is described (Liu et al., 1998), and it is used for calculation total force and moment of cartilage inside the knee during sportsman jumping on the force plate:

$$
\begin{gathered}
m_1\ddot{x}_1 + k_1(x_1 - x_3) + k_2(x_1 - x_2) + c_1(\dot{x}_1 - \dot{x}_3) + c_2(\dot{x}_1 - \dot{x}_2) = m_1 g - F_g \\
m_2\ddot{x}_2 - k_2(x_1 - x_2) + k_3(x_2 - x_3) - c_2(\dot{x}_1 - \dot{x}_2) = m_2 g \\
m_3\ddot{x}_3 - k_1(x_1 - x_3) - k_3(x_2 - x_3) + \\
(k_4 + k_5)(x_3 - x_4) - c_1(\ddot{x}_1 - \ddot{x}_3) + c_4(\ddot{x}_3 - \ddot{x}_4) = m_3 g \\
m_4\ddot{x}_4 - (k_4 + k_5)(x_3 - x_4) - c_4(\dot{x}_3 - \dot{x}_4) = m_4 g
\end{gathered}
\tag{12.3}
$$

Table 12.2 Parameters for ground reaction model

	a	*b*	*c*	*d*	*e*
Soft shoe	1×10^6	1.56	2×10^4	0.73	1.0
Hard shoe	1×10^6	1.38	2×10^4	0.75	1.0

In Eq. (12.3), m_1 is lower body rigid mass and m_2 is wobbling mass; m_3 is upper body rigid mass and m_4 is wobbling mass; k_1 is compressive spring and c_1 is damper that connect the upper and lower rigid bodies; k_3 is spring and k_2-c_2 is spring-damper unit, which connects the lower wobbling mass to the upper and lower rigid bodies; and k_5 is the spring and k_4-c_4 is the spring-damper unit, which connects the upper wobbling mass to the upper rigid mass (Liu et al., 1998). F_g is the vertical contact force, which is defined as $F_g = A_c[ax_1^b + cx_1^d v_1^e]$, and A_c, a, b, c, d, and e are the parameters for ground reaction model, which define deformation of shoes for sportsman during jumping. Some of the different parameters for soft and hard shoe are showed in the Table 12.2.

2.6 Finite element method for numerical calculation

A cartilage inside the knee is considered as a porous deformable body filled with the fluid, occupying the whole pore volume. The physical quantities for this analysis are as follows: the displacement of solid **u**, relative fluid velocity with respect to the solid (Darcy's velocity) **q**, fluid pressure p, swelling pressure p_c, and electrical potential ϕ.

The governing equations for the coupled problem are described as follows: First, we considered the solid equilibrium equation:

$$(1-n)\mathbf{L}^{\mathrm{T}}\boldsymbol{\sigma}_s + (1-n)\rho_s\mathbf{b} + \mathbf{k}^{-1}n\mathbf{q} - (1-n)\rho_s\ddot{\mathbf{u}} = 0 \tag{12.4}$$

where $\boldsymbol{\sigma}_s$ is stress in the solid phase, n is porosity, **k** is permeability matrix, ρ_s is density of solid, **b** is body force per unit mass, **q** is relative velocity of fluid, and $\ddot{\mathbf{u}}$ is acceleration of the solid material. The operator $\mathbf{L}^{\mathbf{T}}$ is

$$\mathbf{L}^{\mathrm{T}} = \begin{bmatrix} \frac{\partial}{\partial x_1} & 0 & 0 & \frac{\partial}{\partial x_2} & 0 & \frac{\partial}{\partial x_3} \\ 0 & \frac{\partial}{\partial x_2} & 0 & \frac{\partial}{\partial x_1} & \frac{\partial}{\partial x_3} & 0 \\ 0 & 0 & \frac{\partial}{\partial x_3} & 0 & \frac{\partial}{\partial x_2} & \frac{\partial}{\partial x_1} \end{bmatrix} \tag{12.5}$$

The equilibrium equation of the fluid phase (no electrokinetic coupling) is

$$-n\nabla p + n\rho_f \mathbf{b} - \mathbf{k}^{-1} n\mathbf{q} - n\rho_f \dot{\mathbf{v}}_f = \mathbf{0} \tag{12.6}$$

where p is pore fluid pressure, ρ_f is fluid density, and $\dot{\mathbf{v}}_f$ is fluid acceleration. This equation is also known as the generalized Darcy's law. Both equilibrium equations are written per unit volume of the mixture. Combining Eqs. (12.4) and (12.6), we obtain

$$\mathbf{L}^{\mathrm{T}}\sigma + \rho\mathbf{b} - \rho\ddot{\mathbf{u}} - \rho_f \dot{\mathbf{q}} = \mathbf{0} \tag{12.7}$$

where $\boldsymbol{\sigma}$ is the total stress that can be expressed in terms of $\boldsymbol{\sigma}_s$ and p, as

$$\boldsymbol{\sigma} = (1-n)\,\boldsymbol{\sigma}_s - n\mathbf{m}p \tag{12.8}$$

and $\rho = (1-n)\rho_s + n\rho_f$ is the mixture density. Here, $\mathbf{m}$ is a constant vector defined as $\mathbf{m}^{\mathrm{T}} = \{1\ 1\ 1\ 0\ 0\ 0\}$ to indicate that the pressure component contributes to the normal stresses only. We have also taken into account that the pressure has positive sign in compression, while tensional stresses and strains are considered positive. In the following analysis, we employ the effective stress, $\boldsymbol{\sigma}'$, defined as

$$\boldsymbol{\sigma}' = \boldsymbol{\sigma} + \mathbf{m}p \tag{12.9}$$

which is relevant for the constitutive relations of the solid. Using the definition of relative velocity $\mathbf{q}$ as the volume of the fluid passing in a unit time through a unit area of the mixture (Darcy's velocity),

$$\mathbf{q} = n\left(\mathbf{v}_f - \dot{\mathbf{u}}\right) \tag{12.10}$$

we transform Eq. (12.6) into

$$-\nabla p + \rho_f \mathbf{b} - \mathbf{k}^{-1}\mathbf{q} - \rho_f \ddot{\mathbf{u}} - \frac{\rho_f}{n}\dot{\mathbf{q}} = \mathbf{0} \tag{12.11}$$

The final continuity equation using the elastic constitutive law and fluid incompressibility is given in the form (Kojic et al., 2001)

$$\nabla^{\mathrm{T}}\mathbf{q} + \left(\mathbf{m}^{\mathrm{T}} - \frac{\mathbf{m}^{\mathrm{T}}\mathbf{C}^{\mathrm{E}}}{3K_s}\right)\dot{\mathbf{e}} + \left(\frac{1-n}{K_s} + \frac{n}{K_f} - \frac{\mathbf{m}^{\mathrm{T}}\mathbf{C}^{\mathrm{E}}\mathbf{m}}{9K_s^{\,2}}\right)\dot{p} = 0 \tag{12.12}$$

The resulting FE system of equations is solved incrementally (Kojic et al., 2001), with time step Δt. We impose the condition that the balance equations are satisfied at the end of each time step ($t+\Delta t$). Hence, we have the following system of equations:

$$\begin{bmatrix} \mathbf{m_{uu}} & 0 & 0 & 0 \\ 0 & 0 & 0 & 0 \\ \mathbf{m}_{qu} & 0 & 0 & 0 \\ 0 & 0 & 0 & 0 \end{bmatrix} \begin{Bmatrix} {}^{t+\Delta t}\ddot{\underline{\mathbf{u}}} \\ {}^{t+\Delta t}\ddot{\underline{\mathbf{p}}} \\ {}^{t+\Delta t}\ddot{\underline{\mathbf{q}}} \\ {}^{t+\Delta t}\ddot{\underline{\varphi}} \end{Bmatrix} + \begin{bmatrix} 0 & 0 & \mathbf{c_{uq}} & 0 \\ \mathbf{c}_{pu} & \mathbf{c}_{pp} & 0 & 0 \\ 0 & 0 & \mathbf{c}_{qq} & 0 \\ 0 & 0 & 0 & 0 \end{bmatrix} \begin{Bmatrix} {}^{t+\Delta t}\dot{\underline{\mathbf{u}}} \\ {}^{t+\Delta t}\dot{\underline{\mathbf{p}}} \\ {}^{t+\Delta t}\dot{\underline{\mathbf{q}}} \\ {}^{t+\Delta t}\dot{\underline{\varphi}} \end{Bmatrix}$$

$$+ \begin{bmatrix} \mathbf{k_{uu}} & \mathbf{k_{u}}_p & 0 & 0 \\ 0 & 0 & \mathbf{k}_{pq} & 0 \\ 0 & \mathbf{k}_{qp} & \mathbf{k}_{qq} & \mathbf{k}_{q\varphi} \\ 0 & \mathbf{k}_{\varphi p} & 0 & \mathbf{k}_{\varphi\varphi} \end{bmatrix} \begin{Bmatrix} \Delta\underline{\mathbf{u}} \\ \Delta\underline{\mathbf{p}} \\ \Delta\underline{\mathbf{q}} \\ \Delta\underline{\varphi} \end{Bmatrix} = \begin{Bmatrix} {}^{t+\Delta t}\mathbf{f_u} \\ {}^{t+\Delta t}\mathbf{f}_p \\ {}^{t+\Delta t}\mathbf{f}_q \\ {}^{t+\Delta t}\mathbf{f}_\varphi \end{Bmatrix} \tag{12.13}$$

2.7 Caption motion system

For experimental jumping of sportsmen, we designed a separate device, a so-called force plate, with force transducer mounted on it. The force measurements were recorded completely in parallel with video analysis, which was controlled with a specific in-house software system (Fig. 12.5). After acquisition of the experimental results, we used our video and image processing software for analysis and preparation of data into numerical calculations. The measurement results were determined for the ankle and knee separately in the form of triplet values: x-coordinate, y-coordinate, and vertical resistance force F.

2.8 Force measurement

Knee motion involves a series of three rotations (flexion/extension, abduction/adduction, and internal/external rotations) and three translations (anterior/posterior, superior/inferior, and medial/lateral translations). Our first goal was to develop a computer vision system for motion tracking and force measurement analysis of sportsman's knee during jumps. These recorded impact forces are used to calculate internal forces in sportsman's ankle and knee. For this purpose, we designed force plate with force transducer mounted on it. Force measurements are recorded completely in parallel with video analysis, which is controlled with specific developed in-house software system (Fig. 12.6).

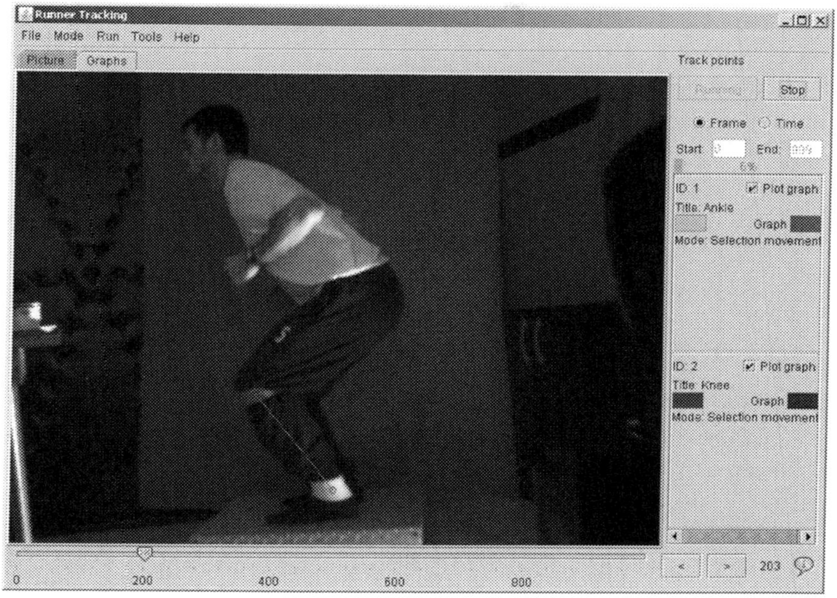

Fig. 12.5 Three-dimensional capture imaging and force plate measurements during sportsman jumping.

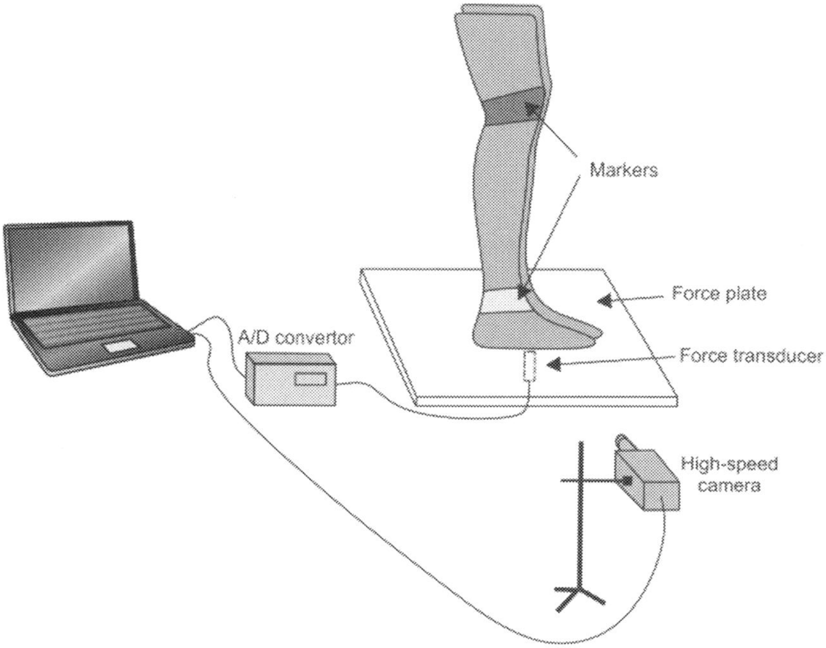

Fig. 12.6 Setup equipment for experimental investigation of jumping impact force and knee force and moment.

After acquisition of experimental results, we used our video and image processing software for analysis and preparation into numerical calculations. The measurement results are determined for the ankle and knee separately in a form of triplet values: x-coordinate, y-coordinate, and vertical resistance force F.

For each individual jump, the following variables are measured: (a) frame rate, (b) total time of experiment, (c) 2D coordinates of the ankle and knee for every time frame, (d) corresponding forces measured on the force plate for each contact between sportsman and force plate, (e) number of jumps, and (f) maximal velocity gained at the moment of sportsman landing to the force plate.

The primary goal of our experiment was to simultaneously record movie of the sportsman jumping on the force plate and to record values of the force transducer mounted on the force plate as well. We used high-speed Basler A602fc camera to record movie in duration of only 10 seconds. It formed a file in AVI format of approximately 1.2 GB, because we did it with the following properties: 100 fps, 656×490 px, YUV 4:2:2 (16 bits/pixel avg.). At the same time (in parallel), we were recording measurements from the force transducer over the serial port. In each case, we also recorded system times, so later, we were able to synchronize results of our measurements.

Additionally, this phase also includes the determination of some parameters that are needed for the following numerical calculations. These are as follows: (a) separation of individual jumps, that is, determination of the start and end jump times; (b) calculating velocities of jump landings; (c) isolating force measurements related only to contact between force plate and sportsman; and (d) filtering force measurements due to inertia of the force plate itself. To fulfill these requirements, we ascertained some criterions. They were mostly influenced by our subjective visual impressions and conclusions about jumps, and they were in accordance with changes in the force measurement diagram. So, firstly, we defined nine phases of the jump: (1) rest (sportsman stands still), (2) preparation for jump (sportsman moves his knees forward), (3) rebound, (4) lateness in air (visual impression due to recording with high-speed camera), (5) fall, (6) toe landing (moment in time when sportsman lands his toes on the force plate), (7) heel landing (similar to 6), (8) preparation for rest (sportsman moves his knees to backward and straightens himself), and (9) rest (sportsman stands still again). The second criteria was related to the determination of landing velocity during jump. There are horizontal and vertical components of velocity during fall, and at the moment sportsman lands his toes on the force plate (sixth phase of the jump), velocity gains its maximum value, which we later used as a parameter for subsequent numerical calculations. This value can be determined by integrating velocities between subsequent frames, starting with the frame where this value was zero. As a starting frame for this, calculations can be considered frame at the very end of the fourth phase of the jump, or the very first frame of the fifth phase of the jump. As the acceleration during fall is approximately constant, maximum velocity at the time sportsman touches force plate can be calculated using next formulae:

$$v_{max}^{px} = \sqrt{v_{\max\_x}^2 + v_{\max\_y}^2} = \frac{\sqrt{(x_m - x_n)^2 + (y_m - y_n)^2}}{m-n} \tag{12.14}$$

where n and m represent indices of starting and ending frames, respectively. In the same sense, x and y values are coordinates of the ankle. Because of that, velocity according to the Eq. (12.1) is expressed in (pixels/frame count). To express maximal velocity in ISO units [m/s], v_{max}^{px} must be multiplied by the coefficient K:

$$v_{max} = v_{max}^{px} \cdot K = v_{max}^{px} \cdot \frac{0.68}{230} \cdot fps \tag{12.15}$$

where 0.68 m is represented by 230 pixels at image frames.

2.9 Motion capture system

For walking analysis, we used a commercial motion capture system OptiTrack. This system consists of six infrared cameras and four retroreflective markers. The markers are 1.5 cm in diameter and are attached at the precise anatomical locations of the participant's leg for unilateral gain analysis (Fig. 12.7). These locations were great trochanter region, femoral lateral epicondyle, tuberosity of the tibia, and the center of the anterior region of ankle joint.

The computerized camera system with accompanying software captures the exact motion of retroreflective markers and thus records their trajectory while the volunteer performs walking over the force plate. The cameras were connected to a computer that collects gait kinematical data. The result of motion tracking is a series of 3D coordinates for each numbered marker. To better understand kinematics and kinetics of gait, we used a three-axial accelerometer.

We used also Sun Small Programmable Object Technology accelerometers (Sun SPOT) to wirelessly detect the middle of the gait stance phase, that is, the moment when the ground reaction force reaches its maximum value. The Sun SPOT has a sensor board that consists of 2G/6G three-axis accelerometer, temperature sensor, and light sensor. In the experimental part, the Sun SPOT LIS3L02AQ accelerometer is used to measure the orientation or motion in three dimensions, X, Y, and Z with the sample rate of 100 Hz. Each

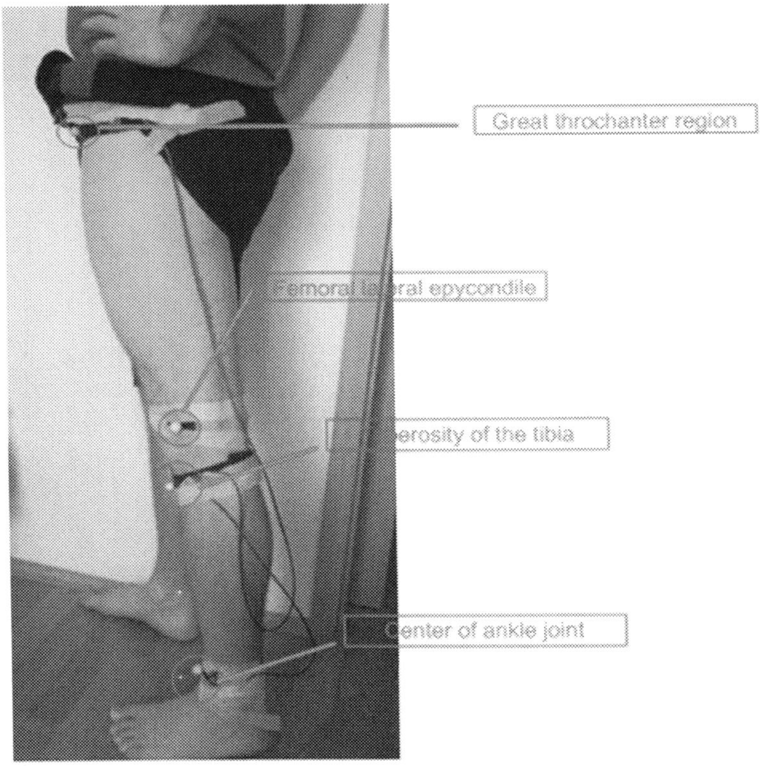

Fig. 12.7 Reflective marker position on the examinee leg.

of these components represents the sum of static accelerations defined by angle of inclination to the corresponding axis and dynamic component stemming from the movement during the walk. The ground reaction force was sampled from the multiaxis AMTI force plate at the rate of 100 Hz. Before the experiment, calibration was conducted to work out the space coordinate system for the camera system field of view. The calibration was performed fully in accordance to the proposed manufacturer's procedure.

2.10 Foot pressure distribution measurement

The foot pressure measurement system was applied for the determination of foot pressure distribution during gate of the subject. This system consists of matrix of capacitor sensor dimension each one 7.6 × 7.6 mm (Fig. 12.8). Range of measurement pressure is from 10 to 1200 kPa. It is possible to

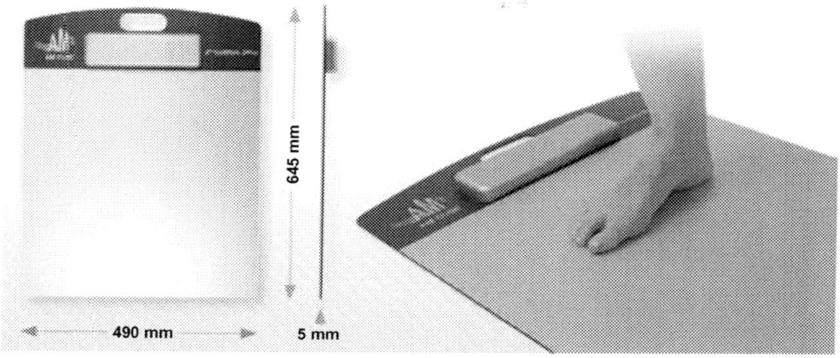

Fig. 12.8 The foot pressure plate.

calculate cumulative force during wade with integration pressure over contact surface. Sample rate of signal acquisition is 100 Hz. This platform is positioned on the pathway of gait and examinee across that by putting one leg on the sensor plate.

2.11 Force plate measurement

We were faced with some serious difficulties, for example, extremely short duration of impact phase or wiggling of motion analysis markers attached to the sportswear. We paid special attention to the second of the previously mentioned problems. Even if we "ideally" position markers relative to the bone joints of the body, they not only considerably shift their position at the impact phase but also started wobbling at the same time. Thus, method of inverse dynamics cannot be accepted without some reserve, especially at the very beginning of the impact, for example, during first 30 ms. Besides, imprecise biomechanical properties of human body represent the other source of erroneous results, so that inverse dynamics must be accepted as a first approximation (Gruber et al., 1998).

Subjects were instructed to perform 10 first (flat) and 10 second (spin) motion, with maximum power. Their motion was recording using Opti-Track motion capture system (six V100:R2 100Hz cameras with ARENA software). Ground reaction force was sampled from the multi-axis AMTI force plate at the rate of 1000 Hz. Before the experiment, calibration was required to work out the space coordinate system of the field of view area. The calibration was performed fully in accordance with manufacturer's procedure. We placed six retroreflective markers on the subjects at the following locations: the fifth metatarsophalangeal joint, the calcaneus, the lateral malleolus, the lateral epicondyle of the femur, the greater trochanter, and the acromion scapulae. The subjects stood in front of the force plate. After the trigger was launched, each subject executed flat and spin serves with maximum power, five times each. The subjects were particularly instructed to execute serve so that they land down on one leg after impact. Rest time between trials was individual, but no longer than 5 min.

2.12 Inverse dynamics

The inverse dynamic solution was employed to calculate the forces, moments, and powers of the supporting leg (Figs. 12.9 and 12.10) (Gruber et al., 1998; Payton and Bartlett, 2008). Body segment parameters, moments of inertia of body segments and their center of masses, were calculated by using literature data (Zatsiorsky et al., 1990; DeLeva, 1996). The free diagram figure of the motion is shown in Fig 12.11. (See Fig 12.12.)

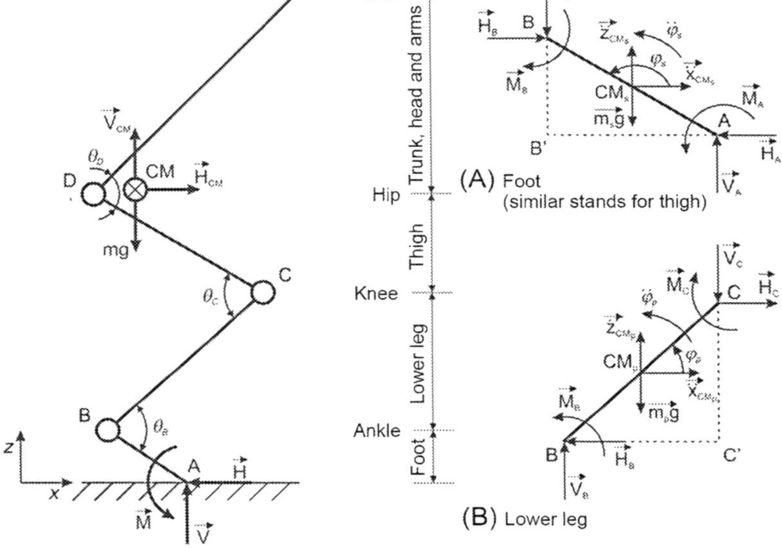

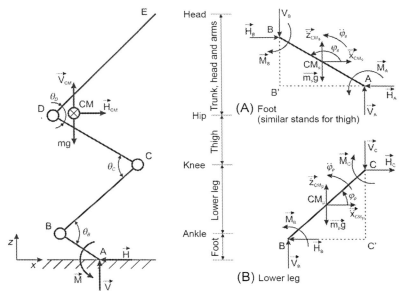

Fig. 12.9 Free body model of the human's body and leg segments.

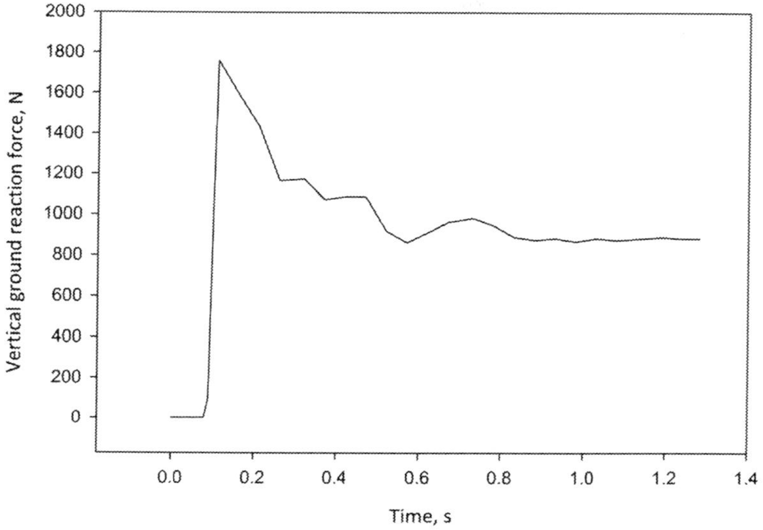

Fig. 12.10 Vertical ground reaction force at impact by landing on one leg.

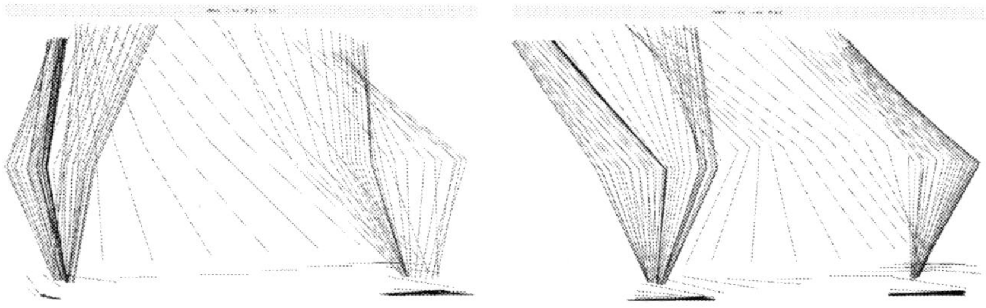

Fig. 12.11 Projections of leg trajectory in *y-z*- and *x-z*-planes during subject motion. Sagittal axis is *y*, and *z* is vertical axis.

3 Results

3.1 Definition of knee geometry from medical images

Knowledge of in vivo joint motion and loading during functional activities is needed to improve our understanding of possible knee joint degeneration and restoration. Internal loadings of knee anatomical structures significantly depend on a lot of factors such as external loads, body weight, ligaments, strengths, and muscle forces. Knowledge about the knee cartilage deformation ratio and the knee cartilage stress distribution is of particular importance in clinical studies due to the fact that these represent some of the basic indicators of cartilage state and that they also provide information about joint

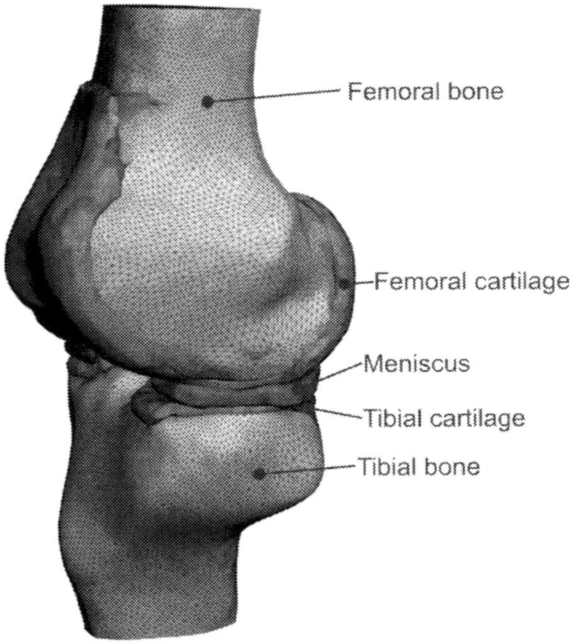

Fig 12.12 Finite element mesh for different knee segments.

cartilage wear so medical doctors can predict when it is necessary to perform surgery on a patient.

In the study, stress on the knee cartilage during kneeling and standing using finite element models is compared. We used magnetic resonance (MR) images of the flexed knee to build a geometrical model. As a computational tool, they used commercial software MIMICS. The results of those studies showed some differences in high-stress regions between kneeling and standing. The conclusion was that the peak von Mises stress and contact pressure on the cartilage were higher in kneeling. The study (Filipovic et al., 2009) used computed tomography (CT) images of knee structures during static loading to determine cartilage strains and meniscal movement in a human knee at different time periods of standing and to compare them with the subject-specific 3D finite element (FE) model.

The initial model is obtained using MRI slice images (Fig. 12.23). The extracted counter at the anatomical landmarks of the lower extremity is detected on MRI 3D reconstruction object.

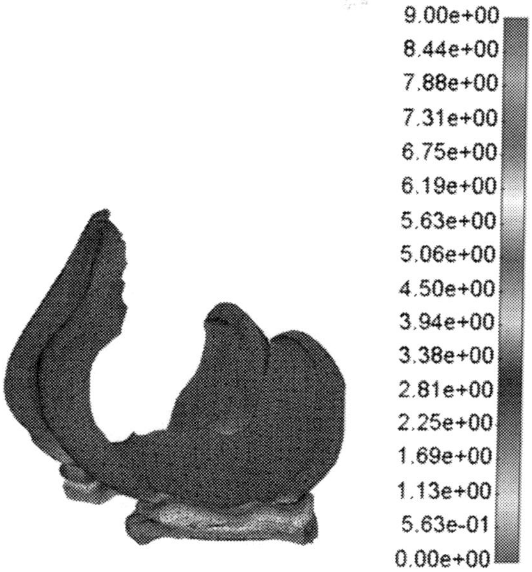

Fig. 12.13 Deformation distribution [%] in the three-dimensional FE model for cartilage during one jump.

3.2 Noninvasive determination of knee cartilage deformation

Deformation distribution in the three-dimensional model for cartilage during one jump has been presented in Fig. 12.13.

More details of this parameter are given in. Performing this system of equation using finite element method and applying the boundary condition for tibia and femur, we obtained mean value of stress distribution during contact period defined in the Fig. 12.13. For the special interest is average value of maximal von Mises stress given by relation:

$$\overline{\sigma} = \frac{1}{T_C} \int_0^{T_c} \sigma_{max}(t)\, dt$$

where T_C is contact period and $\sigma_{max}(t)$ is maximal value of von Mises stress on the knee cartilage in some time moment t (Fig. 12.14).

This procedure is performed on the six subjects.

Effective von Mises stress distribution for patient-specific femoral cartilage, meniscus, and tibial cartilage during one gate cycle has been presented in Fig. 12.15.

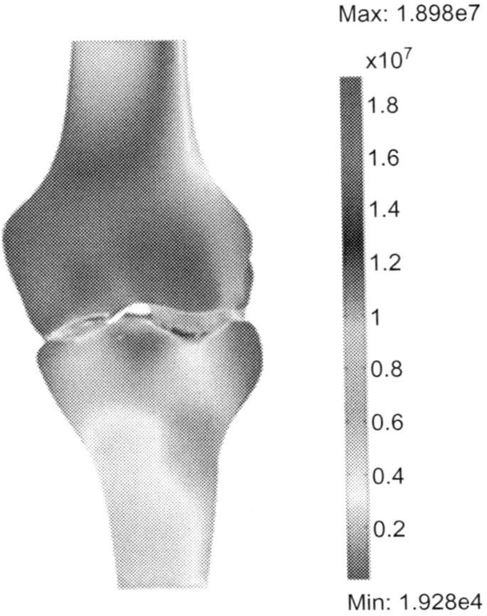

Fig. 12.14 Maximal value of von Mises stress distribution in [Pa].

The volunteer performs walking along 2.5-meter distance path away with his own ordinary velocity and attached infrared marker and accelerometers sensor on the left leg. Results for marker coordinates and corresponding accelerations, measured by the three-axial accelerometer, are presented in Fig. 12.16.

According to Garling et al. (2007), measured values for marker position are influenced by noise due to the wobbling of the participant's skin. The value of this uncertainty is in range of ± 2 mm.

The force plate is positioned in the first half of the walking path. During the experiment, the participants were asked to walk along so the force plate records the value of the ground reaction force (Fig. 12.17).

The force value is zero in the beginning of the walk, and when the participant stands on the force plate, starting with the heel, the force gradually increases and reaches the maximum and then drops to zero again when the foot is detached by the force plate.

3.3 Computer simulation in the jumping force analysis

The participants in experiments followed a standardized warming-up and stretching period. During the measurements, participants wore their own

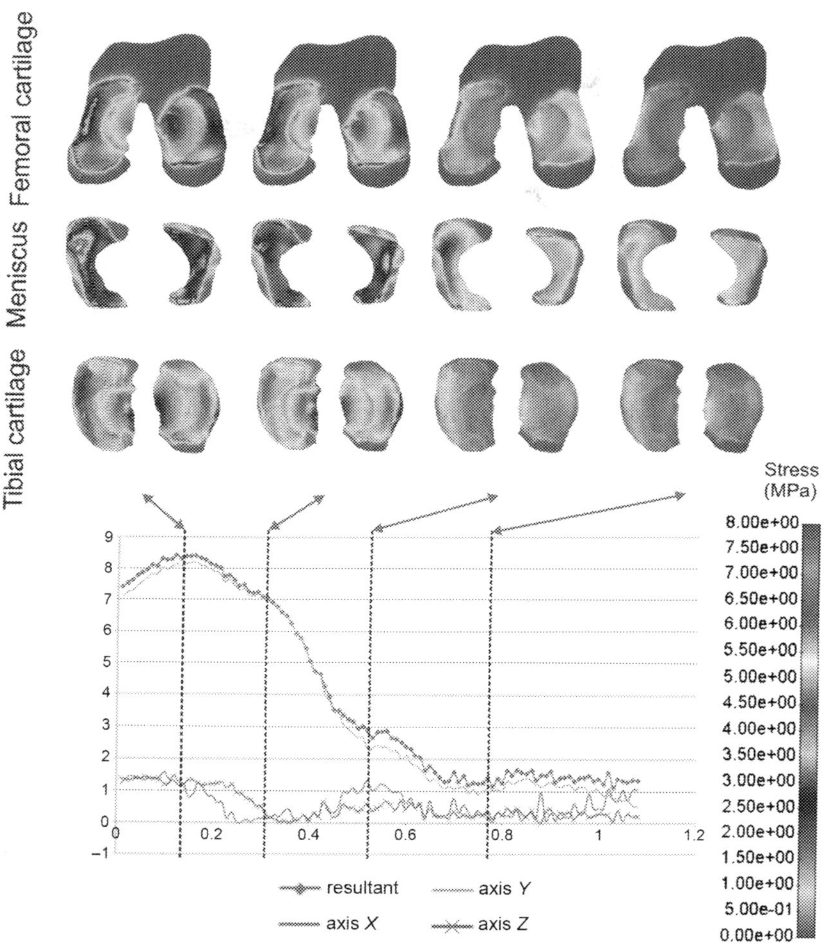

Fig. 12.15 Effective von Mises stress distribution for patient-specific femoral cartilage, meniscus, and tibial cartilage during one gate cycle.

indoor sport shirts, shorts, and sneakers. Participants were instructed to perform four type of jumping: (flywheel jump, jump without flywheel, and jump with left and right leg landing). Their motion was recorded using (innovision Systems Inc.) with two high-speed cameras with frame rate of 200 frames per second. We placed three retroreflective markers on the participants at the following locations: the hip, tuberosity of tibia, and ankle joint (Fig. 12.18).

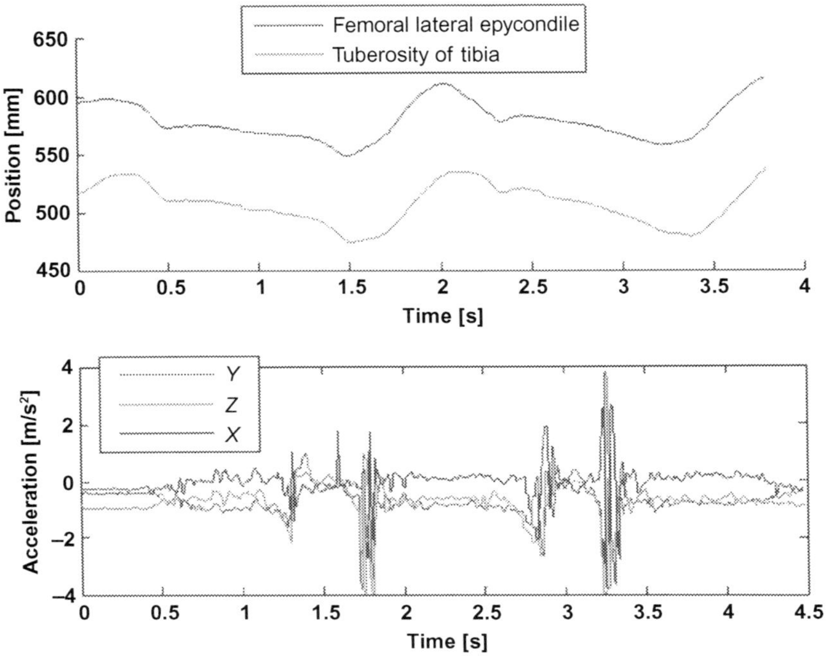

Fig. 12.16 Position of reflective marker during walk and corresponding three-axial acceleration.

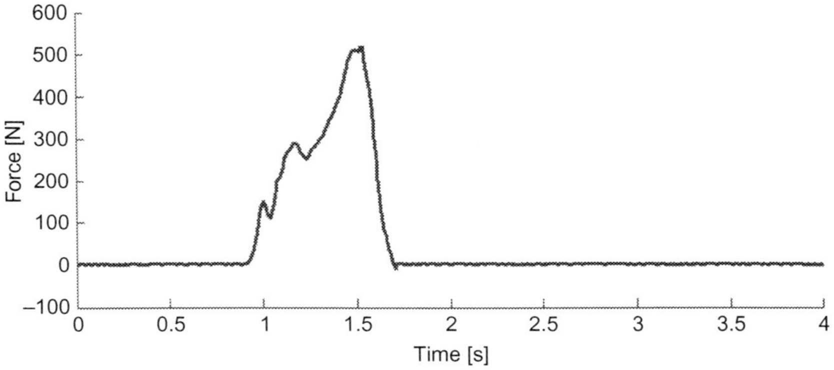

Fig. 12.17 Ground reaction forces during standing on the force plate.

The ground reaction force was sampled from the force plate at the rate of 300 Hz. Before the experiment, calibration was required to work out the space coordinate system of the field of view area. The calibration was performed fully in accordance with manufacturer's procedure.

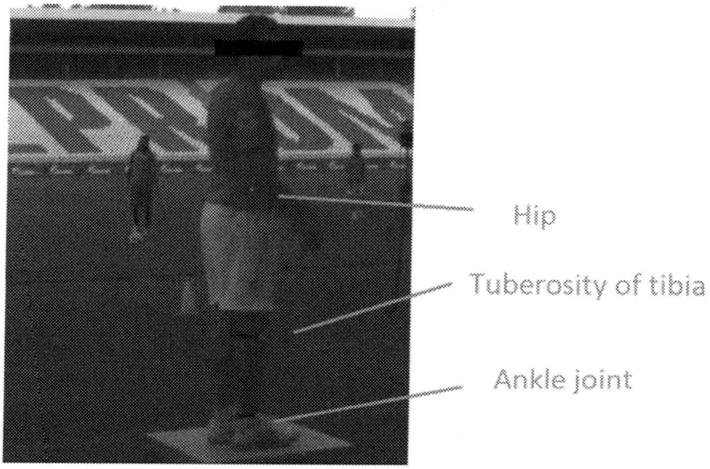

Fig. 12.18 Position of the reflective markers on the participant body.

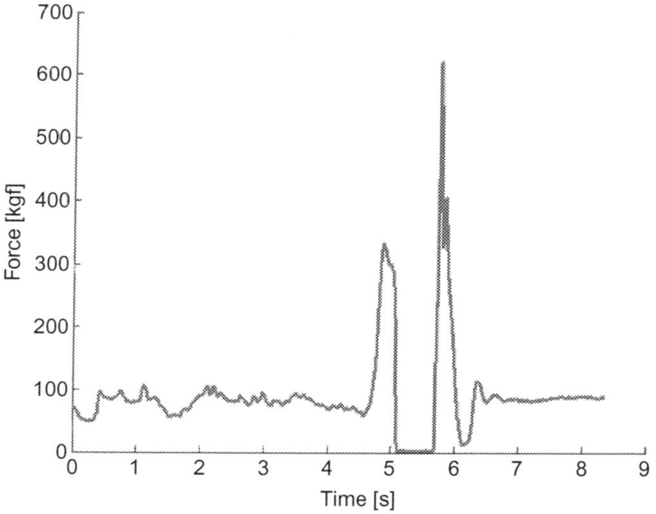

Fig. 12.19 Vertical force value for flywheel jump.

The participants stood in front of the force plate. After the trigger was launched, each participant executed different jumps one by one with break period of around 10 seconds. On Figs. 12.19–12.22, vertical force distributions during time on the force plate for different types of jump are shown, respectively.

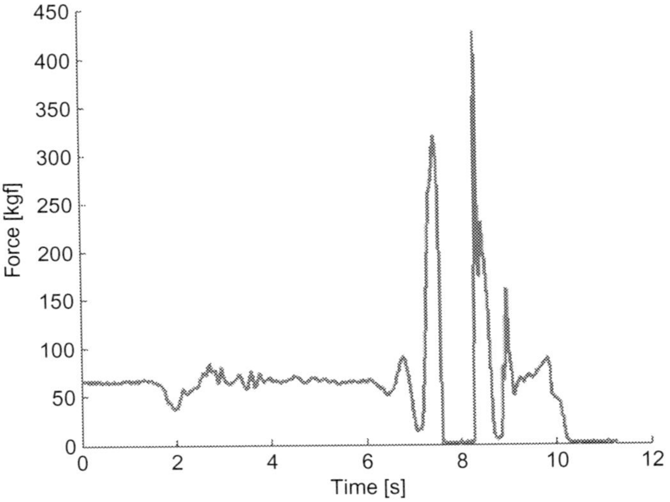

Fig. 12.20 Vertical force value for jump without flywheel.

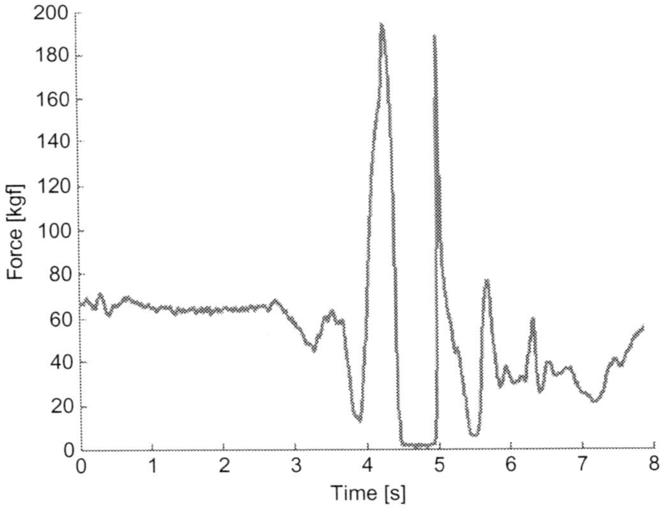

Fig. 12.21 Vertical force value for left leg landing.

The second part of experiment in this study was measure of food pressure distribution for participants. Pressure measurement consists static and dynamic analysis. In the static analysis, participant stand on the foot pressure platform for a 10s. The distribution of foot pressure for static analysis is shown on the Fig. 12.23.

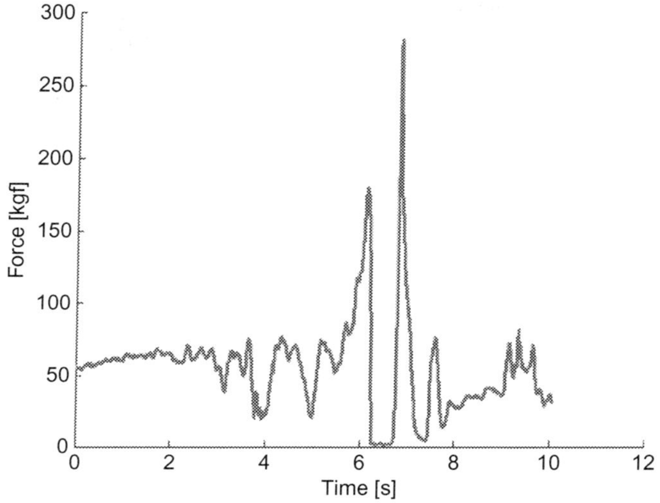

Fig. 12.22 Vertical force value for left right landing.

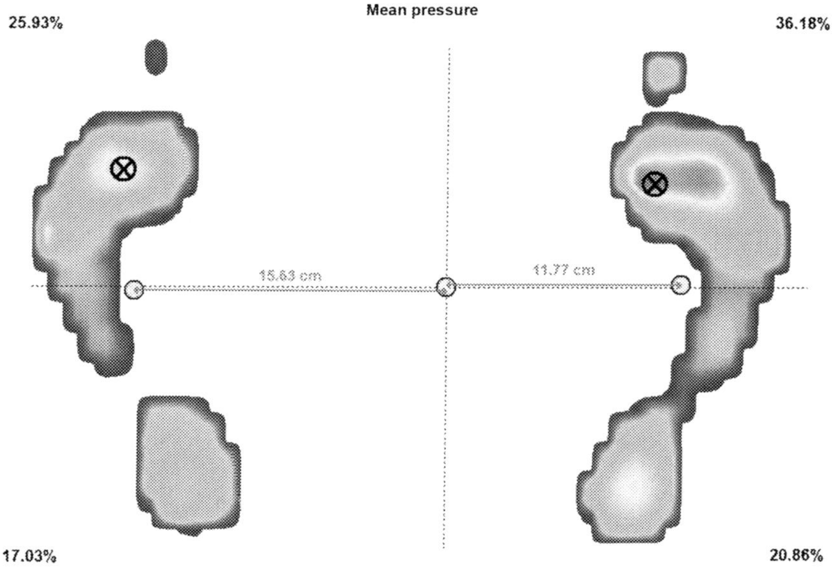

Fig. 12.23 The foot pressure distribution of participant during standing.

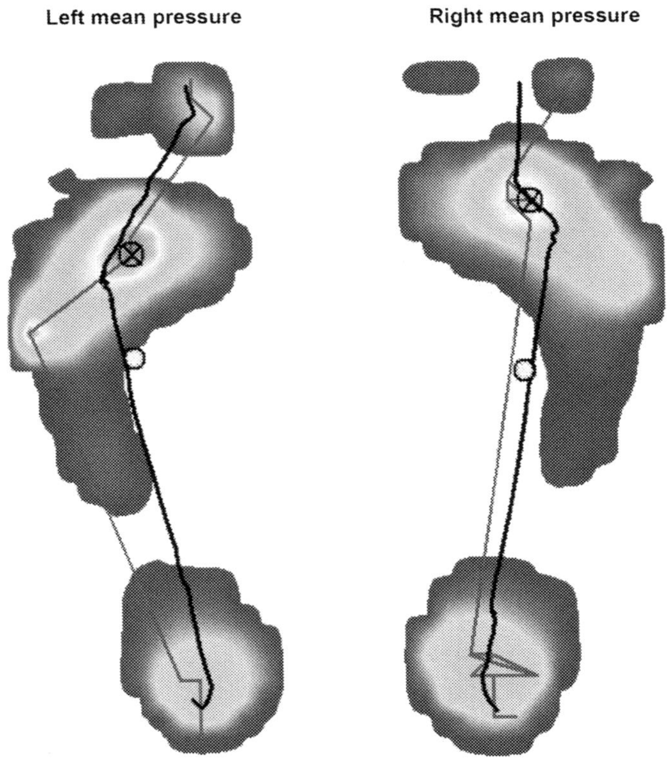

Fig. 12.24 The foot pressure distribution and trajectory of body center mass during walking for both left and right food.

The dynamic analysis is shown on the Fig. 12.24. During this study, participant climbs over foot pressure plate with left and then with right leg.

4 Discussion and Conclusion

In this chapter, result experimental and computational biomechanic analysis of jumping and walking for sportsman were presented. The force plate platform used one basic device for experimental jumping biomechanical analysis. Combing this platform with the high-speed cameras and computer simulation like finite element procedure, it is enabled to understand and analyze different aspects of sport activity and jump as one of them. We presented result of maximal force during jump and average value maximal von Mises stress in the contact period. It is evident that peak of force is not always in correlation with average of maximal von Mises stress due to different contact times and exposure of body by the ground reaction force. We analyzed foot pressure distribution for walking of the subjects. Also, we have connected mechanical parameter of human knee and including influence of foot contact and deformation on the forces and torque in the knee joint.

We have presented a combined image processing-tracking algorithm for the measurement of kinematics and dynamics of jumping on a force plate, for a simplified algorithm of the spring-damper-mass model for determination force and moment on the cartilage inside the knee was developed. Three-dimensional finite element calculation of cartilage deformation during knee motion was implemented.

Experimental and computational techniques could be a good platform for future athlete testing. By using this methodology, coaches may effectively evaluate athletic performance under various competitive conditions and possibly avoid sport injuries during training. A next step would be to include neuromuscular control of human jumping. Also, we might consider the causal mechanism of jumper's knee in sport and possible risk factors in order to better understand them.

Our main goal is to integrate the various steps described previously to produce a practical and diagnostic tool for noninvasively assessing musculoskeletal function on a subject-specific basis. In summary, virtual biomechanical simulation has potential in improving medical science and health care as well as sport trainings. Due to the complexity of medical science and the computer technologies involved in building such simulation systems, there are still a lot of technical challenges in front of us. This work provides solutions to a part of this big problem. Indeed, the 3D reconstruction, motion simulation, and biomechanical visualization techniques presented can be seen as some basic elements of the next generation virtual biomechanical simulation system, a system where all the components of visualization, automated model generation, surgery simulation, and surgery assistance come together.

References

Altmeyer, L., Bartonietz, K., Krieger, D., 1994. Technique and training: the discus throw. Track Field Quart. Rev. 94 (3), 33–34.

Ariel, G., Finch, A., Penny, A., 1997. Biomechanical analysis of discus throwing at the 1996 Atlanta Olympic Games. In: Biomechanics in Sports XV, pp. 365–371.

Baldwin, M.A., Clary, C.W., Fitzpatrick, C.K., Deacy, J.S., Maletsky, L.P., Rullkoetter, P.J., 2012. Dynamic finite element knee simulation for evaluation of knee replacement mechanics. J. Biomech. 45 (3), 474–483.

Bei, Y., Fregly, B., 2004. Multibody dynamic simulation of knee contact mechanics. Med. Eng. Phys. 26 (9), 777–789.
Borelli, K.D., 1989. De motu animalium.pais prima 1680. Springer Verlag, Berlin.
Braune, W., Fischer, O., 1987. The Human Gait. Springer Verlag, Berlin.
Cross, R., 1999. Standing, walking, running, and jumping on a force plate. Am. J. Phys. 67 (4), 304–309.
Dejnabadi, H., Jolles, B.M., Aminian, K., 2005. A new approach to accurate measurement of uniaxial joint angles based on a combination of accelerometers and gyroscopes. IEEE Trans. Biomed. Eng. 52 (8), 1478–1484.
DeLeva, P., 1996. Adjustments to Zatsiorsky-Seluyanov's segment inertia parameters. J. Biomech. 29 (9), 1223–1230.
Djurić-Jovičić, M.D., Jovičić, N.S., Popović, D.B., 2011. Kinematics of gait: new method for angle estimation based on accelerometers. Sensors 11 (11), 10571–10585.
Filipovic, N., 2008. Finite element solver PAK-C. Faculty of Engineering. University of Kragujevac, Serbia.
Filipovic, N., Vulovic, R., Peulic, A., Radakovic, R., Kosanic, D.J., Ristic, B., 2009. Non-invasive determination of knee cartilage deformation during jumping. J. Sports Sci. Med. 8, 584–590.
Finch, A., Ariel, G., Penny, A., 1998. Kinematic comparison of the best and worst of the top men's discus performers at the Atlanta Olympic Games. In: International Symposium on Biomechanics in Sports I, pp. 93–96.
Fitzpatrick, C.K., Rullkoetter, P.J., 2012. Influence of patellofemoral articular geometry and material on mechanics of the unresurfaced patella. J. Biomech. 45 (11), 1909–1915.
Garling, E.H., Kaptein, B.L., Mertens, B., Barendregt, W., Veeger, H.E.J., Nelissen, R.G.H.H., Valstar, E.R., 2007. Soft-tissue artefact assessment during step-up using fluoroscopy and skin-mounted markers. J. Biomech. 40, S18–S24.
Godest, A.C., Beaugonin, M., Haug, E., Taylor, M., Gregson, P.J., 2002. Simulation of a knee joint replacement during a gait cycle using explicit finite element analysis. J. Biomech. 35 (2), 267–275.
Gruber, K., Ruder, H., Denoth, J., Schneider, K., 1998. A comparative study of impact dynamics: wobbling mass model versus rigid body models. J. Biomech. 31, 439–444.
Halloran, J.P., Clary, C.W., Maletsky, L.P., Taylor, M., Petrella, A.J., Rullkoetter, P.J., 2010. Verification of predicted knee replacement kinematics during simulated gait in the Kansas knee simulator. J. Biomech. Eng. 132(8). 081010.
Halloran, J.P., Petrella, A.J., Rullkoetter, P.J., 2005. Explicit finite element modeling of total knee replacement mechanics. J. Biomech. 38 (2), 323–331.
Jósza, L., Kannus, P., 1997. Human Tendons. Human Kinetics, Champaign, IL, p. 576.
Khan, K.M., Maffulli, N., Coleman, B.D., Cook, J.L., Taunton, J.E., 1998. Patellar tendinopathy: some aspects of basic science and clinical management. Br. J. Sports Med. 32, 346–355.
Kojic, M., Filipovic, N., Mijailovic, S., 2001. A large strain finite element analysis of cartilage deformation with electrokinetic coupling. Comput. Methods Appl. Mech. Eng. 190, 2447–2464.
Kutzner, I., Heinlein, B., Graichen, F., Bender, A., Rohlmann, A., Halder, A., Beier, A., Bergmann, G., 2010. Loading of the knee joint during activities of daily living measured in vivo in five subjects. J. Biomech. 43 (11), 2164–2173.
Lee, W., Zhang, M., Jia, X., Cheung, J., 2004. Finite element modeling of the contact interface between trans-tibial residual limb and prosthetic socket. Med. Eng. Phys. 26 (8), 655–662.
Li, G., Van de Velde, S.K., Bingham, J.T., 2008. Validation of a non-invasive fluoroscopic imaging technique for the measurement of dynamic knee joint motion. J. Biomech. 41 (7), 1616–1622.

Liu, W., Benno, M., Nigg, A., 1998. Mechanical model to determine the influence of masses and mass distribution on the impact force during running. J. Biomech. 33, 219–224.
Payton, C.J., Bartlett, R.M., 2008. Biomechanical Evaluation of Movement in Sport and Exercise. Routledge.
Pingali, G.S., Jean, Y., Carlbom, I., 1998. Real time tracking for enhanced tennis broadcasts. In: Proceedings of the IEEE Computer Society Conference on Computer Vision and Pattern Recognition. IEEE Computer Society, pp. 260–265.
Pingali, G.S., Opalach, A., Jean, Y., 2000. Ball tracking and virtual replays for innovative tennis broadcasts. In: IEEE International Conference on Pattern Recognition. vol. 4, pp. 152–156.
Plagenhoef, S., 1968. Computer programs for obtaining kinetic data on human movement. J. Biomech. 1, 221–234.
Reinschmidt, C., van den Bogert, A.J., Lundberg, A., Nigg, B.M., Murphy, N., Stacoff, A., et al., 1997. Tibiofemoral and tibiocalcaneal motion during walking: external vs. skeletal markers. Gait Posture 6 (2), 98–109.
Setuain, I., Martinikorena, J., Gonzalez-Izal, M., Martinez-Ramirez, A., Gómez, M., Alfaro-Adrián, J., Izquierdo, M., 2015. Vertical jumping biomechanical evaluation through the use of an inertial sensor-based technology. J. Sports Sci, 1–9 (ahead-of-print).
Tao, W., Liu, T., Zheng, R., Feng, H., 2012. Gait analysis using wearable sensors. Sensors 12, 2255–2283.
Thambyah, A., Goh, J.C.H., Das De, S., 2005. Contact stresses in the knee joint in deep flexion. Med. Eng. Phys. 27 (4), 329–335.
Wismans, F.V., Janssen, J., Huson, A., Struben, P., 1980. A three-dimensional mathematical model of the knee joint. J. Biomech. 13 (8), 677–685.
Yanga, N.H., Canavanb, P.K., Nayeb-Hashemia, H., Najafic, B., Vaziria, A., 2010. Protocol for constructing subject-specific biomechanical models of knee joint. Comput. Methods Biomech. Biomed. Eng. 13 (5), 589–603.
Zatsiorsky, V., Seluyanov, V.N., Chugunova, L.G., 1990. Methods of determining mass-inertial characteristics of human body segments. In: Chernyi&, G.G., Regirer, S.A. (Eds.), Contemporary Problems of Biomechanics. CRC Press, USA, pp. 272–291.

Further reading

Aiello, F., Fortino, G., Gravina, R., Guerrieri, A., 2011. A java-based agent platform for programming wireless sensor networks. Comput. J. 54 (3), 439–454.
Dabiri, Y., Li, L.P., 2013. Altered knee joint mechanics in simple compression associated with early cartilage degeneration. Comput. Math. Methods Med, 1–11.
Fernandez, J.W., Pandy, M.G., 2006. Integrating modelling and experiments to assess dynamic musculoskeletal function in humans. Exp. Physiol. 91, 371–382.
Frank, E.H., Grodzinsky, A.J., 1987a. Cartilage electromechanics-I. Electrokinetic transduction and the effects of electrolyte pH and ionic strength. J. Biomech. 20, 615–627.
Frank, E.H., Grodzinsky, A.J., 1987b. Cartilage electromechanics-II. Noninvasive determination of the knee deformation. A continuum model of cartilage electrokinetics and correlation with experiments. J. Biomech. 20, 629–639.
Glassner, A., 2001. Fill er up! IEEE Comput. Graph. Appl. 21 (1), 78–85.
Halonen, K.S., Mononen, M.E., Jurvelin, J.S., T yr s, J., Salo, J., Korhonen, R.K., 2014. Deformation of articular cartilage during static loading of a knee joint—experimental and finite element analysis. J. Biomech. 47 (10), 2467–2474.

Hills, A., Hennig, E., McDonald, M., Bar-Or, O., 2001. Plantar pressure differences between obese and non-obese adults: a biomechanical analysis. Int. J. Obes. 25, 1674–1679.

Jurisic Skevin, A., Filipovic, N., Mijailovic, N., Divjak, A., Nurkovic, J., Radakovic, R., Gacic, M., Grbovic, M., 2018. Gait analysis using wearable sensors with multiple sclerosis patients. Tech. Gazette 25 (2), 339–342.

Mark, N., Aguado, A.S., 2002. Feature Extraction and Image Processing. Newnes.

Omkar, S.N., Vanjare, A.M., Suhith, H., Kumar, G.H., 2012. Motion analysis for short and long jump. Int. J. Perform. Anal. Sport 12 (1), 132–143.

Tong, K., Granat, M.H., 1999. A practical gait analysis system using gyroscopes. Med. Eng. Phys. 21 (2), 87–94.

Wang, Y., Fan, Y., Zhang, M., 2014. Comparison of stress on knee cartilage during kneeling and standing using finite element models. Med. Eng. Phys. 36 (4), 439–447.

Young, W.B., Behm, D.G., 2003. Effects of running, static stretching and practice jumps on explosive force production and jumping performance. J. Sports Med. Phys. Fitness 43 (1), 21–27.

Index

Note: Page numbers followed by *f* indicate figures, *t* indicate tables, and *s* indicate schemes.

B

N

O

P

Q

R

S

T

V

W

Z

Printed in the United States
By Bookmasters